Get Onboard With E-Charting

The Complete Reference Guide to Electronic Charting and PC-Based Marine Navigation

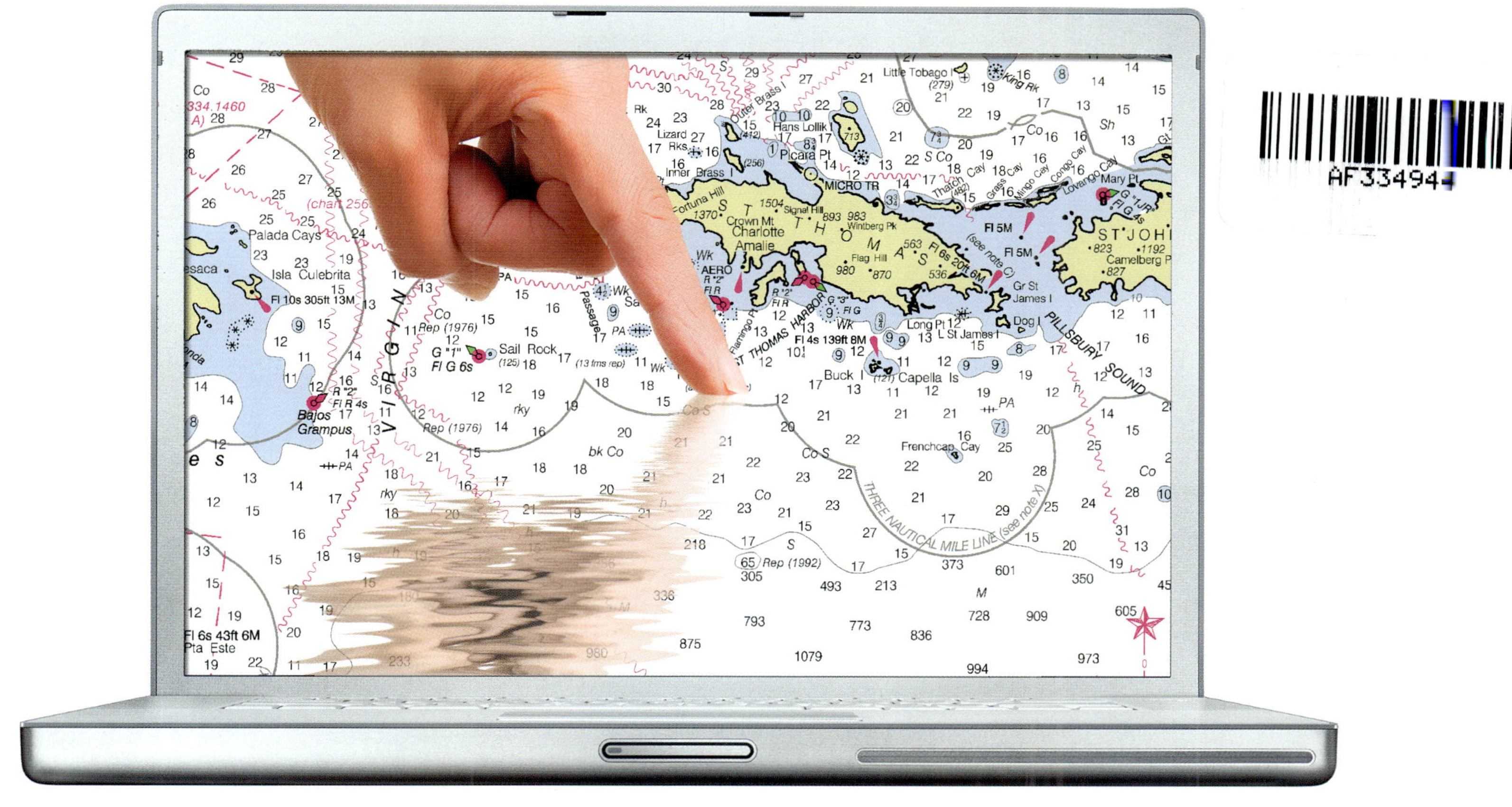

Mark and Diana Doyle, Authors of *Managing the Waterway*

Foreword by Tony Bessinger • Introduction by Glen Justice

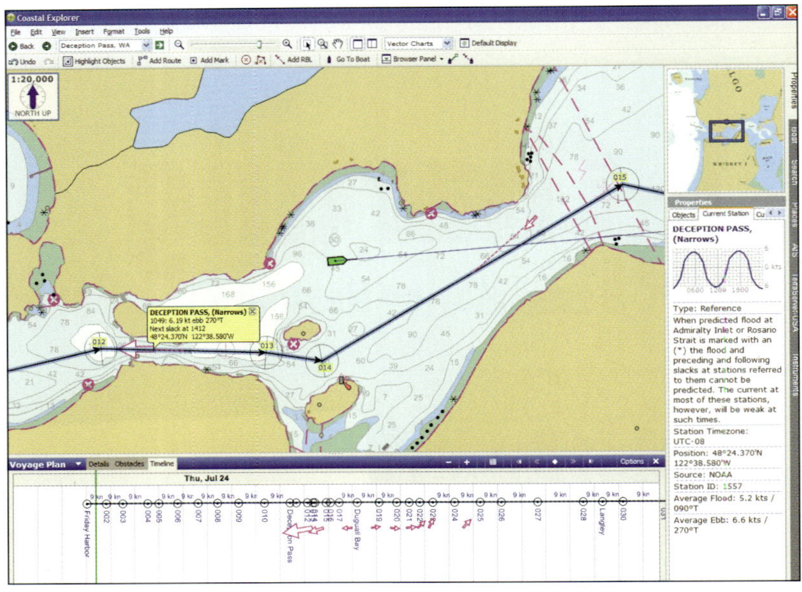

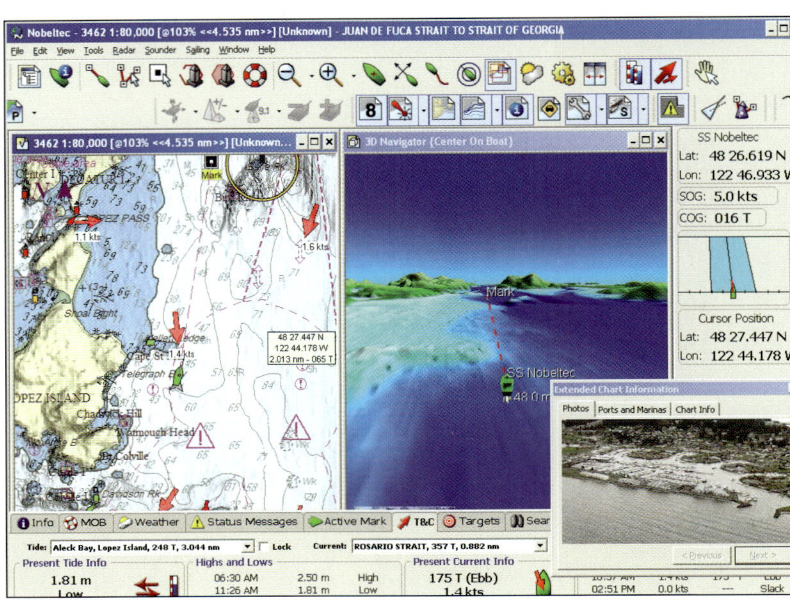

Left: *Coastal Explorer (Chapter 19). Coastal Explorer's unique Timeline feature makes it easy to optimize fuel consumption, departure and arrival times, or tide and current considerations.*

Right: *Nobeltec VNS MAX Pro (Chapter 21). C-Map MAX Pro cartography and supplemental data now fuel Nobeltec VNS MAX Pro and Nobeltec Admiral MAX Pro.*

Get Onboard With E-Charting

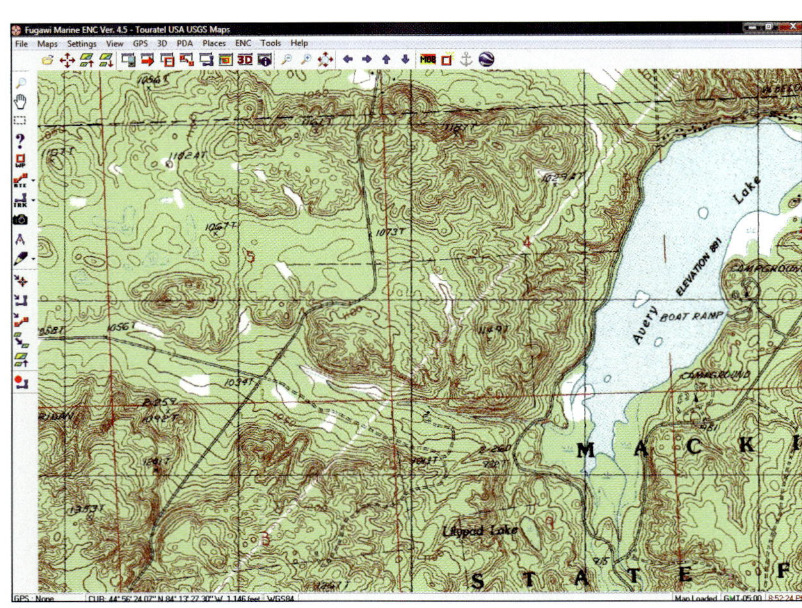

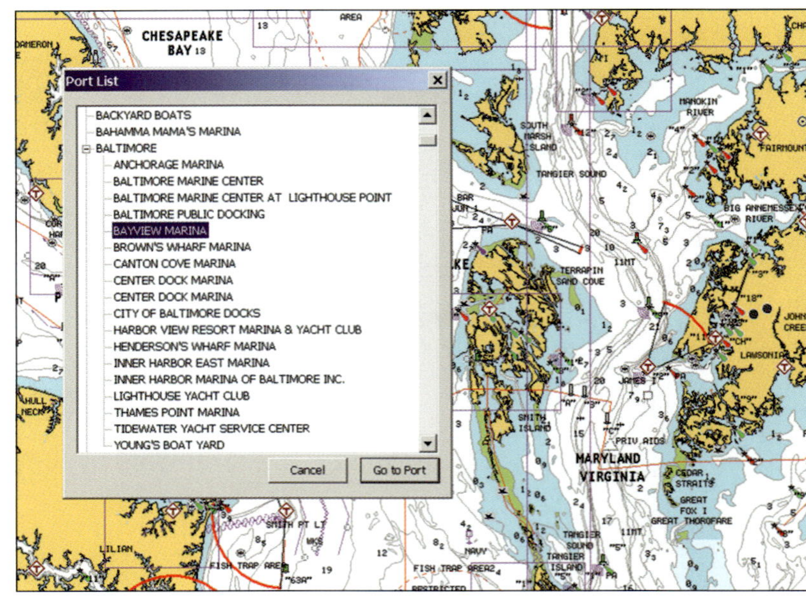

Left: *Fugawi Marine ENC (Chapter 17). Also use Fugawi Marine ENC to plan your next fishing, hunting, hiking, or snowmobiling trip.*

Right: *Navionics NavPlanner (Chapter 7). Locate ports and services with Navionics NavPlanner Go-To-Port functionality.*

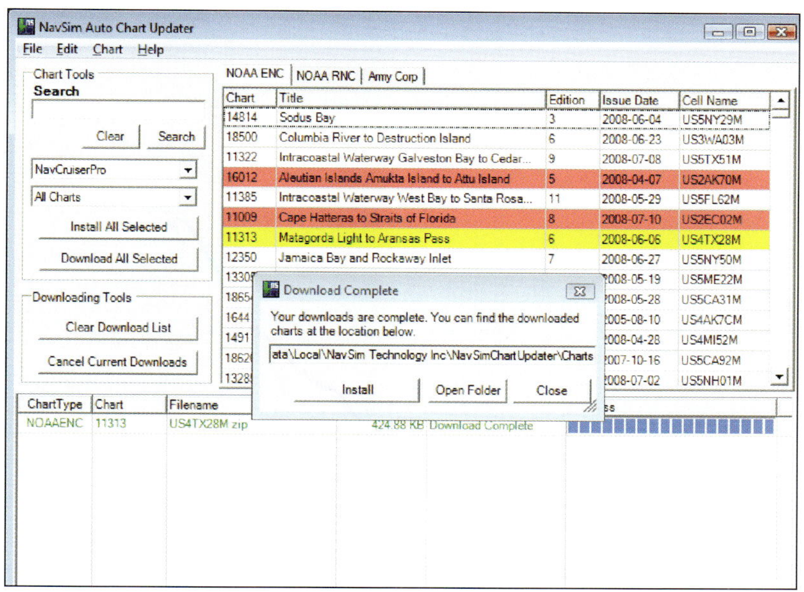

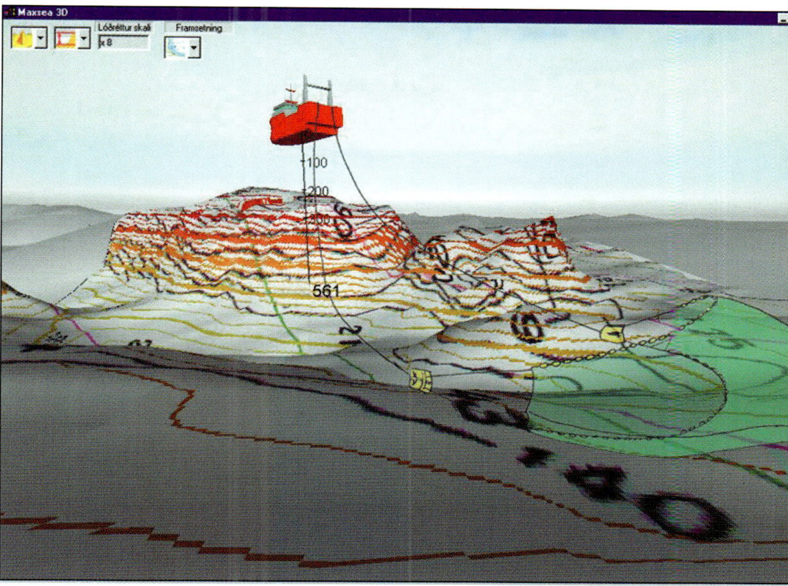

Left: *NavSim BoatCruiser (Chapter 18). NavSim's Chart Updater automatically keeps your NOAA, USACE, and CHS charts current.*

Right: *Furuno MaxSea (Chapter 24). MaxSea has many advanced features, including Trawler Net Tracking.*

The Complete Reference Guide to Electronic Charting and PC-Based Marine Navigation by Mark and Diana Doyle

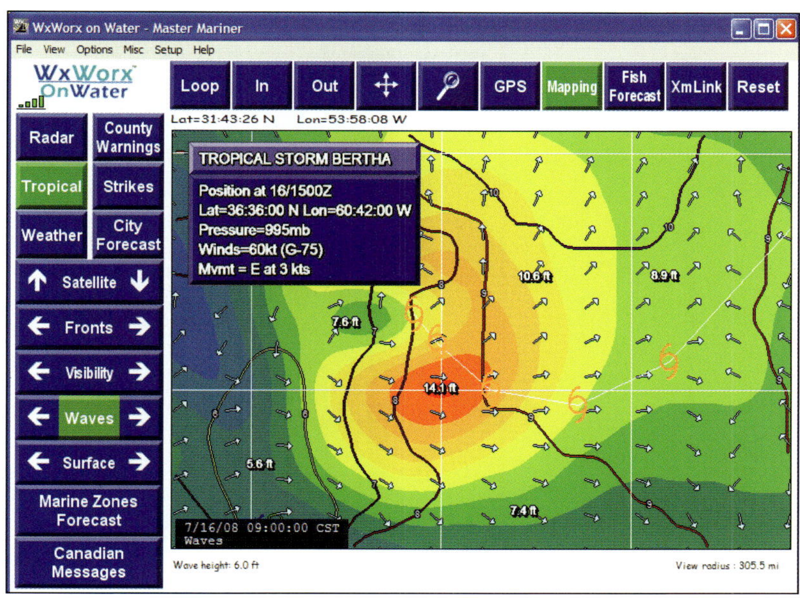

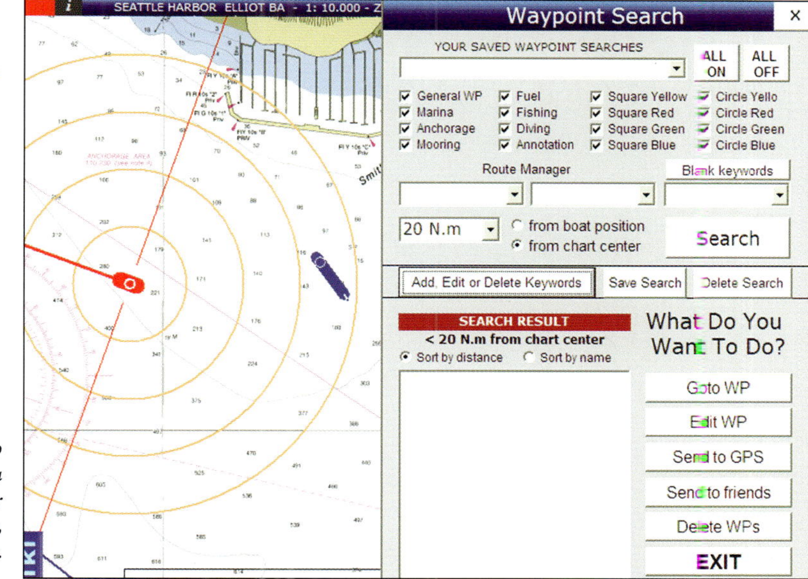

Left: *XM WX Satellite Weather (Chapter 9). Overlaying multiple data products simultaneously, XM Weather displays tropical storm Bertha's wave height and wave direction.*

Right: *TIKI Navigator Pro (Chapter 16). TIKI lets you organize and search your waypoints by type, distance, and keywords.*

Published by semi-local publications LLC
Visit us at www.semi-local.com

Publisher Cataloging-in-Publication Data
Doyle, Mark Stephen, 1956—
Get Onboard with E-Charting: the complete reference guide to electronic charting and PC-based marine navigation / Mark and Diana Doyle.—
1st ed.
p.   cm.
ISBN: 978-0-9758617-2-1
1. Electronics in Navigation. I. Doyle, Diana Richards, 1963.
VK560.D69 2009
623.89´3—dc22   2008904883

For ordering information, including bulk discounts, contact:
semi-local publications LLC
670 West 92nd Street
Bloomington, MN 55420
sales@semi-local.com

Book, Guide, and Chart Updates by Email
Managing the Waterway cruising guides are updated twice yearly through text-only email notices of change and include important information on bridge schedules, marine facilities, anchorages, and shoreside services. In addition, notices of chart errors or substitutions as announced by NOAA or the USACE are included, as are updates to *Get Onboard With E-Charting*.

To receive the updates, subscribe at www.managingthewaterway.com.

Back Cover

Ⓐ TIKI Navigator Pro is able to track buddy boats using AIS or cellular phones, displaying their position on a small side-panel chart.

Ⓑ XM Weather's FishBytes data product creates forecasts of optimal locations by fish species, such as this display for wahoo and white marlin.

Ⓒ Navionics NavPlanner lets you plan ahead and build routes on your PC, then conveniently transfer the waypoint and route information to a chartplotter.

Ⓓ Fugawi Marine ENC includes a Google Earth plug-in that can display a split screen of a chart and a satellite image, useful to double-check a proposed channel.

Ⓔ Furuno MaxSea can display a variety of data layers, such as this image of chlorophyll concentration, useful for sportfishing enthusiasts.

Ⓕ NavSim BoatCruiser allows users to create custom views of charts and instrument data, and to link photos to locations on a chart.

Ⓖ Nobeltec Admiral MAX Pro can integrate data from the radar and a heading sensor to overlay radar scans on a nautical chart.

Ⓗ Coastal Explorer includes a Great Circle Route planning feature that makes it easy to plan long transits or transoceanic voyages.

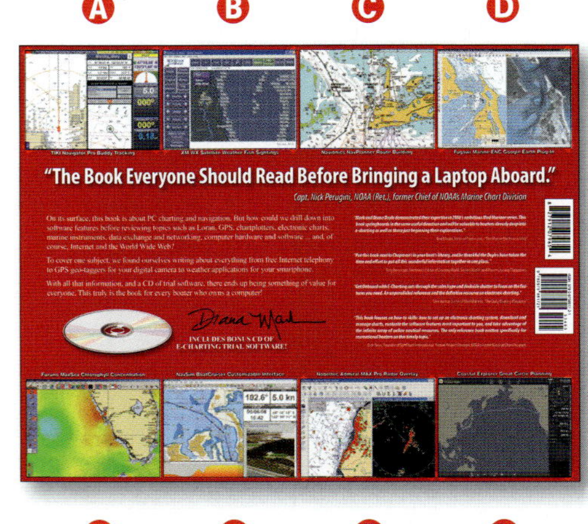

Table of Contents

Part One: Introduction to E-Charting

Part Two: E-Charting System Components

Part Three: E-Charting in Practice

Part Four: Choosing an Application

Part Five: E-Charting Resources

Acknowledgments

Well if it takes a village to raise a child, it seems more than reasonable that it always takes us a gaggle of contributors to produce a book. Each new writing project adds to our posse of subject matter experts, grammar and punctuation masters, and technology wizards who, as luck would have it, become friends. This, in fact, may be our greatest reward in writing.

Special thanks go to Glen Justice for commissioning a series of articles on PC-based e-charting—and convincing us that despite all our other projects, we simply had to do it. What a wonderful experience it was to work with the exciting new *Mad Mariner* e-zine.

We couldn't have undertaken this project without the time and support provided by the e-charting vendors. Thanks to Karen & Jeffrey Siegel of ActiveCaptain; Patrick Cannon at Barco Software; Sheri Flanagan and Julien Barbeau at CARIS; Linda McAndrews and Brandy Power at C-Map; Simon Blundell at DigiBOAT; Virgil Zetterlind at EarthNC; Robin Martel, Amanda Holt, and Kristina Menalo at Fugawi; Jeff Kauzlaric, Peter Prowant, Larry Till, Brice Pryszo, and Iker Pryszo at Furuno; Crystal Friedman, Monica Johnson, and Matt Taylor at Jeppesen Marine; Martin Fox, Craig Cushman, Ian Quarrier, and Mike Bogannam at Maptech; Susan James, Kathy Speers, and Kristina Bacchiocchi at Navionics; Piotr Waclawek, Jaye Gordon, Lisa Piercey, and Eugene Lilly at NavSim Technology; Rich Ray at GPSNavX; Nancy Baumgartner, Chuck Anderson, and Liz Boal (Greenough Communications) at Raymarine; Jeff Hummel, John Cuttitte, and Steve Hodgen at Rose Point Navigation; Olle Soderholm at Spring; Fred Jenssen at TIKI Navigator; and Jenna Shepard and Joe Millard of XM Weather.

The idea for this book came about during the 2007-2008 boat show season, while giving presentations on e-charting and learning from our many sidebar discussions and Q&A sessions with boaters. It was a busy season but special thanks to Raychel Brown of Strictly Sail, Natalie Friton of Trawler Fest, and Nancy Zapf of Seven Seas Cruising Association who gave us multiple opportunities to share our ideas with captive audiences in beautiful waterfront venues. Thanks also to our co-presenters, who selfishly offered great ideas on marine publishing: Kevon Andersen (*Sobe Boatees*), Kim Hess (*Yoga Onboard*), Chuck Kanter (*Cruising Catamaran Communiqué*), Corinne Kanter (*The Cruising K.I.S.S. Cookbook II*), and Kathy Parsons (*Spanish for Cruisers*).

Several people provided invaluable technical advice. Thanks to Dick Davis, founder of SoftChart International and former Project Manager for NOAA's Raster Nautical Chart Program; Alex Heliotis, Deputy Chief of NOAA's Marine Chart Division; Tom Loeper, Chief of NOAA's Coast Pilot Branch; Tony Niles, IENC Project Manager at the U.S. Army Corps; and Nick Perugini, former Chief of NOAA's Marine Chart Division. Thanks to electronics experts Tony Bessinger of *Cruising World, Sailing World, and Power Cruising* magazines and Ben Ellison, editor of www.panbo.com, for their words of wisdom throughout the project.

Some folks kindly agreed to slog through early drafts of the chapters, providing incredibly helpful comments. Thanks to Kevon Andersen of s/v Ionia; John Cuttitte of Rose Point Navigation; Dick Davis of m/v Chart Maker; Sam Densler of s/v Lady of the Lake; Larry & Sharon Duhaime of m/v Lead Free Too; Jeff Hummel of Rose Point Navigation; Fred Jenssen of TIKI Navigator; Glen Justice of m/v Anonymous Source; Capt. Rick Kilmer of s/v The Cat's Aweigh; Capt. Ann Kinner of Seabreeze Nautical Books & Charts; Cliff McKay of s/v Ceilidh; Sylvia Messner; Chuck O'Brien of Bay Books; Capt. Nick Perugini, NOAA (Ret.), former Commanding Officer of NOAA Ship RUDE; Paul Pregent; Hugh Reilly of m/v Westward; Susan Scollay of m/v Sojourner; Capt. Larry Spisak; and Ann Westergard of m/v Galivant.

And last but not least, thanks once again go to our technical friends Paul Pregent, Peter Truskier, and Ben Jenssen who consistently bail us out of the chowder with website, InDesign, and Photoshop issues beyond our meager understanding.

With this august community of friends in place, producing a book in five months can even be fun. Well, sort of ... Again, thank you all.

Foreword

Tony Bessinger, Electronics Editor
Cruising World, Sailing World, and Power Cruising Magazines

I first started using electronic charting when I was a mate on a 98-foot sloop in the late 1980s. It was crude, buggy, and, thanks to the U.S. government's policy of reducing the accuracy of GPS positioning for civilian use, not very accurate. In the nearly 20 years that have passed since then, I've watched with something akin to awe as electronic charting has become so much better, and GPS units far more accurate. As an early adopter and true believer in this technology—and computers in general—I feel as though everything I've ever wished for in digital navigation has become a reality.

While racing in the 2008 Bermuda Race, I used technology that old salts like I used to sail with in the late 70s either screaming for the hills, or drooling with envy. When I'm down below, I'm plotting our basic course with a Raymarine chartplotter. I can bring in huge amounts of weather data via Sirius Satellite Weather, data that includes satellite imagery, water temperature, frontal systems, weather warnings, and I can even listen to the latest news or chill out to some reggae. On the laptop, I've got Expedition tactical racing software, which uses many sources of data, including the wind instruments, GPS position, water temperature, GRIB weather files, and the boat's custom polar data to determine the best course to the islands of Bermuda or the next mark up the race course. To put the icing on the cake, I can even sit on deck with a daylight-viewable, water-resistant touch-screen display, and hike hard while providing solid tactical data to the skipper and the tactician.

Compare that technology and its obvious usefulness to my first Bermuda Race in 1982, when we used a combination of RDF, LORAN (which ran out of signal strength well before you got to Bermuda), and sextant. Our navigator at the time—who claimed to be the last Nantucket Starbuck to have killed a whale (by running it over with his minesweeper during World War II)—came out of the hatch a few hours out of Bermuda, shouting: "Turn around! We've missed the Island!" We as-sured him that we hadn't, as we were surrounded by many similar-sized boats all heading the same direction, and were following the passenger jet contrails heading for the Islands. Any of you who are old enough to recognize the acronyms RDF and LORAN will realize his mistake: a reciprocal bearing generated by the RDF.

It's now possible to determine your position within a meter (and be made aware when the accuracy isn't that good), see exactly where your vessel is on an accurate chart, and integrate that information with real-time and forecasted weather, radar, AIS, autopilot, and depth sounders. You can view charts in traditional NOAA mode, use satellite and aerial photographs, and even see three-dimensional renderings of the waters you're traversing. It's easy to keep those charts up-to-date via the Internet, add your own notations, and add to your chart catalog as your cruising grounds change. Given the prices of good handheld GPS units, laptop computers, navigation software, and even plotters, there's absolutely no excuse to not have some type of digital navigation aboard your boat. And once you make that leap, you'll be amazed at how you ever navigated without it.

I've been lucky in that I've been able to keep up over the years, thanks to a heavy schedule of navigating race boats, working at Newport's Armchair Sailor for many years in the navigation department, and, most recently, as the electronics editor for three marine magazines. Not everyone has had the exposure I've had to this ever-improving technology, and that's why Diana and Mark's latest book is so valuable to anyone who wants to use digital charts and the associated technology. Using clear, easy to comprehend language, terrific illustrations, and an obviously firm grasp of what's available and how it works, the Doyles have distilled what's taken me years to glean into one volume.

From buying the proper laptop, to picking the right software, to planning and plotting, the Doyles have covered every aspect of the digital charting universe and made it easy for anyone to either start from scratch, or simply increase the knowledge they already have. Put this book next to Chapman's in your boat's library, and be thankful the Doyles have taken the time and effort to put all this wonderful information together in one place.

Tony Bessinger

Introduction

Glen Justice, Editor, Mad Mariner
Former Washington Correspondent, The New York Times

To a teenager learning to sail, the nav station on my father's sailboat was captivating. The VHF, weather indicators, and Loran C seemed downright futuristic—to say nothing of the radio direction finder in the corner.

Today, the pilothouse of my trawler makes it all look quaint. Multiple color displays talk to sensors and satellites to provide chartplotter, radar, and fishfinder screens—and that's without the laptop in the corner.

While the changes in marine electronics over a generation are stunning enough, the vast leaps made in the last few years are nothing short of revolutionary. Technology is transforming the way we move over water, putting powerful and affordable navigation tools in the hands of everyday boaters that increase confidence, comfort, and safety. I'm not just talking about the digital toys on display at the local boat show, though they are plenty impressive. The real revolution is in software and online applications.

The average captain, using the laptop he bought last year, instruments already installed on his boat, some free government charts, and an electronic charting program—widely available for less than $300—can now synthesize that mixture into a navigation station that would have required thousands of dollars to replicate even just a few years ago.

That captain can plan a journey in great detail, track his course as he travels, receive warnings of approaching ships, access real-time weather information, and calculate both his time to arrival and the fuel needed to get there—all from that single, modest laptop. Using Internet applications, he can choose an anchorage based on reviews from other boaters, view aerial or satellite photos to help his approach, and consult tide predictions before he digs in for the night.

It is hard to overstate the potential of PC-based navigation systems and electronic charts, and their power to make boating safer and more affordable. Of course, it may take some work to understand these systems and to get them running properly (we are, after all, talking about boating). But that is where this book comes in—it is as valuable to your nav station as your engine manual is to your power plant.

As the Editor of *Mad Mariner*, an online boating magazine, I set out in 2007 to find a writer who was interested in doing deep reporting on navigation software and how boaters could make use of it. My search ended with Mark and Diana Doyle, a pair of experienced cruisers, licensed captains, and cruising guide authors—and they were already writing about electronic charting! The result of our partnership was *Hard Facts on Software*, a series that ran more than 20 weeks and provided a definitive resource for boaters interested in exploring PC-based navigation.

With the publication of *Get Onboard with E-Charting*, Mark and Diana have expanded that resource even further. The authors have spent months conducting in-depth product tests of a type rarely seen in marine journalism. Targeting more than a dozen top navigation programs, they have put software through its paces (including trying to crash it), read manuals and FAQs, wrangled with help desks, deconstructed websites, and interviewed both software developers and the boaters who use their products. There are few mariners on the water today with more detailed knowledge of modern navigation programs.

The information gained from these long efforts has been distilled into *Get Onboard with E-Charting*, a book destined to become an unparalleled reference and a definitive resource. In readable language, the book discusses how to understand the components of an electronic charting system, assess your needs, and make sound purchases.

At the heart of the book are detailed comparison tables that explain the capabilities of more than a dozen top software packages, allowing boaters to cut through the sales hype and dockside chatter and focus on the features that are meaningful. It also describes the full landscape of marine cartography, explaining the utility and limitations of electronic charts available worldwide.

Perhaps most important, the book focuses on skills, such as how to evaluate your computer needs; connect your boat's instrumentation; download and manage charts; and expand the utility of your system using online resources. It is my hope that boaters will find this book before buying an expensive software package or even a chart card. It is required reading for anybody who wants to set up a PC-based navigation system and do the job right.

Preface

Mark & Diana Doyle, Authors
Managing the Waterway Cruising Guides & Electronic Charting

With the shelf of navigational reference books full for decades now, why add a book on electronic charting? It's a good question—and it has a good answer.

Whether you're comfortable with the trend or not, nautical charts have gone digital. Although we believe paper charts will never—and should never—become obsolete, government hydrographic offices are moving to digital charts as their primary format. The commercial shipping industry now navigates using electronic charts as well. It's only a matter of time until personal computers—in whatever form—become as fundamental to recreational navigation as paper charts, parallel rules, and dividers.

Yet as this trend unfolds—with laptops becoming increasingly commonplace, new e-charting companies emerging, and governments worldwide transitioning to digital charts—most boaters are woefully out of the loop. Many already own a laptop for work or email, and probably use a chartplotter aboard their boat. But when it comes to computer-based navigation, they find it difficult to keep up with what's available and what would work best for them.

One of our reviewers wisely pointed out there is more to an electronic charting system than hardware and software—there is a human component also. How true! The user, and by extension the developers of the software, the retailers who sell the products, and the customer support technicians who help solve problems, are integral to e-charting. In fact, you are the linchpin of any electronic charting system!

This book is written for you: the user. It is intended as a practical guide to the background, advantages, weaknesses, resources, and choices in e-charting today. Although our target audience is recreational boaters, its content is also relevant for commercial captains who use navigation software for planning or piloting.

We wrote this book to build from the basics, including sections such as a glossary with more than 460 navigation, electronic charting, and general computing terms. We show novices how to get running with everything they need at very little cost in time or money. If you're considering e-charting, this book clears the forest for you, letting you winnow the choices down to two or three software applications that make sense. With the choices narrowed, you can then download the vendor's trial software to make your final decision.

For boaters familiar with e-charting, we delve into advanced features, such as collision avoidance systems, networking, and data exchange. We also include a chapter that suggests ways to expand your laptop's power as an onboard resource. Experienced e-charters can glean suggestions for additional resources and features, or can use the reviews to evaluate whether they are ready to switch to a more sophisticated modular software package.

By helping you understand e-charting, we hope your entire onboard navigation system becomes stronger and smarter. The best e-charting system centers on an informed user. Of course, the software reviews, and any other content, are simply our opinions based on our own experiences. We recognize this information may be incomplete or subject to disagreement—and so should you. Be sure to check with each vendor for specific details and their latest features, and be sure to take advantage of their trial software.

But why write a book? Shouldn't a digital topic like e-charting be disseminated in a digital format? Isn't the topic of electronic charts and e-charting software outdated immediately when committed to the printed page?

Indeed, portions of this book were originally written for *Mad Mariner*, a web-based electronic boating magazine. Unfortunately, after 25 articles and 85,000 words, we discovered we had only scratched the surface of e-charting. We continued to receive questions by email, through forums, and at boat shows. It was time to write "the complete reference guide to electronic charting and PC-based marine navigation."

All printed material becomes dated to some extent as soon as the ink dries on the paper. But even an emerging technology such as e-charting shows surprising consistency over time. Perhaps the adage, "The more things change, the more they stay the same," applies to e-charting as well. We believe this material will hold up well over time, and that a reference like this provides first-timers with information in a format they well understand—a book.

Although relatively new to most recreational boaters, e-charting is now nearly three decades old. In fact, this evolution of charts, hardware, and software is important to understanding today's e-charting choices. Although devices become smaller, GPS accuracy becomes more precise, screen displays become more vivid, and user interfaces become more convenient, the system itself remains relatively constant. The fundamentals of position finding, personal computers, marine instruments and

Google | Google Search Terms

Always Up-to-Date: *Some aspects of the e-charting and general computer markets evolve at an astonishing pace. Often we're afraid to list a product, technology, or URL knowing it may change before the ink dries. To keep abreast of these trends, we suggest topical Google Search Strings throughout the book.*

The H (Hotel) flag signals *Pilot on Board* and denotes a practical suggestion, power user tip, or speedbump warning culled from personal experience and the many conversations we've had with novice and veteran e-charters.

To receive twice-yearly text-only email updates to this book, subscribe at **www. managingthewaterway.com**. See page iv for details.

sensors, the file formats of electronic charts, and the standards for networking and data exchange change fairly slowly. And these are the topics that comprise the bulk of this book.

The least stable component of e-charting are the software applications, 16 of which are reviewed in Part Four. Developers and vendors continue to add new features and improve their products in a very competitive e-charting software market.

However, even software developers must operate within the constraints of their previous offerings and their customer base. They add new features and functions, or maybe update their graphical user interface, but the base software code is not tossed out as obsolete. This "backward compatibility" will ensure that the information in these reviews has a shelf life and remains useful.

While software vendors strive to out-program their competitors, this book lets you enter the fray as an informed consumer and user. You can ask which features have been added and which weaknesses have been addressed. You can evaluate current offerings based on the history of a software package, its core functionality, and its trajectory in terms of adding and improving chart and instrument support.

For e-charting aspects that change quickly, we incorporate some flexible tools. For example, although current prices and system requirements are listed in a sidebar at the beginning of each software chapter, we also provide the vendor's website. A quick check will provide the latest version, pricing, and hardware requirements.

To prevent broken web links, we include Google search strings. In rapidly-developing product areas, we provide a Google search string that brings up the latest products and vendors. This lets you quickly locate the most relevant websites on a topic.

To address new technology, we include sidebars on emerging topics. The sidebars explain the significance of a new technology, such as the resurgence of eLoran, or cover un-

folding events, such as acquisitions or mergers that may affect future products. In many chapters, What's New, What's Next, and Trends sections cover forthcoming products and versions.

E-charting is fundamentally about making boating safer and more enjoyable. Volumes of data can be accessed in an easy-to-interpret, user-friendly PC format, giving the navigator the ultimate in situational awareness—and that translates into confidence. You probably already have a personal computer on board. This book teaches you how to take full advantage of it!

Diana Monk

Part One

Introduction to E-Charting

First things first. Electronic charting may be a relatively new technology, but its component parts all have rich histories. Before you jump in upgrading computers, purchasing software, or cabling instruments, it's worth taking the time to read a few pages on the big picture of e-charting: what it is and where it came from.

Chapter 1, *What is E-Charting?*, explains e-charting in terms of an electronic charting system. Written for recreational boaters, it explains the difference between recreational and commercial e-charting systems and the advantages of computer-based navigation and piloting systems.

Chapter 2, *The History of Global Position Finding*, takes you through the evolution of position-finding systems such as Loran and GPS, the development of electronic charts, and the early software pioneers. Don't skip this chapter thinking you don't need a history lesson! Chapter 2 explains GPS enhancements such as WAAS-enabled receivers, and previews developing position-finding systems such as eLoran and Galileo. Most importantly, discover how the *limitations* of position-finding systems and electronic chart conversion affect your boating in terms of the accuracy of your electronic chart displays.

Chapter 1
What is E-Charting?

Type "Annapolis" into the search field and a brilliant, full-color nautical chart appears on your laptop's display. Zoom to a nearby bridge and mouse-click on its symbol to check vertical clearance. Scroll the chart image, seamlessly previewing adjacent charts. Select an anchorage and drop a waypoint. Zoom out, string waypoints into a route, and automatically be alerted to an obstacle you hadn't noticed on the chart.

A pop-up window overviews large vessel traffic in your area. Another pop-up shows the positions of your buddy boats. Tides, current, forecasted weather, and sea conditions download automatically to your laptop. A small window at the bottom corner of your screen shows a live video feed from your engine room.

This isn't navigation of the future; it's electronic charting today! Although digital charts have been available through private vendors for over twenty years, the hardware to empower electronic charting—namely an inexpensive personal computer with gigabytes of memory and a lightning-quick graphics card—has only recently become standard boating fare.

Add to this the recent scuttlebutt concerning free government charts and it's easy to see why e-charting has ignited into one of the hottest topics on today's coconut telegraph.

No one can deny that e-charting has come a long way. It's a different e-charting world from the early days of jury-rigged "MacSea" prototypes on world-girdling race boats, or bootlegged copies of The Capn passing dinghy-to-dinghy among Caribbean cruisers. Chapter 2, *The History of Global Position Finding*, covers these staggering developments. With laptops comprising two-thirds of all U.S. computer sales, pocket-sized GPS sensors costing less than $60, and Internet access as simple as inserting a wireless card, e-charting is now easily within reach of the smallest vessel or even the pathologically technophobic.

At a minimum, you most likely already own a computer, and charts for the U.S. are literally free. With free software available, the only cost of entry is time and a little education.

This book explores the new phenomenon of e-charting—displaying charts for planning, navigation, or piloting using a desktop or laptop personal computer. It is written for recreational boaters, whether old salt or newbie, computer geek or technophobe.

So boot up that laptop and download some charts. It's time to move beyond email and digital photos and get onboard with electronic charting!

WHY ADD COMPUTER E-CHARTING?

You may not realize it, but chances are you already have an Electronic Charting System (ECS) onboard.

An Electronic Charting System refers to any system of hardware devices that displays electronic charts and integrates marine electronics. Whether for a large merchant ship or fishing skiff, an ECS has four components: hardware, external sensors, electronic charts, and navigational software. Part Two, *E-Charting System Components*, covers these components in detail, including hardware specifications, where to obtain charts, software choices, and external sensor options.

If you have a dedicated chartplotter or multifunction display, such as a Garmin, Raymarine or Furuno unit, that is one form of an ECS. It displays the chart database contained on a memory chip or cartridge loaded into a hardware device. Most boaters connect one or more marine instruments to their chartplotter. At a minimum, a GPS displays your boat's actual position on the chart.

A desktop or laptop computer can also serve as the hard-

Marinized (salt air and water resistant)
Bright sunlight-viewable display
Small, compact form

PC Advantages

Low cost (you may already own a PC)
Mobile and portable
Large display area
Easy data entry (keyboard and mouse)
Free charts of United States waters
Customized printing capability
Archived nautical references
Unlimited Internet resources

Typical Chartplotter ECS: *Hardware, charts, software, and GPS (not shown)*

ware and display of an electronic charting system. A computer-based ECS uses a standard personal computer combined with a chart database stored on its hard drive. Although the interface, storage, and input devices are different, the concept is the same: the computer displays charts and integrates with a vessel's marine electronics.

If you already own a chartplotter, you may wonder, "Why add computer e-charting?" Your chartplotter displays digital charts, shows your vessel position, stores waypoints and routes, and integrates with electronic instruments such as an autopilot. Plus your chartplotter is weather-proof and designed with a screen bright enough to read out on the water.

You could dismiss chartplotters for their small screens, tiny buttons, and lack of connectivity to the outside world, or casually pigeonhole the laptop as a mere back-up. But that would be selling both short! The biggest bang for boaters lies in embracing laptops and chartplotters in concert, leveraging the advantages of each. Neither is intended to replace the other. Computer-based e-charting brings easily accessible advanced features, improved situational awareness, streamlined data entry, Internet resources, automatic updates, and portability to your onboard electronic charting system.

A laptop's e-charting applications include an extraordinary number of impressive features. At a minimum, you can store and access a complete library of U.S. charts at no cost. Imagine being able to browse and view any chart in the comfort of your desk chair or nav station bench.

E-charting is much more than simply displaying a chart on a computer. Fundamental navigation tasks—such as planning, monitoring, and documenting your transit—are standard features. Electronic charting provides a more robust framework for conventional chart work. On an electronic chart it's a simple matter to draw bearing lines or distance rings, calculate fixes, or add annotations. With a small GPS connected to the computer's USB port, your boat position is displayed directly on the chart image. Waypoints and routes created on an electronic chart can be transferred to a chartplotter or autopilot. Vector charts look the same but the chart attributes become interactive data for the ECS, alerting you of nearby charted hazards such as shallow depths, obstructions, or shipping lanes.

Computer-based e-charting now includes features far beyond the nuts-and-bolts of navigation and piloting. You can link a complete photo library of digital images to your charts by location. You can stream video camera footage from your engine room, child's stateroom, or deep-sea fishing fighting chair. Fishermen can cruise over productive grounds and create their own 3D bathymetric charts of the seafloor. You can monitor buddy boats, and even display a companion window to illustrate their location.

A dynamic computer screen can display multiple types of navigational information in a compact, user-friendly, and flexible format. A mouse-click displays different charts, chart scales, or supplemental data. Split-screen and tabbed windows allow for multiple views, ranging from detailed harbor charts to satellite imagery. Because electronic charts have the ability to store and display data in different layers, navigators can customize their chart display for changing situations, reducing information overload. For example, depths can be shown at a particular threshold; chart notes can be displayed or hidden; and the movement of other ship traffic can be tracked or obscured.

No chartplotter can compete with the user interface of a personal computer. Extensive waypoint and route libraries are much easier to create and manage with the luxury of a full-sized computer screen, keyboard, and mouse. Computer screens are larger and have better resolution and color than chartplotters. A familiar keyboard and mouse are easier for extensive scrolling or data entry than an infrequently-used keypad or touch screen. In fact, most e-charting applications bypass data entry altogether by letting the user visually select waypoint locations using a cursor. Graphical data entry is much less error-prone than typing latitude and longitude number sequences—even for the most meticulous typist.

Computer-based e-charting also provides a portal to a staggering array of planning resources. You can collect folders of reference texts, government publications, and nautical calculators including *Chart No. 1*, the *USCG Navigation Rules (Inland and International)*, or the complete text of *American Practical Navigator* (Bowditch). All these documents are printable, searchable down to the word level, and easily updated or supplemented as needed. In Chapter 9, *Extending the Digital Metaphor*, we spotlight these and many other digital resources.

In this Internet-enabled era, e-charting also goes hand-in-hand with the World Wide Web. Although e-charting began as a way to display graphic images of charts, the Internet is now an integral part of e-charting's added value. With wireless Internet affordable and commonplace, most e-charting applications leverage this phenomenal source of information, integrating features such as weather downloads or waypoint and route overlays on Google Earth satellite imagery. Most recently, updated charts can be downloaded automatically, making obsolete the laborious practice of hand-copying notes from Notices to Mariners.

Instead of entering chartplotter waypoints by hand while tied up at the marina, you can plan your routes and annotate your charts in advance, saving fair weather boating time and reducing the risk of data entry errors. To bring your ready-to-go navigational homework on board, either store them on a portable USB drive or memory card, or take along your laptop. You can bring the laptop from boat to boat for delivery work or chartering. You can exchange assets such as waypoints and chart annotations with cruising club members, buddy boats, or flotilla participants.

You probably already bring a laptop aboard for email, digital camera downloads, or Internet surfing. The advantage of a computer-based charting system is that you can, quite literally, take it anywhere with you. With a laptop-based system you have the ultimate in portability, bringing charts (and valuable assets such as waypoints, routes, and annotations) from home to helm. Chapter 8, *Networking and Data Exchange*, discusses this topic in detail.

From Kayaks to Mega-Yachts

E-charting choices are available for every budget, level of computer expertise, and operating system. Although you can spend thousands of dollars for sophisticated networked systems, there are also excellent choices for less than $150—including a very robust free e-charting package. There are options for beginners who simply want to view charts, and for computer-savvy circumnavigators who want to scan and geo-reference their own paper charts. Even boaters on Macintosh or Linux operating systems now have an array of e-charting options. Part Four, *Choosing an Application*, includes detailed reviews of 16 popular packages.

E-charting can be as simple as viewing charts on your home computer, or as sophisticated as an interfaced network of onboard computers and marine instruments. Some boaters only use e-charting at home, studying charts or planning routes from the comfort of their desk during the off-season. For others, an onboard personal computer is the cornerstone of their vessel's navigation system. At boat shows we've been approached by kayakers as well as mega-yacht owners, each interested in getting started with e-charting.

At the simplest level, *viewer* software only displays an electronic chart on a computer screen. Viewer applications are perfect for boaters who do not plan routes or navigate on their personal computer. Viewers let you display electronic charts to study and/or print snapshots. Even if you don't use e-charting for creating waypoints and routes, these simple applications take advantage of the large screen display, and the comfort and convenience of a personal computer.

For example, canoeists, kayakers, and flats-boat fishermen can use their personal computer to create custom chart-books of their local waters. We were approached at the Miami International Boat Show by a retiree preparing to kayak the entire Mississippi River. We advised him to create and

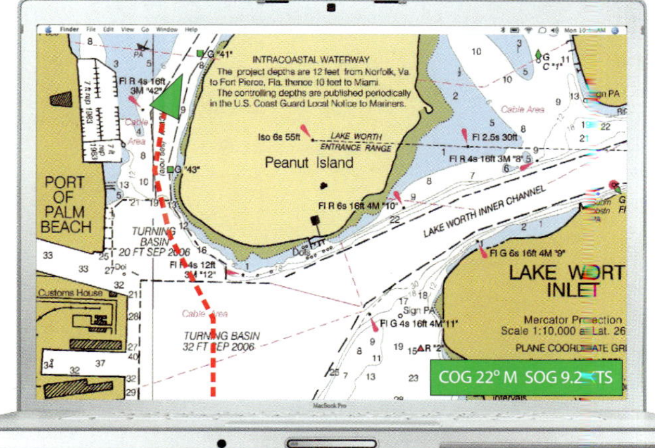

Typical Laptop ECS: *Hardware, charts, software, and GPS (not shown)*

Three Types of E-Charting Software

There are three types of e-charting software, each progressively more capable, and more expensive.

Viewers
Chart Display

Planners
Chart Display
+ Waypoints/Routes
+ Data Transfer

Full-Featured Applications
Chart Display
Waypoints/Routes
Data Transfer
+ Boat Position
+ Piloting Features
+ Instrument Connectivity

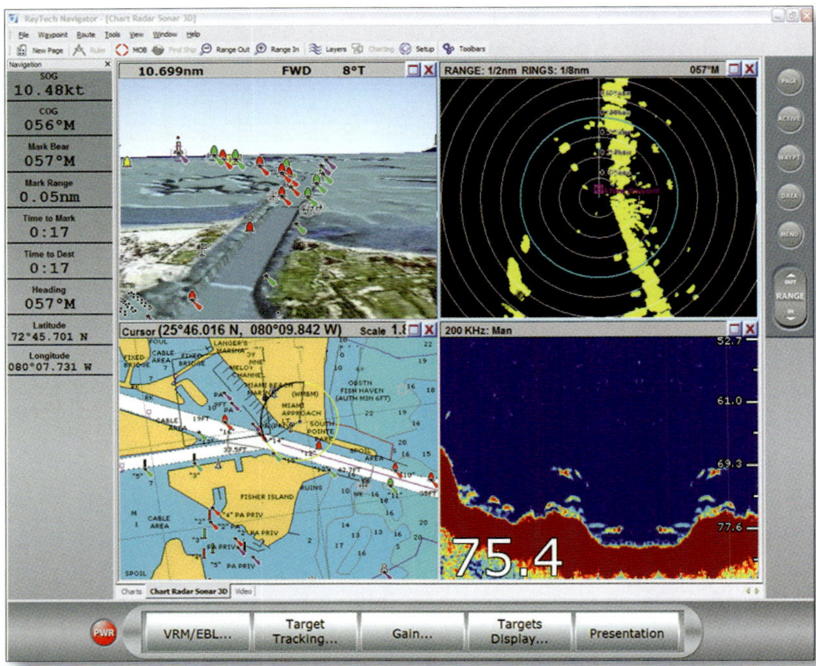

Split Screen: *Several full-featured e-charting applications can display multiple panes of information simultaneously, such as 3D land and sea, radar, sonar, and a vector chart (clockwise from top left).*

then laminate custom chartkits at home, using free U.S. Army Corps of Engineers charts and a free viewer. These chartkits include locks, dams, campsites, and provisioning stops on his route.

Planner software adds the ability to diagram your transit by creating and storing waypoints and routes. These assets can then be exported (either by removable media or via a data cable) to a dedicated marine device such as a GPS or chartplotter.

Planning software is perfect for boaters who want to complete their navigational homework in advance in the comfort and convenience of their home or office. Navigation assets can be brought aboard on a thumb drive or memory card for quick transfer to a chartplotter or onboard computer. Sailboaters and weekenders can use e-charting to plan routes in advance, saving scarce race evenings and weekends for on-the-water time. In small sailboats or sport cruisers, the cockpit typically provides little space or protection for safe laptop operation. In these cases, computer-based charting works best as a planning tool down in the saloon or at home the evening before. You can view and study charts, print chartlettes, or create waypoints and routes to transfer to a chartplotter.

Full-featured e-charting applications are designed for recreational use but now rival the electronic charting systems found on large commercial vessels. Although these packages support advanced features such as ship traffic alerts, automated route optimization, and extensive networking, the clincher for most boaters is simply the support of a GPS sensor. Regardless of whether one uses all the sophisticated piloting and instrument options, a full-featured application adds the ability to view the boat's current position on a chart display. These robust applications take advantage of the staggering memory, speed, Internet connectivity, and graphic display options available with today's portable personal computers.

Boaters using a full-featured application typically install the software on a laptop to be brought aboard. Trawlers, cabin cruisers, and catamarans have the luxury of an enclosed helm or navigation station with space for an accessible laptop that can be used underway. Offshore or ocean racing yachts take advantage of the sophisticated sail optimization and weather route planning algorithms included in several high-end packages. The laptop serves as a central node of a navigation system, displaying boat position, piloting routes, and monitoring vessel and environmental conditions.

Even mega-yachts aren't too big for a personal computer-based electronic charting system. On the contrary, many large luxury vessels connect multiple computers in an onboard network, taking advantage of advanced features to monitor the engine via live video feed or to track expensive tenders with theft sensor technology.

In Chapter 10, *Putting It All Together*, we outline seven e-charting scenarios—different ways boaters can and have put e-charting into practice. Although no single scenario will match you and your vessel perfectly, the scenarios illustrate the variety of ways e-charting and an onboard laptop compliment traditional navigation.

Chapter 2
The History of Global Position Finding

In the era before electronics, navigating essentially meant figuring out where you had been. This sounds odd, but rather than pinpointing your current position, visual bearings or fixes gave an approximate triangle of where you were at the time you took the sightings. The logic was, if you could figure out where you were at a particular time, then you could extrapolate to make an inference about where you were most likely to be at the present time. Navigators "deduced" their current position based upon a previously determined position, advancing that position using presumed speed, elapsed time, and course. Navigation was all about "deduced" reckoning, a phrase that became *dead reckoning* or simply DR.

Technology, in the form of global position finding systems, changed the foundations of navigation. Navigators can now determine their position at any moment within incredible accuracy anywhere on the globe. Regular citizens take it for granted, whether boating or driving or hiking, to know where they *are* to within a few feet.

But this era of instant and accurate global position finding took decades to develop, spurred by WWII research in radio-navigation. It began with Loran, then early satellite systems such as Transit and NAVSTAR, then Differential GPS, to today's WAAS-capable GPS. These systems, even seemingly-ancient Loran, are relevant to understanding global position finding today.

Global position finding could never have developed without the parallel development of electronic cartography. Of course, this resource also did not spring from a blank slate; its rich history impacts the ability to place your vessel's icon on a electronic chart. Viewing your position on a chart requires a mutual synthesis of *position accuracy* and *chart accuracy*.

Yet most boaters are shocked to learn that much of the data viewed on their state-of-the-art e-charting system is stale. The problem is not that their charts are old (although sometimes this is the case), but that even the newest charts may contain survey data that is several decades old. Understanding the evolution of electronic cartography helps you understand why some charts are only available in particular formats, or only from particular sources, or unavailable at all. If you understand the source of electronic charts, you also understand the potential for chart anomalies, making you a more vigilant and prudent e-charting navigator.

THE DEVELOPMENT OF POSITION-FINDING SYSTEMS

Shop the used boat market and you still see Loran receivers mounted at some navigation stations. Walk the docks and some masts still carry SatNav receivers, precursors to today's GPS system. It wasn't too long ago that Loran and SatNav provided state-of-the-art position finding—a phenomenal improvement over land- or celestial-based fixes.

From Loran to GPS, each technological turn-of-the-crank resulted in quantum leaps for electronic position finding. With each new technology, position fixes became more accurate, more frequent, and most recently with the integration of electronic cartography, more visual.

Accuracy has improved by more than two orders of magnitude in only a couple of decades. Loran, one of the first electronic position-finding systems, was hailed for its "plus or minus one-quarter mile (95% of the time)" accuracy. A contemporary WAAS-enabled GPS receiver—now a standard consumer item—provides accuracy to within three meters (ten feet). That's about the length of most cruising dinghies!

Each newer system also calculated positions more frequently. NAVSTAR, the precursor to GPS, could only obtain a fix when satellites passed overhead, which occurred every ninety minutes. Yet compared to the alternatives, a precise

Chapter Terms

BSB4, BSB5: Proprietary versions of BSB files requiring a user license.

CRADA: Cooperative Research and Development Agreement. An agreement between a private company and a government agency to work together on a project in order to speed the commercialization of a federally-developed technology.

GPS: Global Positioning System. A receiver that uses a constellation of satellites that transmit microwave signals to determine the receiver's location, speed, and direction.

NOAA: National Oceanic and Atmospheric Administration. The federal agency responsible for oceanic and atmospheric data that includes the National Ocean Service and Office of Coast Survey.

WAAS: Wide Area Augmentation System. A feature on GPS receivers to improve position accuracy, typically to within three meters, using a system of approximately 25 ground reference stations positioned across the U.S.

Chapter Questions

- *Is Loran obsolete? What about eLoran?*
- *How did GPS develop?*
- *What is DGPS and WAAS?*
- *Who produces electronic charts?*
- *Why did the U.S. release its electronic charts for free? Why aren't other countries' charts free?*
- *How accurate are electronic charts?*
- *What were the first e-charting software applications?*

Accurately projecting a three-dimensional object onto a two-dimensional plane is difficult.

Similar to cross streets on a map, locations on Earth are recorded using hypothetical lines called *latitude* and *longitude*.

Lines of latitude run horizontally, with 0° at the equator to 90° North or 90° South at each pole. Lines of longitude run vertically, with 0° longitude at the prime meridian and proceeding 180° east or west to a line opposite the prime meridian, called the International Date Line. The prime meridian, which is also the time reference for Coordinated Universal Time (UTC), passes through Greenwich, England. The International Date Line runs down the middle of the Pacific.

Latitude and longitude can be expressed in degrees, minutes, and seconds; or more recently in degrees and minutes with decimal notation. Below, San Juan Harbor's latitude is noted as N 18° 27'06" or N 18 27.100 (because one-tenth of a degree is six minutes).

One minute of latitude always equals one nautical mile. However, because lines of longitude run closer together as they near the poles, this relationship does not hold true for one minute of longitude.

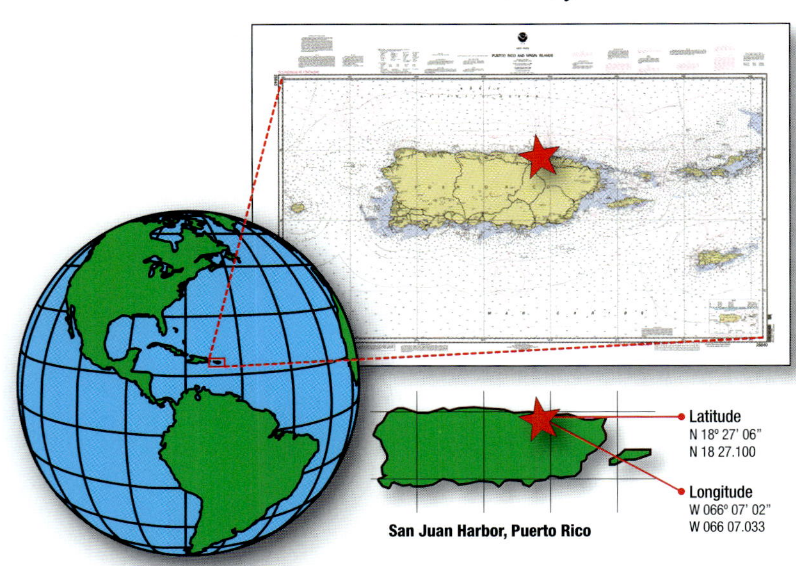

San Juan Harbor, Puerto Rico

global position sixteen times a day was a Godsend. Contemporary GPS receivers are always able to pick up at least four signals—the minimum required to compute a fix—a result of many more satellites with more sophisticated orbit placements. Positions are now calculated so frequently they are near instantaneous.

To display position fix information, early systems were reminiscent of old digital clock radios, providing simple digital numeric read-outs. Loran receivers initially only provided TDs (Time Delays or Time Differences) that allowed navigators to determine their position on a chart overprinted with TD lines. Later Loran units made these calculations automatically, displaying a digital read-out of latitude and longitude, which was easier to plot on any chart.

The development of electronic cartography allowed for a graphical display of position data. Position data could be shown directly on a chart image, saving the time and risk of error in transcribing from instrument to chart. GPS data now typically integrates with color chart displays on handheld GPS units, chartplotters, PDAs, or personal computers, displaying real-time position as a dynamic visual icon.

Loran

Before you denounce Loran radionavigation as obsolete, realize that the U.S. Coast Guard still maintains about 25 Loran stations. The Loran system, originally developed during World War II and made available for recreational use in the late 1950s, is still used by some recreational boaters as a back-up to their GPS. In fact, Loran is currently receiving an upgrade, motivated by national security concerns.

Loran, short for *Long Range Radio Navigation*, is a medium-range navigation system that covers primarily the U.S. coasts, Canadian coasts, and Great Lakes. The system provides navigational fixes within 50 nautical miles from shore, or out to the 100-fathom curve, whichever is greater.

Loran works using a chain of shore-based radio transmitters. A Loran chain consists of one master transmitter and between two and four secondary transmitters. Each transmitter sends out pulsed radio signals.

A Loran receiver onboard a vessel picks up these pulsed signals and accurately measures the time difference. Unlike a GPS, which measures the absolute distance to a transmitter, a Loran unit measures the difference in time for two signals in a transmitter chain to reach the receiver. This information locates the receiver on a hyperbolic line-of-position. Immediately repeating the process produces a second line-of-position, establishing a fix.

Depending on the Loran unit, navigators could either use this TD data, or the converted latitude/longitude data, to plot their position on a paper chart. Charts were (and many still are) overprinted with Loran TD lines in addition to the latitude and longitude grid.

At its heyday, Loran produced positions that were accurate to about 200 meters, updated approximately once every second. Even by today's standards, knowing your position to 656 feet is not bad if your GPS quits.

Global Positioning System (GPS)

The first global positioning system was revolutionary because it extended position finding beyond the constraints of shore-based transmitters. Moving transmitters into the sky literally launched the potential for a truly global position finding system.

The U.S. Navy Navigation Satellite System, also called *Transit*, was the first satellite navigation system, conceived in the late 1950s and deployed in the mid-1960s. Old-timers may remember this system under the name Transit SatNav.

Transit achieved near-global coverage with a constellation of only five satellites. If conditions were favorable, it was accurate to 35 to 100 meters. But because of limited satellites, it was dependent on the timing of the orbits, providing a navigational fix about every 90 minutes. Mariners had to interpolate their position between satellite passes.

In the mid-1970s, the U.S. Department of Defense developed the NAVSTAR system, for *Navigation Satellite Timing And Ranging*. A constellation of satellites were placed at much higher orbits, ensuring that at least four satellites were in view from any place on Earth at any time. NAVSTAR could provide fixes more frequently and more precisely.

NAVSTAR eventually became the GPS system used by civilians today. Although NAVSTAR GPS began as a military system, in 1983 President Reagan issued a directive making the system available for civilian use. By 1991 the system contained enough satellites for mariners to reliably determine their position nearly anywhere in the world. The final satellite was launched in 1994, completing the system.

Technically GPS only refers to the satellite-based radio-navigation system operated by the United States. Most U.S. citizens do not realize that other countries have global positioning systems. Russia has its GLONASS, which is incomplete as of 2008 and currently without applications for the recreational boater. China has the COMPASS navigation system and India has IRNSS. The European Union is developing the Galileo Positioning System, which is slated to provide civilian accuracy to within one meter. It is expected to be operational by late 2008. Unlike GPS, Galileo would not degrade signals for civilian or foreign government use.

Differential GPS

Although the decision to turn off selective availability (see Selective Availability) greatly improved GPS accuracy for civilian use, position-finding was still distorted by atmospheric effects. Differential GPS, called DGPS, is a network operated by the U.S. Coast Guard that improves the accuracy of GPS readings. Originally intended for commercial vessels, it is also widely used by recreational boaters. DGPS is particularly important for near-coastal navigation, broadcasting its data corrections up to 200 miles offshore.

DGPS improves position accuracy by compensating for known error differences. Using a network of land-based reference stations, each station compares its position using satellite GPS to its precisely-known land location. This local error correction is then broadcast and assimilated by DGPS-capable receivers. With accuracies of five meters (16 feet), DGPS provides great improvements for coastal mariners, allowing for meaningful calculations of a vessel's course and speed over the ground.

DGPS coverage now extends to all U.S. coastal waters including the Great Lakes, Alaska, Hawaii, and Puerto Rico. Many other countries, including Canada, broadcast a similar differential correction.

WAAS Corrections

Most contemporary GPS units are now WAAS-capable, meaning they are designed to incorporate a newer GPS correction system.

WAAS, short for *Wide Area Augmentation System*, was originally developed by the U.S. Federal Aviation Authority for aeronautical use, but is now popular among mariners. WAAS-capable GPS receivers have up to five times better accuracy than non-WAAS units, placing your vessel within three meters (ten feet) about 95% of the time. Unlike DGPS with its coastal coverage, WAAS spans North America.

WAAS works by supplementing the satellite GPS system with approximately 25 ground reference stations positioned across the United States. A master ground station on each coast collects data from the reference stations to create a GPS correction message, compensating for GPS satellite orbit, clock drift, and signal delays caused by the atmosphere and ionosphere.

Graphical Displays

The ability to know your precise position anywhere in the world at any time is an incredible feat. But what does N39.16.880 W076.36.090 really mean to you? The numbers alone do not impart a sense of place to most people unless associated with a chart or map. The numbers convey special meaning only when overlaid on a chart, placing those coordinates at Inner Harbor East Marina in Baltimore Harbor, Maryland, USA.

GPS position fixes have evolved from a list of abstract numbers, beginning with TDs, and then latitude/longitude read-outs, to graphical displays. Today, color LCD screens display your boat icon on a digital chart image. The error-prone step of transferring and plotting latitude and longitude

Google | GPS Loran Receiver

Combined GPS and Loran: *To stay abreast of the developing market for combined GPS and eLoran systems, Google "GPS Loran Receiver."*

It looks like the Loran-C system will continue operation, providing a backup for boaters with older Loran receivers. For the latest information on the status of Loran-C, visit www.navcen.uscg.gov/loran/LORAN_C_status.htm.

For an explanation of GPS waypoint and route transfer utility software, such as EasyGPS or GPSBabel, see **GPS Helper Utilities** in Chapter 8.

Multifunction Display: *Furuno's MaxSea Time Zero MFDs incorporate accelerated graphics technology, allowing data-intensive images to display, scale, zoom, and pan instantly.*

to a paper chart is bypassed by outputting GPS data directly to a chartplotter, multifunction display or personal computer.

Chartplotters

A chartplotter displays a chart, integrates with a GPS receiver to show boat position, and plots waypoints and routes. In addition, some models may display optional numeric course or simple instrument data if connected to devices such as a compass, autopilot, or depth sounder.

Chartplotters are sold by manufacturers such as Furuno, Garmin, Interphase, Lowrance, Raymarine, and Standard Horizon. Electronic chart data is typically loaded by inserting a data card of proprietary charts, such as those purchased from C-Map, Garmin, or Navionics.

Multifunction Displays

A multifunction display is a chartplotter on steroids. These high-end chartplotters are characterized by a networked configuration that allows the unit to simultaneously display multiple views of graphical information. A multifunction display (MFD) is intended as the command center of a network of marine instruments and sensors—your visual portal to all your marine data.

Multifunction displays include all the functions of a chartplotter: an electronic chart, boat position, and waypoints and routes. But a multifunction display can also display graphical images of any data received as part of a network. For example, an MFD has the capability to show color windows for 3D bathymetric charts, digital radar displays, fishfinder sonar views, sea surface temperature overlays, weather forecasts or weather fax, remote video cameras, and DVD or satellite television. Many MFDs can show these graphics in split windows, letting you simultaneously see a 3D chart and fishfinder sonar, or weather and digital radar on a single color screen.

The most popular multifunction displays are made by Furuno, Garmin, and Raymarine. Furuno sells MFDs under the brand name *NavNet*, consisting of a network connected with a simple Ethernet cable and integrating with its MaxSea-NavNet PC software. Garmin's multifunction displays include the 4000- and 5000-series units. Raymarine was one of the early leaders in multifunction displays beginning with its A-Series. In 2008, Raymarine launched its most recent successor, the G-Series.

Computer-Based Systems

Chartplotters and multifunction displays are both dedicated marine instruments, manufactured for the specific task of showing GPS position data on an electronic chart image. A multifunction display simply adds more processor and graphics power to handle diverse graphical information received over a network.

Displays, processors, graphics, networks—these terms sound a lot like the components of standard personal computers. Why not leverage the rapid technology improvements in personal computers for marine applications? The increasing popularity of PC-based charting is a natural outcome of the incredible advances in personal computing. The past ten years have brought revolutionary changes in consumer personal computers. They now sport large bright color screens, powerful graphic accelerators, better user interfaces, and increasing portability. A contemporary laptop has more computing power than an early Cray supercomputer and weighs only a few pounds. It's logical that a laptop would become a standard item in most boaters' gear bags.

Like a multifunction display, an onboard personal computer can display diverse types of information on an easy-to-read graphical interface. Charts, weather patterns, satellite images, aerial photos, and bathymetric data can all be displayed in full color and can be accessed quickly by flipping through layers of data windows.

Personal computers are also designed to be networked. In fact, the most sophisticated MFDs use Ethernet cables for networking, a plug-and-play standard borrowed from the computer industry. The networking configuration of personal computers allows them to serve either as the head of a network of marine instruments or simply as a mirror monitor to display an instrument's data.

FROM PAPER CHARTS TO DIGITAL FILES

In order to display a position fix on a digital chart, you need a library of electronic chart files. Unfortunately, the creation of every single electronic chart file is a huge undertaking, beginning with survey data and ending with an electronic database. In fact, the task of "charting the world" is so large a project it remains incomplete. Any world chart portfolio is an amalgam of data contributed by government hydrographic offices and private cartography companies.

Government Hydrographic Offices

The need for accurate nautical charts has always been driven by commercial and military needs. It's no surprise that the most important producers of worldwide charts are the governments of past or present superpowers: Great Britain, the former Soviet Union, and the United States.

Each national hydrographic office (HO) is responsible for conducting hydrographic surveys of its waters to ensure safety of navigation. These government agencies issue charts of national waters, extending coverage as needed for its shipping and strategic needs. The HO of the United States is the Office of Coast Survey within the National Oceanic and Atmospheric Administration (NOAA). The National Geospatial-Intelligence Agency (NGA) also maintains a worldwide chart suite.

The U.K., Russia, and the U.S. have been the leaders in global chart coverage. The British Admiralty, Britain's HO, developed an extensive chart portfolio to support its international maritime shipping. With coverage of former colonial areas in Asia and Africa, the British Admiralty remains one of the most important sources of worldwide raster charts. Similarly, the Russian Head Department Navigation and Oceanography (HDNO) maintains a near-global chart portfolio. The United States still produces charts for areas important during World War II, including coverage of some Pacific islands.

However, all these HOs remained islands of survey data and cartography, motivated by their own national commercial and military needs. No single government has the resources or motivation to chart the world. In fact, many lesser-developed countries (including most African countries and countries along the Mediterranean) don't maintain hydrographic offices, resulting in missing or unreliable charts for some regions. It took the entry of private companies—purchasing licenses for government and private survey data to produce *proprietary* charts—to pull together a near-global portfolio.

Proprietary Charts

A Russian company, Transas Marine, was the first entity to attempt world chart coverage. Known as "the granddaddy of nautical data," Transas consolidated nautical surveys from other agencies with the goal of a worldwide charting system. However, Transas was still motivated by commercial demands and their chart coverage reflected that economic bias.

Private companies in Europe and the U.S. also emerged to meet the demands for world cartography. Many of these companies are now familiar names in electronic charts, such as C-Map, Garmin, MapMedia, Maptech, Navionics, and Nobeltec.

These private companies generally do not conduct hydrographic surveys—it's simply too expensive. This task is largely left to government hydrographic offices. The Bahamas are a rare exception of private rather than government surveying. *Explorer Charts* of the Bahamas were surveyed and originally hand-drawn by Sara and Monty Lewis. *Wavey Line Charts*, also covering the Bahamas, were surveyed by sailors Bob Gascoine and Jane Minty.

Typically, private cartography companies obtain their data by purchasing government or private licenses for chart assets. For example, one of the most famous licensing agreements was Nobeltec's acquisition of the Transas data, resulting in Nobeltec Passport Charts. In the U.S., Maptech

How GPS Works

GPS uses a constellation of 24 Earth orbiting satellites, currently spaced in six orbital planes with four satellites each. Each satellite orbits the Earth every 12 hours. The orbits are arranged so that at least six satellites are always within line of sight from almost anywhere on Earth.

Within its *view*, each GPS satellite casts an equal-timed signal on the Earth's rounded surface. This signal defines a series of possible location points, called a *circle of position*.

Three satellites can provide an accurate two-dimensional electronic fix. Below, where three circles intersect is the vessel's position.

Most current GPS units receive signals from about a dozen satellites, selecting the four best-positioned to obtain a 3D fix. As the satellites and your boat move, the receiver monitors and substitutes better-positioned satellites. You can typically view a status window on your GPS to see the number of satellites in view with their identification, position, and signal strength.

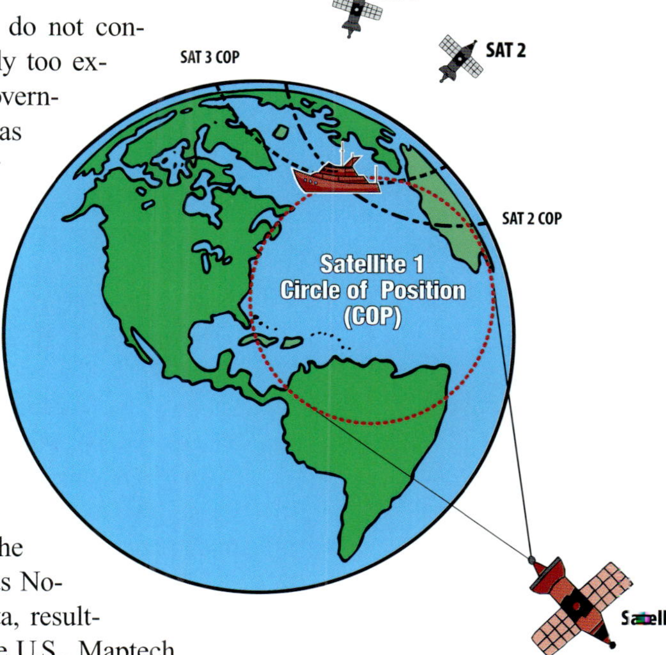

cut a deal with NOAA to be the exclusive producer of U.S. electronic raster charts. More recently, Explorer Charts were licensed for inclusion in Nobeltec Passport and C-Map charts. In 2008, EarthNC (a Google-Earth-based charting package) licensed the use of Wavey Line Bahamas charts. Proprietary chart companies continue to seek out new opportunities to buy the rights to incorporate new and updated chart data.

The fact that no single agency has charted the world, resulting in chart portfolios that are an amalgam of chart sources, accounts for the incomplete coverage and varying quality of today's nautical charts. In the end, each cartography company has high quality charts for some regions and weaker coverage in others. Depending on their licensing opportunities, a company may be forced to accept substandard charts of a region rather than suffer a gap in coverage. Each company's evolving chart portfolio depends on the original source of the data and the licensing deals they have won over time as they attempt to build a worldwide collection of the best possible charts.

The Story Behind Free U.S. Charts

The licensing agreement between the government agency NOAA and the private company Maptech had the greatest impact on U.S. cartography. In 1995, when NOAA originally introduced Raster Navigational Charts (RNCs), which are paper charts scanned and reproduced into digital files, they did so as part of a public-private *Cooperative Research and Development Agreement* (CRADA). After considering 37 bids, the winner was BSB Electronic Charts, a company in Bangor, Maine, who produced spiral-bound collections of paper charts called ChartKits.

If you think BSB Electronic Charts sounds a lot like Maptech, you're correct: BSB Electronic Charts was soon acquired by Maptech. The partnership between NOAA and Maptech made Maptech the only official source of NOAA Raster Navigational Charts. Maptech distributed the electronic files in a format known as "BSB," which remains a trademark of Maptech.

Although Maptech owned the license to work with NOAA, other companies circumvented Maptech's exclusive license by scanning NOAA's paper charts and creating files in other formats. For example, SoftChart sold CD-ROMs of charts in the GEO/NOS format, which was supported by early charting software such as The Capn and Nobeltec. SoftCharts were the alternative to Maptech's exclusive agreement with NOAA. When Maptech acquired SoftChart in 2006, the BSB format effectively became the U.S. raster chart standard.

Under the CRADA arrangement, electronic charts were available, but expensive. With the high price of admission, many recreational boaters used old charts year after year, creating a navigational safety concern. NOAA and Power Squadron sources estimated that most charts on pleasure vessels were an average of four years out of date. All NOAA's efforts to maintain and publish accurate charts were for naught unless boaters carried up-to-date editions.

In August of 2005, NOAA's agreement with Maptech expired. According to Tom Loeper, then of NOAA's Marine Chart Division, the agency initiated a new two-step strategy to get electronic charts into the hands of more boaters.

First, NOAA announced that its entire raster chart library, covering all U.S. waters with about 1000 charts, would be available as free downloads. Second, NOAA established a process whereby value-added distributors were authorized to package electronic chart files and make them available conveniently and affordably. Any Certified Raster Navigational Chart Distributor could sell chart sets, providing a service to boaters who do not want to download their own charts, are unsure which charts they need, or simply want the convenience of a CD-ROM format.

You don't need an advanced degree in economics to realize that when NOAA released its charts, prices for electronic cartography plummeted. Instead of hundreds of dollars per region, electronic chart catalogs for the entire U.S. could be purchased for under $50. Boaters who already used electronic charts could update their charts more frequently at lower cost. Best of all, the chart shake-out lowered the price of admission for boaters who had never tried electronic charting.

Unfortunately, the revolution of free charts only covers United States waters, including U.S. possessions around the world such as Puerto Rico, U.S. Virgin Islands, and assorted smaller Caribbean and Pacific Islands. Coverage also exists for Hawaii, Alaska, and the U.S. Great Lakes.

Popular cruising areas outside of the U.S.—including the Bahamas, Mexico, Canadian ports on the Great Lakes, or the Inside Passage to Alaska through British Columbia—are not covered. For cruising outside U.S. waters, electronic charts must be purchased through companies licensed by other countries' hydrographic offices. Unfortunately, each country produces charts in its own raster format, such as British Admiralty ARCS or Australia's Seafarer charts. There currently is no international file format standard for raster charts.

An International Vector Chart Standard

As electronic charts shifted from scanned paper charts (raster format files) to electronic chart data (vector format files), the advantage of an international standard became apparent. Vector charts were (and still are) the format used by commercial shipping, which is under the jurisdiction of the International Convention for the Safety of Life at Sea (SOLAS). SOLAS may be familiar to recreational boaters as the international agency that approves your flares.

All vector charts produced by national hydrographic offices now abide by an international standard, called "S-57," established by SOLAS. This common file format guarantees that compliant chart files will properly display and transition in all electronic charting systems.

However, each government has the option to encrypt these files, in effect requiring the purchase of a user's license to unlock the files. Unlike the U.S., encryption is the norm for other hydrographic offices. Currently the U.S. is the only major HO to issue its charts for free. Canada recently had the opportunity to follow the U.S. lead when its exclusive arrangement with the private company NDI ended in 2007. But Canada decided to continue to sell its charts, available through appointed licensed chart distributors.

The existence of a chart standard means that HOs can work together to create a common electronic chart database. For example, the U.S. is currently collaborating with Mexico to develop updated electronic charts of adjacent waters such as the Sea of Cortez. Although more than 20 countries have issued vector electronic charts of their waters, there are over 180 countries, so worldwide coverage remains spotty. But recent international cooperation, combined with the acceptance of electronic charts on commercial vessels, suggests a future with truly worldwide electronic chart coverage.

How Accurate Are Electronic Charts?

Despite their high-tech format, electronic charts are no more precise nor accurate than a chart printed on paper. Since electronic charts are the same data transcribed from ink to bytes, they are only as accurate as the paper charts from which they were created.

Extraordinary chart accuracy didn't matter until recently. Charts were used as representations of the approximate locations of hazards such as reefs or rocks. Mariners would note a charted hazard and give it a wide berth.

But with GPS providing fixes to within three meters—and placing an icon for position on an electronic chart—the accuracy of a chart becomes paramount. Yet a chart is only as ac-

Unlike a radar which displays real-world images, an electronic chart is only a representation using (likely dated) survey data. When planning routes, place waypoints away from charted aids to navigation. When piloting, drive the physical rather than virtual aids. Never assume the real world corresponds to what an electronic chart shows. Even if the GPS position is correct, chart accuracy may produce an error.

Ship on the Pier Problem: *Many factors contribute to an inaccurate representation of a vessel's position on an electronic chart. In Kentucky Lake's narrow, recreational Barkley Canal, it would not be unusual to see your boat's icon on land (red circle), 80 yards northwest of your route and "known" position.*

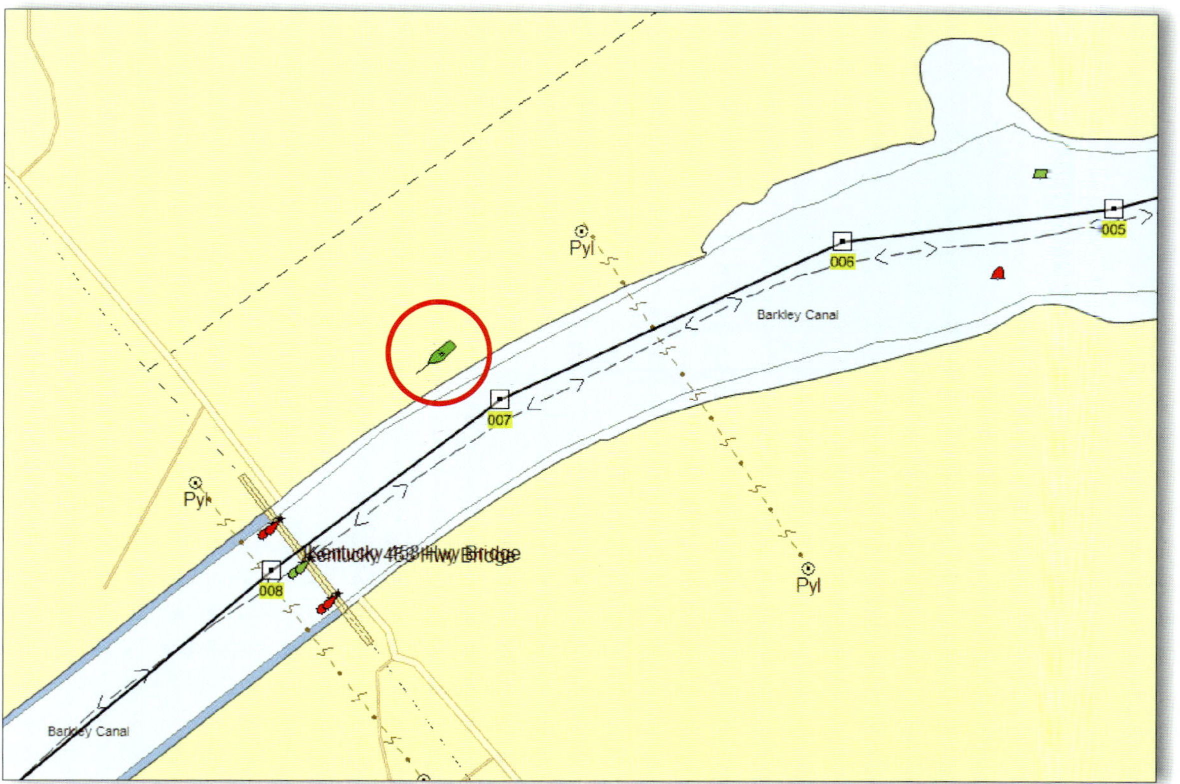

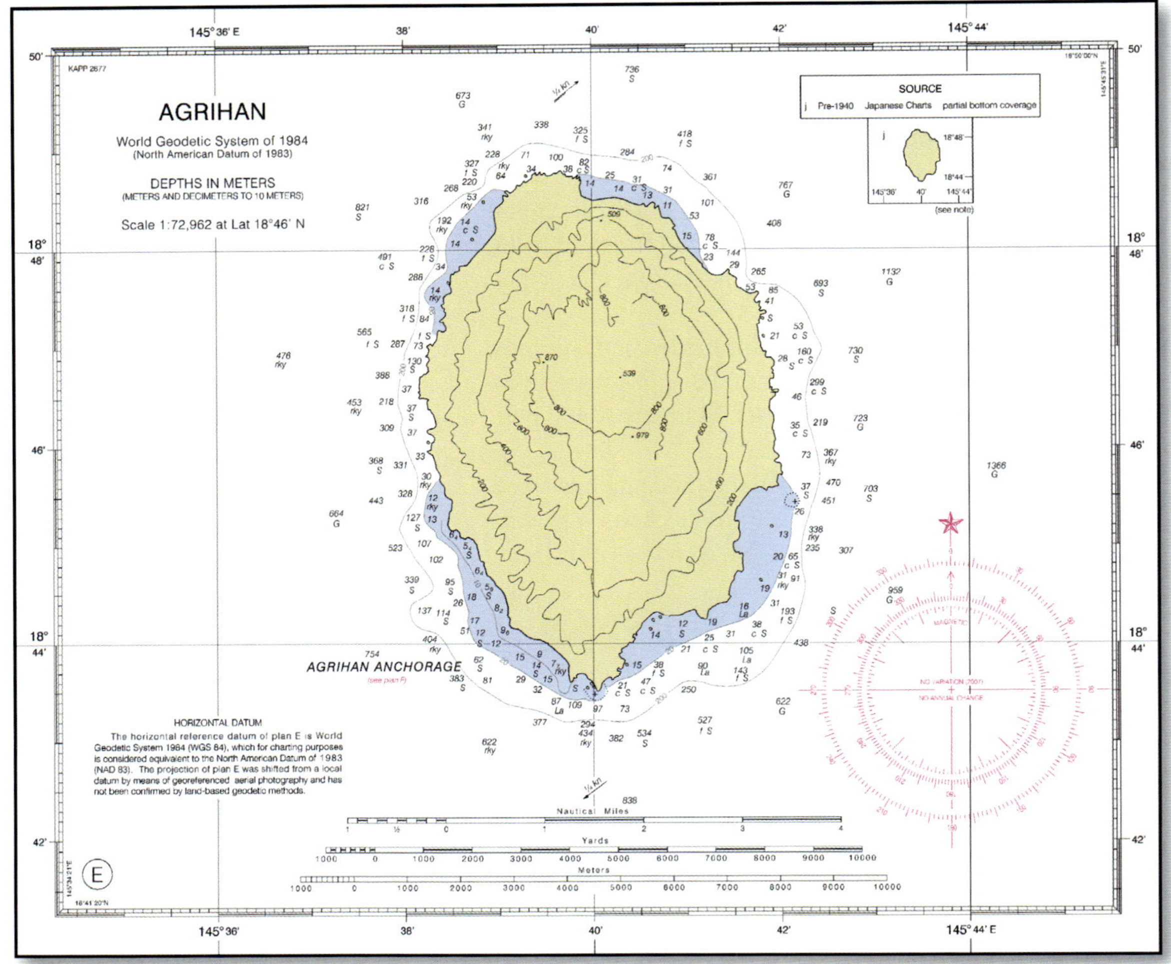

Chart Datum Matters

This Pacific island chart summarizes a navigator's datum fears. Although the datum is technically World Geodetic System 1984 (WGS-84), the survey data is from pre-1940 Japanese charts. The note also states the projection has been shifted by means of aerial photography and has not been confirmed. In all likelihood, this island is *not* where it is charted.

curate as its surveys, image resolution, and choice of a mapping reference system called a geodetic datum.

Chart Surveys

Never think that a chart issued in 2008 represents survey data collected in 2007. Nothing could be further from the truth. Because collecting accurate chart information still typically involves a steel ship with surveying equipment, most charts are produced using historical data supplemented with local updates for high-priority areas.

Faced with an overwhelming task and limited resources, government hydrographic offices are forced to use whatever historical survey data they already own. In some cases, this means survey information originally conducted with lead lines and horizontal sextant angles.

For example, some U.S. charts are based on survey data collected during World War II. Much of Australia's cartography is based on surveys from the beginning of the 19th century—in some cases last surveyed by Captain James Cook using lead lines for depth readings and celestial observations for position. Concerned about this extremely dated information, the Australian HO began a project in the 1990s to update all its marine surveys. Similarly, Norway's charts on its entrance to the North Sea were based on surveys from the early 1900s. Because the entrance is deep with low risk of grounding, it was not a priority in the larger task of surveying Norway's complicated 1,200-mile coastline.

Once an HO collects survey data and issues a paper chart, it still must be converted to an electronic chart file. Paper to digital is relatively straightforward in the case of raster chart files, which are simply computer-scanned images of a paper document. Because of the relative ease of this task, the U.S. promptly issued complete raster coverage for all its waters.

However, converting a paper chart to a vector chart file is much more difficult. Creating a vector chart requires a vector database (see Chapter 6, *Charts and Supplemental Data*), an extremely labor-intensive project. Each HO manages thousands of charts and is hampered by the simultaneous demands of maintaining and updating their existing chart library. Alexandra Heliotis, Deputy Chief of NOAA's Marine Chart Division, points out that an average year produces 5,900 individual updates that result in corrections to over 16,000 chart products.

Because time and resources are limited, the priority for conversion is driven by commercial ship traffic. In 2001, NOAA initially identified 40 major ports for vector chart coverage by 2003. When these were completed, another 100-plus ports were identified for conversion. NOAA's priorities began with major ports, followed by the seaward connections between those ports for commercial shipping traffic. For example, the Intracoastal Waterway between the ports of Norfolk and Charleston is not a priority, whereas the seaward connection between these two ports, including Cape Henry and Cape Hatteras, is a priority.

Currently about 60 percent of NOAA's projected chart cells are available in vector format, or about 600 of its roughly 1000 projected vector charts.

Chart Resolution

Even if the survey data is up-to-date, computers tempt more accuracy than is warranted. Consider a bearing or fix done by pencil on a paper chart. If your fix came down to the width of a pencil line, you'd be proud of your accuracy. But a pencil line on a 1:80,000 scale chart is 320 yards—over one-sixth of a mile. Your laptop may display several decimal places and millions of pixels, but the original chart files do not.

Nowhere is this significant digit problem more apparent than with GPS. A WAAS-enabled GPS receiver shows your position to within ten feet—one hundred times more accurate than the width of your pencil line on the paper chart! Although a GPS position typically displays minutes up to three decimal places, only the second decimal place (representing a granularity of 60 feet) really has meaning in practice.

GPS technology has leapfrogged electronic chart accuracy. In the past, a fix using visual bearings, radar ranges, or Loran gave an approximate triangle of position. This position estimate was good enough, particularly given the survey accuracy of the charts. Today, although you can get a precise GPS fix of your vessel's position, that position may be plotted on charts with outdated hydrographic survey data.

As a recreational boater, this discrepancy of accuracy between the chart image and your GPS fix can result in some strange screen displays. Boaters who zoom in and expect to see their vessel in more detail on their screen may experience what is known as the *ship-on-the-pier problem*: your vessel icon is placed on land, but you're in plenty of water.

Even a well-charted and recently-surveyed area such as Boston Harbor may not have the accuracy to display your vessel represented correctly alongside a pier. The discrepancy is even more apparent in narrow winding channels such as the Intracoastal Waterway through Georgia's marshes. A relatively small mismatch of 300 feet in those narrow channels easily places your boat icon on brown land rather than

blue channel. Nothing is wrong with your GPS, your electronic cartography, or your software—it's simply the discrepancy of accuracies across paper and electronic formats.

The World is Not Round

Columbus was wrong, the Earth is not round. That doesn't mean the Earth is flat, but it isn't a round ball either. It's actually a *rotational ellipsoid*. Because of its rotation and the effect of centrifugal force, it has a slightly larger diameter at the equator than its pole-to-pole axis. Actually, it's even worse: the Earth is also uneven due to an irregular gravitational field.

The Earth's shape matters for electronic charting because the chart image (whether on paper or computer screen) is flat. A GPS is trying to project a position on something that isn't perfectly round. All this wreaks havoc for geodesists, the people who need to relate all measurements to a coordinate system so the location can be entered on a chart at the correct position.

They accomplish this mapping using a reference system called a *geodetic datum*, which is the choice of an ellipsoid and its center point in relation to the center point of the Earth. In other words, this datum is the reference system or conversion scheme from numbers, such as GPS data, to a chart.

With today's impressive accuracy of GPS, the choice of a datum really matters. Unfortunately, originally each country established its own datum, eventually spawning over 100 different standards for charting. In 1950, Europe coordinated to use the same datum system.

Today most of the world uses the World Geodetic System (WGS-84). Russia still uses its own datum PE90 for GLONASS (its GPS equivalent), but PE90 and WGS-84 only differ by a few feet. North America also uses NAD-83, which is the same as WGS-84 for all practical purposes. Before WGS-84, positions using different country's datum systems could differ by over a mile.

Moore's Law in Practice: *Not that long ago, global position finding was a "certain times of the day, some places on the planet, you were roughly here" experience. Today, a $50 20-channel WAAS-enabled GPS sensor fits in the palm of your hand and tells you where you are on the planet, anytime, typically within ten feet.*

If you use older charts (such as with NAD-27 datum) or charts with foreign chart datums, you may experience *significant* errors (from hundreds of feet, up to a mile).

Although it's interesting to know what "WGS-84" means, the take-away message for recreational boaters relates to the match between GPS accuracy and the digital chart image. An old datum means your GPS position may not map correctly onto the charted image.

THE SOFTWARE PIONEERS

Government agencies led the conversion of paper charts to digital chart files, but private developers had to step in to design user-friendly ways for these files to be displayed on a personal computer. Despite the rapid development of personal computers though the 1970s and 1980s, no mainstream software was specifically designed for mariners. On both sides of the Atlantic, boaters with a technological bent had ideas on how to merge the new electronic charts with the rising phenomena of personal computers. MacSea and CAPN were the two pioneering software applications for e-charting.

MacSea

Over twenty years ago, a group of engineers in France realized they could bring a computer onto a boat if they could re-engineer the system to run on 12 volts. They began with hacksaws, soldering irons, and a Macintosh computer.

At that time, an Apple Macintosh was the best choice for do-it-yourself engineers, particularly those contemplating a personal computer on a boat. Macs generated their own monitor video frequency, allowing easier conversion from AC to DC power. They also used rigid plastic-encased 3.5" diskettes instead of the moisture-vulnerable 5.25" floppy disks of PCs.

The system was named *MacSea* after this first onboard Macintosh computer. The founder and current CEO of MaxSea (as the company is now known), Brice Pryszo, used this Mac in 1984 as navigator during the Quebec Saint-Malo Transatlantic Race.

Today MaxSea, recently purchased by Furuno, still produces some of the most sophisticated e-charting software in the industry. (Chapter 24 contains an in-depth review of the latest version of MaxSea.) With an official software release in 1985, MaxSea is technically the oldest e-charting software company and application.

CAPN

Meanwhile, on the other side of the Atlantic, Dennis Mills, a boater and ex-journalist living on the coast of Maine, wanted to create a computer program to help mariners calculate and display lines of position from celestial sights.

Although most people didn't own a personal computer, Mills foresaw a time when it could be a powerful calculator for navigation. His project began as a computerized version of the *American Practical Navigator*, so he called it CAPN, short for Computerized American Practical Navigator.

At the same time, NOAA was beginning to experiment with electronic versions of their paper charts. With charts now available in electronic format, Mills added a revolutionary feature to his CAPN program. He included the ability to plot a vessel's position directly on an electronic chart display. His company, Nautical Technologies, was one of the first companies to take advantage of NOAA's electronic charts. Now called The Capn and sold by Maptech, the original CAPN was the first commercial Windows-based charting and navigation program. (Chapter 20 contains an in-depth review of The Capn.)

Part Two

E-Charting System Components

Before you can understand a system, you need to understand its component parts. Electronic charting relies on an integrated *system* of components. Like any system, each component must be sufficient to its task and able to seamlessly integrate and interact with each other. In this sense, the computer must be capable of handling large graphical chart files, instruments and sensors must be able to communicate with the computer, and the software must work with electronic charts, supplemental data, and instrument input.

Chapter 3, *Four Components of an E-Charting System*, overviews the four "in-the-box" components: hardware, instruments and sensors, chart database, and application software.

Chapter 4, *E-Charting Hardware*, looks at the computing component of an e-charting system: the CPU, display, operating system, input devices, drives, and ports. This chapter also deals with practical questions such as the speed or memory needed for e-charting, the issues of powering a PC, and how to get onboard Internet access.

Chapter 5, *Instruments and Sensors*, outlines many of the marine devices you can connect to an electronic charting system, ranging from a basic GPS sensor to advanced collision avoidance systems such as AIS.

Chapter 6, *Charts and Supplemental Data*, delves into the nitty-gritty of electronic chart formats: the acronyms of RNCs, ENCs, DNCs, BSBs, and so on. This chapter explains the difference between raster and vector format charts and why it matters. We explain how to obtain free charts of U.S. waters and where to source international charts. We also overview the optional supplemental data included with private cartography, such as tide and current predictions, Coast Pilots, satellite images, aerial photos, topographic maps, and bathymetric data.

Chapter 7, *E-Charting Software*, explains the different types of software available for e-charting on a personal computer. This chapter compares the different software options at a high level in terms of ease of getting started, user interface, and basic and advanced features.

Chapter 3
Four Components of an E-Charting System

Electronic charts are rapidly replacing paper charts, the mainstay of marine navigation for centuries. In fact, use of electronic charts on commercial vessels, including the international shipping industry, is now widespread.

Of course, international regulations on safe use had to be put into place before digital cartography could be widely adopted. Beginning in 1996, the International Convention for the Safety of Life at Sea (SOLAS) became steward for electronic chart standards in commercial shipping. Specifically, SOLAS regulates what constitutes an approved "Electronic Chart Display and Information System," also called an ECDIS (pronounced eck-diss).

Among other requirements, an approved ECDIS must use IMO-sanctioned charts. In addition, it must be able to query the chart's objects to retrieve information. This information is used to perform important safety functions in navigation, such as monitoring courses or sounding warning alarms. In effect, these requirements dictate the use of internationally-approved vector format charts, such as the S-57 and S-63 charts produced by national hydrographic offices.

Recreational boaters are not under the jurisdiction of SOLAS regarding the use of electronic charts, giving them flexibility in both system configuration and chart formats. To capture this "official-unofficial" distinction, a non-SOLAS approved e-charting system, such as might be found on a recreational boat, is generically called an "Electronic Charting System" (ECS) versus an "ECDIS.".

An ECS is an electronic charting system that does *not* meet official IMO standards for a commercial ECDIS. For example, a recreational ECS would fail the ECDIS standard if it only displayed raster charts or had limited functionality in terms of piloting and navigation. On the other hand, some full-featured e-charting packages, when loaded with official vector cartography, satisfy the requirements of an ECDIS.

Regardless, an electronic charting system, whether for a large merchant ship or Jon boat, has four components: hardware, external sensors, electronic charts, and navigational software. This chapter provides an overview of ECS components before detailing each in the following four chapters.

HARDWARE

Although electronic charting systems have taken different forms over the past twenty years—and will undoubtedly incorporate new and unforeseen technology in the future—the fundamental hardware components remain the same. Whether a chartplotter or personal computer, there must always be a central processing unit, a display, a human interface, and methods for data input and output. These components may change in size, weight, or specifications across systems or over time, but each function is necessary to complete the hardware requirements of an ECS. Chapter 4, *E-Charting Hardware*, covers the specifics of choosing a computer and peripherals for e-charting.

Most importantly, an ECS must have a central processing unit (CPU). The CPU is the brain of any computer system. The CPU processes tasks at the system level and manages memory and data storage. For example, a chartplotter provides a small built-in CPU with the ability to process commands and store data. Personal computers have more powerful CPUs, distinguished by their processor type (e.g., "Intel Inside"), processor speed, and memory demands.

An ECS must also include a screen display to show charts, menus, and information panes. The screen may be large or small, monochromatic or color. A display may be integrated with the CPU, as with a laptop or chartplotter, or may be freestanding such as a desktop monitor or dedicated helm-mount-

Chapter Terms

CPU: Central Processing Unit. The "brains" of the computer that executes programs. Also called the processor.

ECDIS: Electronic Chart Display and Information System. An electronic charting system satisfying international standards for commercial navigation.

ECS: Electronic Charting System. A generic term for a system comprised of hardware devices integrating a personal computer or dedicated chartplotter with navigation software, chart data files, and additional marine electronics.

Graphical User Interface (GUI): The addition of icons and images, complimenting text, allowing a person means to interact with a computer.

International Maritime Organization (IMO): The specialized agency of the United Nations responsible for measures to improve the safety of international shipping and to prevent marine pollution from ships.

LAN: Local Area Network. A computer network serving a small area, such as a home, office, school, or boat.

Software Application: A computer program that uses the capabilities of a computer to perform certain tasks.

Chapter Questions

- *What is the difference between an ECDIS and an ECS?*

- *What components comprise an Electronic Charting System?*

- *What are some of the "gotchas" when getting started with e-charting?*

ed marinized multi-function display.

An ECS also requires a human interface, namely a way for the user to interact with the device through software applications to perform tasks. This is commonly know as a GUI (pronounced goo-ee) or Graphical User Interface. Decades ago, the human-computer interface consisted of connecting jumper wires or entering punch cards. Today's interfaces are closely coupled with the graphic display. For example, chartplotters typically include buttons on the device that interface with the display. These context-sensitive "softkeys" provide more command flexibility with fewer hardware buttons. Personal computers use a keyboard and mouse as input devices to enter text- or cursor-driven commands. More recently, both

Computer Components: *A computer includes a* Ⓐ *CPU,* Ⓑ *display,* Ⓒ *graphical user interface, and an assortment of input and output devices (such as a* Ⓓ *mouse and keyboard).*

chartplotters and personal computers are incorporating touch screens and voice-activated commands.

Finally, the hardware of an ECS requires a method for data input and output (I/O)—both to receive chart data and to communicate with other electronic instruments. Data exchange technology has also changed radically over the past few decades, from floppy disks to today's DVDs, flash cards, and Ethernet networks. A DVD can store 4.7 GB of data on a tough, lightweight, relatively waterproof disc. A flash card (in CF, SD, or MM format) can store the same amount of data on a tiny memory chip. Chart companies now sell mega-collections of charts and supplemental data, storing everything on a few DVDs or flash cards. Chartplotters nearly always read charts by inserting a card into the chartplotter. A computer-based ECS can read chart files either by DVD (or CD) or by card. DVDs or CDs are read by the computer's disc drive. Cards may be read using a slot-enabled card reader or an external multimedia card reader connected to the computer's USB port.

An ECS communicates with other devices, ranging from a keyboard to a GPS sensor, through physical cables, or more recently, through wireless connections. Any data exchange requires a compatible *physical* (often called *electronic*) connection and a data communication standard. The most important data standard for ECS instrument connectivity is NMEA 0183 (or the newer NMEA 2000), allowing an ECS to communicate with any NMEA-compatible marine instrument.

Common physical connections include Ethernet, FireWire, USB, or serial connectors. Depending on the manufacturer and model, all PCs include some combination and number of communication connections. Cable adapters are available to accommodate all sorts of mismatched options. USB ports are now the most common universal connection on personal computers, replacing older serial (PC) and FireWire (Mac) ports. Many marine sensors now come in USB versions, such as the USGlobalsat and Garmin GPS sensors.

Using cables or wireless interfaces, an ECS can also be created over a network, whether it is your own local area network (LAN) or a proprietary marine manufacturer-based system. A network of components, such as processors, laptops, standard or marinized monitors, keyboards, and wireless routers can also satisfy the hardware requirements of an ECS. Wireless data exchange often occurs over a Bluetooth network, such as between a wireless mouse or keyboard and a computer or a wireless marine instrument and a computer.

INSTRUMENTS AND SENSORS

The real power of an electronic charting system comes when you link data from your boat to your laptop. To do this, you need real-time sensor input. As with anything electronic, there is "Buzz Lightyear" potential here—including external sensors for sonar, radar, or even video cameras to monitor your engine from the helm.

However, the most common and useful external device is a GPS sensor. These small devices, which retail for less than $100, connect to your computer through its USB port. With a GPS sensor connected to your computer, you watch your boat's icon glide across the chart.

A slew of other external sensors or marine instruments can be connected to an electronic charting system. For example, an autopilot or an Automatic Identification System (AIS) receiver is common. Additionally, a wind instrument, depth sounder, water temperature gauge, radar, heading sensor, satellite radio, weather receiver, or video camera can all be potentially linked to your ECS. Each of these sensors is covered in detail in Chapter 5, *Instruments and Sensors*.

CHART DATABASE

An electronic charting system must have a database of charts. These chart files come in many formats—which unfortunately do not look or behave alike. In addition, e-charting software applications may only be able to display a small selection of the superset of chart formats.

Just as photo image files can be stored in many formats, such as TIFF or JPEG files, chart images are stored in many formats. Some of these formats may sound familiar to you, such as BSB, SoftChart, or S-57.

However, unlike photo files, which can be re-saved in another format (such as saving a TIFF file as a JPEG), chart file formats are integral to how the files were created. Electronic charts are created in a format and largely speaking—except for proprietary conversions—they stay in that format.

Also unlike photo files, the format of chart files has huge ramifications on the chart display and chart-related features. For example, a JPEG and a TIFF photo file look the same on your computer screen. Not so with electronic charts. BSB, SoftChart, S-57, and Navionics charts each display with different chart shading, coloring, line quality, and detail levels.

More importantly, different chart formats support very different e-charting functions. For now, the important point is that the format of a chart file affects how the chart looks on the screen, how it works, and what one can do with it. This topic is covered in detail in Chapter 6, *Charts and Supplemental Data*.

APPLICATION SOFTWARE

In order to view an electronic chart, your computer must be loaded with a charting and navigation application. It seems obvious, but a surprising number of new e-charters double-click on a chart file icon and are disappointed when the file fails to miraculously open and display a chart.

Opening and viewing a chart file is similar to opening an email attachment. You can't open an XLS spreadsheet file without Microsoft Excel (or a similar spreadsheet application) loaded on your computer. You can't open a PDF file without Adobe's Acrobat Reader. And you can't open a DOC file without Microsoft Word. The same logic applies to an elec-

It's important that the dongle is not inserted into the USB port prior to software installation. Wait until the software installer prompts you to connect the dongle.

Hardware Dongle: *One method of thwarting software piracy is a hardware key. An advantage is the ability to load the software on multiple computers and move the dongle from computer to computer.*

What's a Dongle?

All contemporary computer applications now incorporate user licenses to prevent *software piracy* (unauthorized duplication of the software or cartography).

A user license is administered in the form of an activation key. This key is an alphanumeric code distributed with the software, or by registering over the Internet.

E-charting software vendors also typically use this common method. However, in two cases (Nobeltec and MaxSea), the user license is administered in the form of a hardware key, commonly called a *dongle*.

Arguably, dongles are a safer and more flexible form of hardware activation in an e-charting environment. A dongle enables the software to be installed on multiple computers for back-up or planning purposes. Furthermore, a dongle allows for restarts without contacting the manufacturer or depending on an Internet connection in the event of a catastrophic computer crash at sea.

Detractors point out that dongles are an inconvenience and a single point of failure.

Instrument Connectivity

Today, many boaters use e-charting for planning purposes only. A growing number are choosing to pilot their vessel with the simple addition of a GPS sensor. In fact, all onboard electronics may be interfaced, including:

- GPS sensor
- Autopilot
- AIS receiver/transponder
- Wind instrument
- Depth sounder
- Water temperature sensor
- Radar
- Heading sensor
- Video camera
- Engine sensor
- Bathymetric recorder
- Satellite radio
- Weather receiver

If you don't already own an e-charting application, download a trial version from the manufacturer's website. This *demoware* lets you open chart files and try out the software's e-charting features. Trial software either limits functionality (such as restricting the number of waypoints or routes) or expires in a given time (such as a 30-day trial period).

tronic chart: you can't open a BSB file without a charting and navigation application that supports the BSB chart file format.

A software application is necessary to control the data management and presentation of the chart files. The software sports its own graphical user interface, which "paints" the screen you see and determines how you interact with the program. More important, an e-charting application not only allows you to open and view a chart file, it is the source of all navigation functions, such as creating waypoints, saving routes, or setting alarms. Chapter 7, *E-Charting Software*, covers the three types of e-charting software packages (viewers, planners, and full-featured applications).

Scores of charting and navigation applications are available through retailers or on the Web. Appendix B lists nearly 30 e-charting vendors, many of whom produce several software packages at different price and feature levels. In Part Four, *Choosing an Application*, we review the 16 most popular e-charting applications.

SOME GETTING STARTED POINTERS

As boaters set out upon the e-charting waters, there absolutely is a list of "gotchas"—problems that consistently rear their ugly heads. The following are lessons we've learned helping others get started with electronic charting:

- Start with a decent computer! This means better than average CPU speed and memory. Can your laptop handle large photos and action games? If the answer is no, it won't be able to display and manipulate large chart images.
- Don't purchase a DVD of charts if your computer only has a CD drive! Many older models only have a CD drive. Check this before ordering charts.
- Always *copy* the charts and chart folders to your hard drive first. Then *load* them into the charting and navigation application. We've witnessed many software applications crash while attempting to load this much data from a CD or DVD drive.
- Although hard disk space is seemingly infinite, only copy charts or chart regions you intend to use in the short term. This prevents filling up your hard drive. If you later re-

quire another chart or region, simply copy it from the CD or DVD to your hard drive and load according to your application's instructions.

- You cannot simply click on the chart data files and have them open! You must have a charting and navigation application active and the charts properly loaded.
- Every charting and navigation application loads or "points" to the charts differently. Follow your software manufacturer's instructions *precisely*.
- Although file naming and structure conventions may seem arcane, never rename or nest chart folders. Charts can easily become "lost" to the application.
- Don't load only raster BSB or KAP files thinking you are saving space. They are a pair. You need both to display all chart features properly.

Chapter 4
E-Charting Hardware

An important advantage of computer-based electronic charting is that you probably already own the hardware. Most boaters (after all, you're able to afford a boat!) own a laptop—either as their sole computer or as a supplement for travel. In fact, it's expected that over half of American households will own a laptop by 2010.

Laptops are already routinely brought aboard, typically for email, Internet access, or camera downloads. Today's laptops have more than enough oomph to handle these less demanding tasks while also serving as the hub of an e-charting system. Larger vessels may opt for an onboard desktop computer, optionally combined with laptop docking stations. If you don't own a laptop—or prefer not to bring it aboard—desktop computers work well for planning on shore.

This chapter details some hardware options to consider as you choose whether to bring your existing laptop aboard, upgrade its specifications or operating system, or confine your e-charting to your desktop computer.

COMPUTER SPECIFICATIONS

It's no secret that displaying and manipulating chart files takes lots of memory. Understandably, most boaters also want to stockpile many charts on their hard drive, leading them to worry that their existing computer might not handle the demands of e-charting.

Actually, your computer's memory constraints arise less from the size of chart files than from the manipulation of chart images. Admit it, you expect to zoom, pan, rotate, and switch charts without delay or image blurs. In effect, you're asking for instantaneous dynamic image display—much like a computer gamer. So the rule-of-thumb on e-charting and a personal computer is: Any personal computer sufficient for graphic-intensive action games can display and manipulate large chart images. Alternatively, if your personal computer can display and manipulate large photos (such as in Photoshop), it should be sufficient for e-charting.

The CPU

No one likes waiting for a file to open; we expect charts to open promptly without experiencing hourglasses, flying sheets of paper, or spinning beach ball icons. Not surprisingly, the specifications of a personal computer do impact the performance of an e-charting program. Although MacENC, the most popular Mac e-charting application, works fine using a Macintosh PowerPC, charts display and pan noticeably faster on the newer Intel Macs. E-charting applications that include 3D features, such as Fugawi Marine ENC or Furuno MaxSea Time Zero, work best with a graphics accelerator video card.

Nearly all full-featured e-charting software applications specify a personal computer with an Intel Pentium processor (or the equivalent for Macs). Although these processors are standard in new personal computers, they are sold at different speeds. Processor speed determines how fast the computer can perform calculations. Speed impacts panning the charts, keeping track of instrument data, and updating boat position.

Although faster processors are obviously better for e-charting, many of the lower- or mid-priced e-charting applications suffice with a 500 MHz processor. More full-featured applications, such as RayTech RNS or Furuno MaxSea, recommend a 2 GHz processor. Nobeltec Admiral recommends a 3 GHz Pentium 4 or better processor.

Memory also affects your computer's performance. When dealing with bulky chart files, more memory is always a good thing. There are two kinds of memory: random access memory (RAM), which can be thought of as the short-term memory used by the processor; and "available hard disk space," which can be thought of as long-term storage and memory.

Chapter Terms

Air Card: A hardware device that connects a computer to a cellular network for Internet access.

Cellular Network: A system of base stations providing mobile telephone support.

Operating System (OS): The software component of a computer system that is responsible for the management and coordination of activities and the sharing of the resources of the computer. Examples include Windows XP, Vista, OS X, and Linux.

Rugged Laptop: Portable computers designed for harsher environmental conditions including temperature change, impact, and humidity.

WiFi: Wireless Fidelity. A group of technical standards enabling the transmission of data over wireless networks.

WiFi Card: A hardware device that connects a computer to a wireless network.

Chapter Questions

- *Is my computer sufficient for e-charting?*

- *Which computer specifications really matter for e-charting?*

- *Will my operating system run my e-charting software—even Windows Vista?*

- *What are some options for onboard Internet access?*

- *How do I manage onboard computer power use and charging?*

- *What are the important trends in e-charting hardware?*

Don't anthropomorphize computers; they hate it.

Rugged Computer: *Several computer manufacturers produce specialized laptops for harsh environmental conditions such as high temperatures, humidity, or impact. This ARMOR C12 operates from -20ºC to 50ºC, is shock and vibration protected, and boasts sealed ports to eliminate water or dust intrusion.*

E-charting applications differ widely in the amount of memory they expect or need. At one extreme, SeaClear II specifies a paltry 64 MB of RAM and 10 MB of available hard disk space. Specifications like these read like a Tandy ad from the 1980s.

Most e-charting applications specify modest memory needs for contemporary personal computers. For example, 512 MB of RAM is the norm. The amount of recommended available hard disk space varies more widely across applications. Some only specify the space required for the software code; others include recommended hard disk space for chart folders. In general, about 250 MB of available disk space is minimal for most e-charting applications, recognizing that more hard disk space is always better.

As with processor requirements, more expensive and full-featured packages from Raymarine, Nobeltec, and Furuno demand more memory. These packages expect RAM and hard disk space measured in gigabytes rather than megabytes. A standard recommendation in this class would include at least 1 GB of RAM and between 10 GB and 40 GB of available hard disk space. These higher-end packages not only require more space for their full-featured code, they are designed for long-distance cruisers who require an extensive library of electronic chart and supplemental data files.

These CPU specifications should not scare you. Most likely, any recently-purchased consumer laptop or desktop is sufficient for e-charting. However, if you intend to use one of the higher-end e-charting applications, be sure to invest in faster processor and higher memory options.

If you'd like to e-chart on an older laptop or desktop, look to the freeware SeaClear II or one of the applications with less computing demands, such as TIKI Navigator Pro, NavSim BoatCruiser, or DigiBOAT Software-On-Board.

Ruggedize It?

Several companies have recently introduced *rugged* laptops, such as the ARMOR C12, Dell Latitude, Itronix GoBook, Motorola Rugged Notebook, and Panasonic Toughbook. These models are specially designed to handle tougher conditions such as drops, vibration, spills, humidity, and extreme temperature changes.

In addition to a tougher casing and less movement-sensitive drives, ruggedized models also typically adjust their screen features. For example, the Panasonic Toughbook has a brighter screen for outdoor readability. The Dell Latitude ATG includes a red task light to illuminate the keyboard at night. Some models add other outdoor-oriented features, such as the Rugged Notebook's hard drive heater for starting up in freezing temperatures.

Realistically, most of us recreational boaters aren't that rugged ourselves. It's analogous to buying a Land Rover to drive to the grocery store. Think hard about whether you really need a laptop that can sit out in the rain or can boot up when there's ice on the deck. Before you decide that a ruggedized laptop is necessary to e-chart, consider your likely use and the price difference.

A ruggedized laptop is generally lower performance, significantly heavier, and more expensive. For example, purchasing a Dell Latitude (their rugged version) adds $1000 over the equivalent standard Dell laptop. In addition, some of the high-end rugged features, such as heaters or brighter screens, use more battery power (see Powering Your PC).

Most mainstream laptops—designed for the harried and oftentimes careless business traveler—are engineered for a fair amount of abuse. They can handle moderate vibration, tropical humidity, and normal bumps. Most boaters have little trouble with today's laptops, provided they are used in an enclosed saloon or helm station. Even a rugged laptop does not take kindly to salt spray.

To care for your onboard personal computer, consider these three vulnerabilities: temperature differences, pounding movement, and salinity.

For example, never start up the computer if its temperature differs greatly from the environmental temperature. Never start up a warm laptop in a cold cockpit, or conversely, a cold laptop from outdoors brought into your overheated saloon. Leave the laptop off until it reaches the temperature of its surroundings. Moderating your laptop's temperature changes will minimize internal moisture condensation.

If sea conditions are exceptionally rough, cushion the CPU. This can be done permanently, such as with a cushioned mount installation, or as simply as placing the laptop on a cushion down below. Even coastal waters, such as the wild rides through Cape Cod Canal, Baltimore's Inner Harbor, or the standing waves of the Great Lakes, can cause your boat and your laptop hard drive to pound. We once had to temporarily place our computers on a berth mattress until we passed through the worst of the waves bucking a strong current.

Laptops also do not like moisture. Although the normal humidity of the subtropics or tropics is fine, any spill, rain, or spray—especially if salty—is disastrous. Keep the laptop away from any sudden moisture source and consider investing in a water-resistant, drop-protection laptop case. We use several products from OtterBox to protect our laptops, digital phones, and PDAs (www.otterbox.com). At less than $200, a waterproof case is much less expensive than a rugged laptop.

The Display

Despite the heavy graphics of e-charting applications, a standard color computer monitor is sufficient. A display monitor may be part of an integrated unit, such as a laptop or unibody desktop, or may be added as a peripheral device. A typical computer monitor—as specified by most e-charting applications—provides 1024 x 768 resolution with 32-bit color. Unless your monitor is exceptionally old, it should fit the bill.

However, if your e-charting needs call for 3D displays of the seafloor or rotating 3D topographic maps, you've entered the realm of interactive games. Mainstream computer companies sell to this gaming market, and the graphics tools needed for 3D action games also work excellently for rotating chart images.

Several e-charting applications, most notably Furuno MaxSea and Fugawi Marine ENC, include excellent 3D display capabilities, and therefore require the hardware to support these features. For example, MaxSea requires a DirectX9 compatible video card (the same card used for gaming) and Fugawi Marine ENC requires a graphics accelerator for its 3D viewing features.

Most boaters rightly worry more about their CPU specifications than their monitor specifications. However, bringing a laptop onboard does entail an important monitor issue: screen brightness.

Laptop screens vary in their level of brightness, sometimes compromising screen brightness in order to manage battery needs. In fairness, standard laptops are designed for office or home use—very different lighting conditions than bright outdoor sunlight. In contrast, a chartplotter is specifically designed for varying lighting conditions, ranging from bright sunlight to nighttime viewing. Some chartplotters have a *transreflective* display for bright light, where the display becomes brighter as the light gets brighter.

Nighttime viewing is less of a problem. Nearly all full-featured e-charting software have different display options, such as for night, twilight, and daytime. By altering chart colors and intensity, these options help—but are constrained at the upper limit by the brightness of your screen.

Fortunately, personal computers continue to improve screen technology. As battery life improves, more power can be allocated for a brighter screen. Manufacturers recognize the need to respond to outdoor laptop use. Today's laptop users consider the outdoors—whether at a sidewalk coffee shop or the lawn of a college quad—as a routine setting for computing.

In the interim, there are several ways to compensate for the poor sunlight viewing of a computer screen: keep the computer down below, bring it above decks but shade it from sunlight, or add a peripheral marine monitor.

The simplest solution is to keep the navigational computer out of bright light, namely by placing it down below. For many boaters, especially those with exposed cockpits, the elements force a laptop to stay below long before the sunlight does. A laptop down below is used for planning, waypoint and route creation, periodic viewing of in-

If you're shopping for a new laptop, especially if you are new to computers, consider a Mac. Nineteen percent of all laptops sold in the U.S. are now Apples and many are purchased by so-called "PC switchers." Although Mac laptops have a price premium and some software choices are more limited, the Mac operating system is significantly easier to set up and use compared to dealing with wizards and ports on a Windows computer. If you want to run a PC package, you can still do so with Parallels Desktop or VMware Fusion (see **Your Intel Mac is a PC** and **Parallels Desktop in Action** in Chapter 26).

Waterproof Laptop Cases: *Keeping your laptop protected in a waterproof case is a wise investment, avoiding expensive repairs or replacements.*

formation such as tides and currents, or by a second person navigating while the other takes the helm.

For those who want their laptop helm-side but have problems with glare, a display hood may solve the problem. We've heard of all sorts of solutions, from simple pieces of canvas to custom-built shields or hoods. You can purchase a computer screen glare shield for less than $50, or fashion one yourself to fit your helm station.

The most expensive option is to purchase a marine daylight-viewable monitor and connect it to your computer as a peripheral device. Today, a few thousand dollars will get you a large ruggedized and weatherproof monitor, designed with an extremely bright LCD, typically with anti-reflective glass. Both Raymarine and Furuno produce marine-grade color LCDs, at price tags that start at $4,000 and go up to $7,500 depending on the screen size.

User Interface

Don't take that keyboard and mouse for granted. Although all e-charting applications currently rely on a keyboard and mouse interface, developers are thinking ahead to alternative user interfaces. In fact, some e-charting applications are already designed for new technologies, retaining keyboard and mouse options in the interim.

As with many onboard tools, ranging from image stabilization binoculars to gimbaled stovetops, keyboard and mouse use must be adapted for use at sea. Typing on a keyboard or selecting a waypoint with a mouse often proves the ultimate challenge. To mitigate this, most e-charting applications adapt their interface specifically for marine use.

For example, some e-charting applications prefer the mouse, but adapt the graphical user interface for at-sea mousing. TIKI Navigator includes large menus and buttons for easier mouse targeting. It also shows only context-relevant menus and constrains mouse movement to lock into the target choices.

Other applications take the opposite approach, relying on a more stable and less sensitive keyboard interface. They consider a mouse to be too "slippery" when used underway—

overly difficult to click on small icons or drag-and-drop objects in a moving environment. DigiBOAT Software-On-Board is designed with no menus and few actions that require dragging a mouse. It includes mouse support, but considers it additional functionality until touch screens become more mainstream.

The touch screen is the latest innovation, where a user can touch the screen with a stylus or fingertip to initiate actions. Touch screens are just entering the consumer market, most notably with the Apple iPhone which is entirely touch-based. Garmin recently introduced its GPSMAP series of touch screen chartplotters. Several notebook computers, all launched recently, use a touch screen interface, such as the ARMOR X10, Canova Dual Touch Screen Laptop, Fujitsu LifeBook, and HP Pavilion Entertainment Notebook.

Drives and Ports

Accessing different forms of data—ranging from nautical charts to weather forecasts—is a real benefit. Because today's personal computers assume networked data exchange, they easily serve as a flexible command center for other marine instruments. But this requires a computer with the appropriate data input/output drives and ports.

Ideally, an e-charting computer should have a DVD drive, which typically reads both DVD-ROM and CD-ROM discs. Many e-charting packages include sizable cartographic collections, most notably Maptech Chart Navigator Pro's 15 DVDs of resources. A DVD drive is necessary in order to load these assets. A CD drive cannot read DVDs. However, it is not necessary to have a DVD-RW, or so-called *superdrive,* which can *Read-Write* blank DVDs.

Some chart distributors, including C-Map, Navionics, and Garmin, also sell chart collections on compact cartridges, commonly called "cards." Although chartplotters include a cartridge slot for loading cards, many personal computers do not. In this case, reading a card requires a card reader, a small peripheral device costing less than $100 that plugs into the computer's USB port.

A personal computer connects to peripheral devices, such as card readers, additional monitors, GPS sensors, or autopilots, via its *ports.* Ports are the connections on the side of the

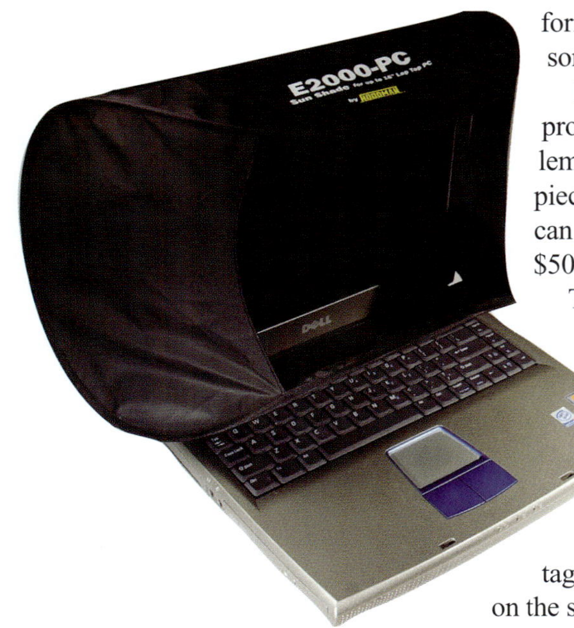

Daylight Viewing: *An inexpensive hood reduces glare and improves daylight viewability. Hoodman Corporation (www.hoodmanusa.com) sells Mac and PC products such as the E2000-PC above ($39).*

Peripheral USB ports, such as on your monitor or keyboard, are not always "powered." Use the ports on your computer to directly power your GPS or other USB devices.

Don't forget Bluetooth, a wireless connectivity option for cell phone headsets and computer peripherals such as mice, keyboards, and GPS sensors.

computer, exchanging information across devices. Every personal computer includes ports for connecting peripherals or networking with other computers, although the number and specifications vary. Your computer may have one or more serial, FireWire, USB, or Ethernet ports.

Older PCs typically have serial ports. Older Macs typically have FireWire ports (400 or 800), but this standard was not widely adopted. Fortunately, the computer industry has finally converged on the USB standard for peripheral devices, either USB 1.0 or the new, faster USB 2.0. If you are purchasing a computer or peripherals, look for USB 2.0. An existing computer that only supports USB 1.0 will work fine, but be aware that any task done across the USB connection, such as reading charts from a card reader, will be slower.

New personal computers now include several USB ports, typically adding extra USB ports on monitors or keyboards. In addition, the two most common e-charting peripheral devices: GPS sensors and multicard readers, are both available in versions that can be plugged directly into your computer's USB port.

Ethernet is the standard for computer networking. Most personal computers include an Ethernet port, the connection that looks like a large phone jack. Some marine manufacturers, such as Raymarine, have adopted Ethernet for easy computer/marine instrument networking.

If your computer's ports don't match the cables on your peripheral devices or marine instruments, don't panic. You can purchase an adapter for any connector combination. You may think them pricey—often $30 for a short length of cable with two connector ends—but compare this with the cost of a new device before discounting this option.

Internet Access

The Internet is one of the most important portals for receiving data to a personal computer. Charting software, electronic charts, nautical references, and live weather are all available if you equip your onboard computer with Internet access.

In Chapter 6, *Charts and Supplemental Data*, we cover the logistics of downloading free U.S. charts. Chapter 9, *Extending the Digital Metaphor*, details how to use your computer to access a myriad of marine assets, including satellite images, street and point-of-interest maps, tourism information, weather data, and reference publications. But first you must have the hardware components necessary for onboard Internet access.

WiFi

The Internet has changed boating. Only a few years ago, boaters hauled laptops ashore to beg use of a phone jack for dial-up connection to an AOL account. "High-tech" boaties carried small modem-enabled devices, such as a PocketMail Composer, to access email using a pay phone receiver.

Dial-up and pay phones seem like antiquated relics now. Within a span of three years, everything changed and boaters can now effortlessly stay in touch from their vessels. Marinas, waterfront areas, and entire towns (such as Annapolis, Maryland) became *hot spots*, broadcasting wireless Internet service that can be picked up by anyone with a computer and a WiFi card. Although some facilities charge a daily fee, more and more free signals are available. Free Internet access has spread from Starbucks Coffee to parks, libraries, and even McDonald's fast-food restaurants.

A WiFi card, which enables a computer to talk to a WiFi base station, is a standard built-in item on today's laptops. If your computer is an older model, or you purchased it without a WiFi card, you can probably still add the card. The card can be installed inexpensively at the service desk of a consumer electronics store. For more flexibility, you can purchase a WiFi card that inserts into your computer's PCMCIA or Express slot (the slot on the side that takes cards for different functions). WiFi cards are also available that connect to your computer's USB port if you do not have a card slot or if it is already occupied. Most people retrofit an older computer by buying a slot-based WiFi card. Both slot-inserted and USB-connected WiFi cards can be swapped and used on multiple computers.

Although WiFi cards are all similar in "power," they vary in signal strength due to differences in antenna quality. A slot-inserted or USB WiFi card that includes a tiny external flip-up antenna typically has better reception than an installed card, which must rely on the computer's default antenna. Although personal computers do include built-in antennas, like any an-

Making the Connection: *Luckily, there are as many adapter designs as there are physical connection options. This Keyspan USA-19HS connects an RS-232 serial device to the USB port of a PC or Mac.*

Multimedia Reader: *This C-Map USB card reader accommodates C-, SD-, and CF-formatted cards and may be used with DigiBOAT Software-On-Board, NavSim BoatCruiser, and Furuno MaxSea to read C-Map cartography.*

Cables & Ports & Cards: *Oh my! There are a confusing number of computer physical connection types. Shown above are (1) common network and peripheral connectors* **A** *USB A* **B** *USB B* **C** *FireWire 400* **D** *FireWire 800* **E** *Ethernet* **F** *serial (male and female); (2) video connectors* **G** *VGA* **H** *DVI; and (3) computer cards* **I** *Express* **J** *PCMCIA.*

tenna, design matters. The newest laptops build an antenna into the entire screen frame, providing much better reception.

Cellular Network

Although free WiFi may seem like free candy falling from the sky, there are some practical glitches. Quite literally, things get in the way. Signals are not always available or strong enough—even in areas that should be bombarded with WiFi signals.

WiFi is short range and *line of sight*. Objects such as buildings can block reception. To illustrate the ramifications of line-of-sight WiFi, we were unable to pick up a signal on the downtown Boston waterfront for literally half of each day— low tide put us down behind a stone pier! WiFi is also subject to interference from devices that operate at the same frequencies, including other WiFi networks, cordless phones, and microwave ovens.

In order to pick up a good WiFi signal, many boaters bring their laptop ashore to *sniff* for a strong WiFi signal or to work in an Internet cafe. But this strategy is inconvenient, takes time, and risks damage or theft of your laptop. Often, it also entails hidden costs we call the "bagel-and-coffee surcharge." Although the Internet is "free" at a Bruegger's Bagels or Starbucks Coffee, the five dollar food cost of admission adds up on a daily basis. For some users, a monthly plan using an air card and a cell phone provider makes better economic sense.

Cellular Internet access plans, available for about $60 per month through companies such as Verizon, Sprint, T-Mobile, and AT&T, provide unlimited Internet access anywhere you can pick up a cell phone signal. With the rapid sprouting of local cell phone towers, coverage now extends to even seemingly remote locations. The speed varies by location, but is typically equivalent to broadband rates. This system, officially called Evolution-Data Optimized or Evolution-Data Only (abbreviated as EVDO) is a telecommunications standard for the wireless transmission of data through radio signals for broadband Internet access.

Unlike many marina or public WiFi networks, a cellular network is secure, which is a big plus for boaters conducting financial business. With encryption built-in by the cellular provider, you can securely log in to your financial institution account and make credit card payments and purchases.

A cellular Internet plan works by connecting your computer to a cellular network. The hardware connection consists of an air card, which differs from a WiFi card. Simply put, an air card is a cellular modem connecting to a cellular network; a WiFi card connects to a WiFi computer network. As part of the setup, the cell phone company provides an air card compatible with your computer. The air card is the wireless modem that connects your computer to your Internet account.

There are three options for an air card. An air card can be inserted into your laptop's computer slot. Alternatively, if your computer does not have a slot, you can purchase a USB air card which plugs into your computer's USB port. Better yet, an air card can be plugged into a wireless router so all computers in the LAN have Internet access.

Signal Strength Improvement

Can you really pick up Internet service on the water? The answer depends on your Internet source (WiFi or cellular), your antenna, and your location. Although there are no technological constraints to having onboard telephone and Internet service literally anywhere in the world, you must configure the hardware set-up for your boating confines.

We've already mentioned that in-port WiFi may be spotty due to line-of-sight obstructions or signal interference. Even without these hindrances, WiFi signals only transmit about 300 feet (outdoors) from the wireless access point, and often much less. However, if you're committed to using WiFi rather than a cellular network (the choice of many boaters since WiFi is often free), there are ways to boost your WiFi reception.

Most likely, the problem lies in your computer's built-in antenna. We've already mentioned the advantage of a slot or USB WiFi card, which include tiny, but effective, flip-up antennas. Several companies, including Linksys, D-Link, IOGear, Buffalo, Hawking, and hField Technologies make USB WiFi adapters or boosters, which generally sell for less than $80. These devices connect to your computer via its USB port and either suction a WiFi antenna unit onto your computer or sit separately at the end of a relatively long USB cable. Some boaters even temporarily place the device outside on deck and run the USB cable down below to the computer. Many boosters claim ranges over 1000 feet, enough to increase your odds of picking up a signal while on a mooring or in an anchorage.

Boaters are no strangers to antennas, in fact you probably have one or more antennas mounted for your VHF or SSB radio. Several companies have adapted these standard marine antennas to include WiFi reception. For example, Radiolabs sells a WaveRV marine antenna for about $170 that fits into the standard marine antenna fitting (www.radiolabs.com). It connects with, and is powered by, a single wire to your computer's USB port. The company claims a one-mile range from the WiFi access point. Syrens Onboard Wi-Fi offers several antenna/booster solutions (www.syrens-at-sea.com). Syrens' powered Ethernet amplifier/antenna option reportedly has a range of several miles. Similarly, a Port Networks system links an outdoor antenna and radio to your computer's Ethernet port (www.portnetworks.com).

Although you will pay the dreaded marine surcharge, there are advantages to choosing a marine WiFi booster system (commonly called a "marine wireless bridge") rather than a standard consumer WiFi product. The antennas, wiring, and devices made by Radiolabs, Syrens, and Port Networks are designed for a marine environment, with attention to weatherproof seals, power consumption, and compatibility with marine antenna base fittings.

WiFi boosters, whether designed for consumer or marine use, improve signal reception but have nothing to do with providing the signal. They assume you are picking up a free or subscriber WiFi signal. Another option is to combine your antenna purchase with an Internet plan. For example, Shakespeare, an established manufacturer of marine antennas, sells a mobile broadband system called CruiseNet (www.shakespeare-marine.com). CruiseNet combines a marine cell phone router with comparably-priced mobile phone plans through several popular cellular companies. Shakespeare's CruiseNet cell phone router, combined with a 2-foot to 18-foot antenna, is a marine alternative to a laptop-installed air card and your own Internet plan.

In contrast to WiFi networks, which are broadcast from a

Google   USB WiFi adapter booster

WiFi Improvement: *To find currently available signal boosters, Google "USB WiFi adapter booster.".*

New Partnerships: *Some marine manufacturers are leveraging their equipment and marine customer base, partnering with cellular companies to provide "always-on" Internet for boaters. An example is Shakespeare's CruiseNet.*

Cellular Data Plans: *To find currently available data plans, Google the cellular carriers.*

If your computer is connected to the Internet, you need to protect against viruses, worms, trojan horses, adware, spyware, and other Internet hazards. Purchase and install security software such as ESET NOD32, Norton AntiVirus, or McAfee Internet Security Suite. Also consider privacy software such as Ad-Aware or Spybot. Once installed, be sure to update the virus definitions frequently, or whenever the software prompts you for a new release. No software can remain effective against rapidly-evolving viruses and attacks unless you keep it up-to-date!

local wireless access point, cellular networks transmit from tall cellular towers. That means anywhere you can pick up a cellular telephone signal, you have cellular Internet access. This range extends offshore (usually several miles at least), as well as to seemingly remote locations. Cell phone towers have been sprouting up like weeds, providing service to low-population areas. We've had near-broadband rates in the flats of Chincoteague Island on Virginia's sparsely populated Eastern Shore and in the middle of the Florida Everglades.

If you do feel you need more signal strength, most likely because of a poorly-designed computer antenna, you can purchase cellular boosters. Digital Antenna's cellular signal amplifiers or wireless cellular signal repeater systems can boost your cell phone or air card signal up to 50 miles (www.digital-antenna.com).

For truly offshore passages, you need a satellite voice and data service. Several companies maintain their own private satellite communication systems. Although it will cost you (with per-minutes usage charges), most provide phone, fax, and Internet service anywhere in the world. Inmarsat is one of the oldest network providers, dating back to the early 1980s (www.inmarsat.com). A 62-country internationally-owned cooperative, it was originally formed to facilitate communication to the commercial shipping and fishing industries. Globalstar (www.globalstar.com), Iridium (www.iridium.com), and iDirect (www.idirect.net) are three well-known companies with near-global coverage. Iridium's coverage even includes the polar regions. Thuraya mobile satellite systems focuses on Europe, North and Central Africa, the Middle East, and Central Asia (www.thuraya.com).

Operating Systems

An operating system (OS) is the system software responsible for the direct control and management of the computer hardware. E-charting software runs "on top" of this system software, adding its own graphical user interface and command feature set. Since an e-charting application is written to be compatible with the system software, an e-charting application may only run on particular operating systems.

Most e-charting applications run on Windows 2000 or the newer Windows XP operating systems, although several applications have already phased out Windows 2000 support. If you have a old machine running Windows 98, NavSim Boat-Cruiser still supports this legacy operating system. Microsoft's newest Windows operating system, Vista, is discussed in the next section.

Macintosh users must use OS X, the operating system that completely replaced OS 9 in 2001 (or use Windows through virtualization software). Generally speaking, all Mac e-charting applications expect at least OS X 10.3, also known as "Panther." All are compatible with 10.4 (Tiger) and the newest 10.5 (Leopard) release. Any of these versions, Panther, Tiger, or Leopard, are sufficient for e-charting—an upgrade is not necessary.

Is E-Charting Vista Ready?

When it comes to software, everyone asks, "Is it Vista ready?" However, as the joke goes, Microsoft wasn't quite Vista ready. So don't be surprised if your e-charting software stumbles a bit or lacks certain peripheral connectivity in Vista.

Unfortunately, there is rarely a simple yes-or-no answer on an application's Vista-readiness. Companies are in various stages of adoption, compliance, and support. Some packages, such as Coastal Explorer or Chart Navigator Pro, were genuinely Vista ready. Others, such as Nobeltec VNS and Admiral, initially appeared ready to run on Vista but continued to troubleshoot Vista-specific glitches until the recent release of their MAX Pro versions. Still others, such as The Capn, have chosen to let the dust settle on Vista and are best run on an XP partition.

What comprises "Vista capable" is a bit slippery. Although e-charting applications are written in languages ranging from Visual Basic to C++, they all should boot up in Vista. But simply booting up shouldn't count for full Vista compliance. If key features don't work—such as connecting a GPS—then we don't consider it Vista ready. This criterion is especially relevant since Vista's mantra is "plug-and-play."

The current issue with Vista is the lack of drivers for external devices, including card readers and GPS sensors. Pundits are jokingly calling Vista's handling of peripherals in general as "plug-and-pray" instead of "plug-and-play." E-charting companies are scrambling to deal with issues as they are identified, posting solutions and new drivers on their website support pages.

If you intend to e-chart using the Vista operating system, begin by checking the software vendor's website for Vista support. Explore the vendor's support pages and forums to see the true extent of Vista compatibility across all features and peripheral devices.

Your Mac is a PC

Although Macintosh users are safely removed from the Vista shake-out, they face another choice: Intel or not Intel?

If you own a newer-model Mac with an Intel processor, don't forget that now your Mac can also be a PC, letting you run any charting and navigation application.

In the old days, Mac users had to run an *on-the-top* emulation application such as Virtual PC. But it was slow and expensive. Now that Macs have switched to Intel processors, they can run other operating systems including Windows, Vista, Linux, and BSD Unix.

Apple's free dual-boot system, called Boot Camp, lets you boot up your Mac in either Windows or OS X. Two better solutions, Parallels Desktop (www.parallels.com) and VMware Fusion (www.vmware.com), let Mac users run Windows applications *side-by-side* by partitioning the processor and running multiple operating systems concurrently. These virtualization tools also create shared folders (handy for chart files) that can be accessed from either operating system.

Powering Your PC

Desktop or laptop, your onboard computer requires electrical power. Fortunately, laptops have become extremely energy efficient, making them a feasible appliance on many boats. But don't rule out a desktop choice, now that CPUs are smaller than a lunch box.

For most boaters, a laptop is still the best onboard choice. With built-in batteries, today's laptops easily run for several hours without an electrical connection. Most boaters can simply charge their laptop at home at the end of the day or when connected to a marina's 30- or 50-amp service. In addition, since they are designed for travel, laptops are more resistant to impact, vibration, and moisture. Regardless, they should always be cushioned and secured, with movement minimized during hard drive and disc drive operation.

For extended voyages (away from shore power), you can charge a laptop either directly from your boat's 12-volt DC system (with the proper optional "cigarette-lighter" charging cable) or with an inverter if your boat has a 120-volt AC set-up. Regardless of how you charge your laptop, always use the correct batteries and cables and be alert to "unclean" power sources. A laptop, storing all your valuable files, is a fragile device. If plugged into an AC system, particularly with marina dock power sources, use a surge protector.

Desktop computers are built for 120-volt AC power input. Ironically, the computer circuits actually run on DC power, but since households are wired with AC systems, the included power plug is a transformer that steps down 120-volt AC to the appropriate DC voltage. The best way to power an onboard desktop is by plugging it into your boat's 120-volt system (if you have one) and letting the computer transform the voltage to its specifications. This two-step approach may not seem efficient, but it keeps the power consistent with the manufacturer's engineering.

Although laptops are now more popular than desktops, a desktop remains an option for e-charting on vessels with 120-volt electrical systems. We're not talking about a behemoth under-desk CPU, but one of the new miniaturized models. For instance, a Mac mini is a six-inch by six-inch by two-inch CPU that weighs less than three pounds. These stackable flexible CPUs can be combined with your own monitor, keyboard, or other peripherals. Mac minis double as storage for music or photo libraries, a DVD player, or a backup drive.

Hardware Trends

By using mainstream hardware and software, you leverage the cumulative developments of the entire computer industry. This rapid development cycle is known as Moore's Law: computer process-

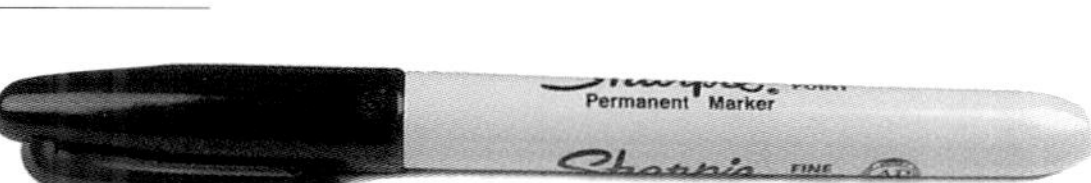

12-Volt Computers: *Boaters can take advantage of the mobile computing market explosion. This $500 12V Logic Supply unit comes with Windows XP and has standard VGA, USB, and Ethernet connectors.*

Pint-Sized Powerhouse: *Apple offers an Intel Core 2 Duo-based ultra small computer (6.5 x 6.5 x 2") called the Mac mini. With Parallels Desktop, you can run any PC or Macintosh e-charting application.*

ing speed, memory capacity, and even the resolution of digital cameras improve exponentially, doubling about every two years. And while computing power increases, computer prices stay constant or even fall.

Computers are not only moving toward more power, they are undergoing radical changes in form and user interfaces. In the near future, look for onboard options that include miniaturized computers, UMPCs (Ultra Mobile Personal Computers), subnotebooks, and tablets.

Many industries already use these newer technologies. Miniaturized dash-mounted computers are standard equipment in police vehicles and ambulances. Parcel companies log deliveries en route using tablet computers.

The Mac mini is an example of a miniature computer, a tiny CPU the size of a lunch box. For PC users, Mini-ITX and mini-ATX systems provide a miniaturized motherboard that can be installed with standard operating systems, a CD-ROM drive, housed in a small casing, and connected to an LCD monitor. The Mini-ITX is the new trend for those who want a computer in their car.

Some mini-computer manufacturers are: mp3Car (www.mp3car.com), Logic Supply (www.logicsupply.com), and Simplified Innovation (www.simplifiedinnovation.com).

For an integrated computer, the new subnotebooks or mini laptops incorporate a keyboard and screen into a streamlined package. Apple just released its MacBook Air, a full-powered laptop that is a thin as your index finger. Although it's light and slim, it includes a 13-inch widescreen display, a full-size keyboard, and a multi-touch trackpad.

Subnotebooks are even smaller, pushing for a streamlined but integrated device using tablet or swivel screen designs. Panasonic, Fujitsu, Flybook, Sony and Samsung all have new subnotebook models. If these miniaturized laptops seem too small for marine viewing, you can always mount a standard monitor, or even upgrade to a marine monitor.

Tablet computers, based on a touch screen interface, are also new on the market. The ARMOR X10, Motion Tablet, and Sony Vaio series are leaders in consumer tablets. Despite their compact size, touch tablets are full-fledged computers. The CPU is built into the tablet design.

Tablet computers recently entered the marine industry. Maptech began with an InMotion PC tablet running Maptech navigation software. More recently, Maptech and Nobeltec teamed up with Faria Instruments to market a marine-specific tablet computer (www.faria-instruments.com). Faria's Maestro Marine runs Maptech Navigator (a software program specifically developed for touch screen hardware) or Nobeltec charting and navigation software. The Maestro, specifically designed for marine use, also includes a GPS antenna, depth transducer, and USB port for uploading charts from a thumb drive. For Internet access, it includes integrated WiFi and a PCMCIA slot for cellular air cards.

Touch tablets have several advantages. Instead of a keyboard and mouse interface, commands are simply entered by touching the screen. Touch the screen to create a route, view and change charts, get tide predictions, or control the autopilot. The peripheral mouse and keyboard devices are gone, replaced by a large LCD screen. The Maestro's screen is waterproof, daylight readable, and weather resistant.

Although a marine tablet computer will set you back several thousand dollars, Moore's Law is in your favor. Expect even more power and features, and for prices to drop as touch screens, tablets, and subnotebooks continue to enter the mainstream market.

Chapter 5
Instruments and Sensors

Most boats incorporate networked marine electronics, allowing data from a GPS sensor to appear on a chartplotter, or a VHF radio to transmit position in an emergency, or an autopilot to steer routes created on a chartplotter. These marine devices *talk* and *listen* to each other via NMEA data strings transmitted over the network.

If you bring aboard a personal computer equipped with the appropriate e-charting application, you also have the option of connecting your marine instruments. Your computer can receive and display data, such as GPS position, depth soundings, wind speed and direction, AIS ship traffic, and even radar scans. Likewise, your computer can transmit data, such as sending route information to an autopilot.

Instrument support is strong and relatively uniform across today's e-charting choices. Even the lowest-priced packages support an impressive number of instruments. For example, DigiBOAT's Software-On-Board, at only $53, can connect to every instrument with the exception of a video camera. Sea-Clear II, the freeware choice, links to every instrument except water temperature, radar, and video.

In order to decipher data from marine instruments, e-charting programs are written to accept NMEA 0183 data *strings*, the standard used for marine instrument communication. However, the potential to understand a data string is not the same as the ability to display the data. The *Software Comparison Tables* in this chapter summarize instrument and sensor support for different e-charting applications.

For instance, in order to display wind information, an e-charting application must be engineered with an interface (window or data field) to display wind speed and direction. Any given application, although able to accept NMEA data, may not support the display of a particular instrument's output. As another example, some applications display extremely detailed AIS information, including a vessel's type, home port, destination, call sign, MMSI, speed over ground, turning indications, and so on. Other e-charting applications receive the same data strings, but are designed to only use some of the data, such as closest point of approach (CPA) and time to closest point of approach (TCPA).

Before you start drilling through bulkheads to run cables, or rush out to purchase the e-charting software that supports the most instruments, ask yourself what instrument data you really need to see on a laptop while underway. Your answer will depend on your helm situation and your type of boating.

The location of your onboard desktop computer or laptop influences which instruments make sense. Collision avoidance displays, such as a window of AIS data, are of little value on a laptop buried down in a saloon. Unfortunately, because most laptops are not marinized, and even the brightest screens are difficult to see in sunlight, boats with open or exposed cockpits, such as sailboats or small powerboats, may need to keep the laptop down below. Trawlers or cabin cruisers with enclosed and spacious helm stations can view a laptop or computer screen while underway. Thus, it only makes sense to specify advanced instrument features if the computer is easily visible and actively used at the helm.

In addition, not all boaters need all instrument data. For example, do you really need water temperature displayed on your laptop? If you are a Gulf Stream racer or an offshore fisherman, the answer is yes. But for most boaters, this one bit of data probably isn't worth the cabling hassle. Ask yourself what instrument data you really need to network to your computer. And keep in mind that an e-charting laptop can legitimately be an *island of information*. It still accomplishes important tasks such as creating waypoints, planning routes, and obtaining additional information such as weather— even

Chapter Terms

AIS: Automatic Identification System. A system for identifying, locating, and communicating between vessels by integrating an electronic navigation system with a GPS and a VHF transceiver system.

ARPA: Automatic Radar Plotting Aid. A radar that is ARPA-enabled can create tracks using radar contacts—calculating the tracked object's course, speed, and closest point of approach (CPA)—in order to warn of a potential collision.

Driver: A computer program allowing higher-level computer programs to interact with a device. Also called a device driver or software driver.

Global Positioning System (GPS): A global navigation system using satellites to determine position, maintained by the United States.

MARPA: Mini Automatic Radar Plotting Aid. A radar that is MARPA-equipped monitors a selected target to provide Closest Point of Approach (CPA) and Time to Closest Point of Approach (TCPA) data. Used for collision avoidance.

NMEA 0183: An industry standard for electrical and data specifications, set by the National Marine Electronics Association (NMEA), for communication between marine electronic devices.

NMEA 2000: A new data communications protocol that allows up to 50 devices to be connected together.

Chapter Questions

- *What instruments can be connected to a computer-based electronic charting system?*

- *How do the instruments and computer communicate for data exchange?*

- *Which instruments make sense for me?*

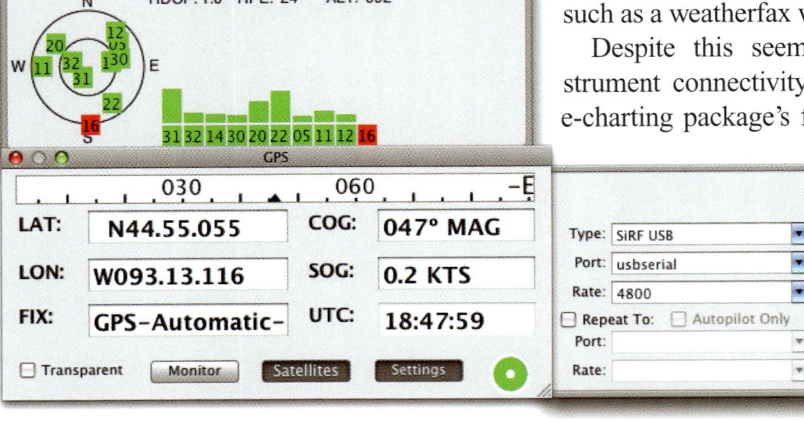

Satellites and Settings: *GPS port settings and satellite constellations can be monitored through separate control windows.*

Typical GPS Sensor: *Maximize your laptop and application software by attaching external sensors. The most common choice is an inexpensive USB GPS sensor. With a GPS sensor, you not only see a chart, you watch your vessel transit the chart in real time.*

if it is stowed and supplemented by marine devices such as a weatherfax while underway.

Despite this seemingly skeptical comment, instrument connectivity is a major component of an e-charting package's features. Although few boaters will connect every instrument, nearly all boaters will connect at least one. The most popular devices, in order of priority, are a GPS sensor, autopilot, and an AIS receiver. Additional instruments, such as wind, water temperature, radar, electronic compass, or video, are either more specialized or more complicated to implement.

GPS

The real allure of e-charting is watching your boat icon move over a digital chart on your computer. Without a GPS, even the most sophisticated e-charting application becomes a crippled viewer or planner. A GPS sensor is the most important instrument to connect to an e-charting computer.

Fortunately, connecting a GPS doesn't require difficult or inconvenient cabling from your computer to your onboard GPS receiver. A much easier solution is to purchase a USB GPS sensor for less than $100, such as the USGlobalSat BU-353 or Garmin GPS 18. These units are smaller than a hockey puck and plug directly into a computer's USB port. In addition, they provide an important backup to the vessel's main GPS receiver.

AUTOPILOT

A truly full-featured charting and navigation application supports autopilot connectivity. Without autopilot support, assets such as waypoints and routes can only be used for active navigation if transferred to a chartplotter. Although many boaters choose this option, an e-charting laptop can also function like a chart-plotter, sending route information directly to an autopilot. All full-featured applications reviewed in Part Four, *Choosing an Application*, connect with a vessel's autopilot.

Much like a chartplotter, a helmside laptop can be used actively for piloting, auto-steering to routes created and stored in your computer. It also integrates transit features such as cross-track error, boundary or depth alarms, route steering, and fuel or transit calculators.

However, connecting an autopilot is not as easy as connecting a USB GPS sensor. Although autopilots are NMEA 0183 compliant—ready to communicate with your computer's e-charting software—the two devices must be physically connected. The autopilot's cable connects to a computer, using a serial or USB adapter if needed. On some vessels, particularly where a computer can be located at a navigation station while cabled to the autopilot, this is an easy connection. In other cases, it may require long or inconveniently-run cables.

AIS

The Automatic Identification System (AIS) was originally developed for commercial ships, providing information to identify vessels and avoid collisions. The IMO requires all commercial vessels over 300 tons and all passenger vessels to be outfitted with AIS. Because AIS identifies a vessel, the U.S. Department of Homeland Security recently proposed that *all* vessels be required to carry AIS. However, AIS is currently not required for recreational boats.

AIS has become a must-have feature in e-charting software. Except for RayTech RNS, all full-featured applications reviewed in Part Four include AIS support. In fact, two of the less-expensive packages, SeaClear II (free) and TIKI Navigator Pro, both have extremely robust AIS functionality.

A complete AIS unit contains a GPS receiver, a VHF transmitter, two VHF receivers, and an additional DSC (Digital Selective Calling) receiver. Because AIS uses VHF frequencies, a VHF antenna splitter allows a single VHF antenna to be used for VHF and AIS. Less expensive receive-only AIS units are available, which do not include the transmitter. Receive-only units provide data about other vessels, but your vessel does not appear on their collision systems. In other words, you see them, but they don't see you. The *transceiver* (transmits and receives) option is also required to use an e-charting software's AIS buddy boat features.

With an AIS receiver (and optional transmitter), vessels electronically exchange ship data. Each e-charting software limits which information to display, but commonly includes the vessel's name, position, course, and speed. Some applications display more of the AIS data stream, displaying draft, length, turn rate, MMSI identifier, call sign, destination port, and type of vessel (such as marine patrol or ferry).

E-charting applications not only show ship data, they also typically display a chart overlay of ship positions. Vessels in the area are shown with an icon, which can be mouse-clicked to open the window of detailed AIS information. Turning on the AIS layer lets you see all the bogies in your proximity; simply turn off the layer to return to an uncluttered chart. These features provide what is called *situational awareness*. More importantly, e-charting software uses the AIS data to provide warnings of potential collisions. These *collision avoidance* features calculate each vessel's closest point of approach (CPA) and time to closest point of approach (TCPA), and can be set to sound warning alarms.

Because AIS alerts of potential collisions, it is an important safety supplement for many types of boaters. AIS is most relevant in areas of heavy commercial traffic, whether cargo ships, ferries, or commercial fishing. This includes the Chesapeake Bay and Puget Sound. AIS also adds a safety layer in fog or during offshore night passages. By setting collision alarms, AIS helps off-watch crew members sleep more soundly, knowing there is an electronic watch to backup a sleepy helmsman.

Although e-charting software integrates sophisticated AIS features, it's also possible to display AIS data on a personal computer without a full-featured e-charting application. Stand-alone packages, such as ShipPlotter (www.shipplotter. com), decode AIS signals using the sound card in your personal computer. You need a suitable VHF radio receiver tuned to one of the two AIS channels. ShipPlotter decodes the received data and displays it on a personal computer.

RADAR

Many boaters mistakenly assume that radar is an easy-to-implement feature in e-charting software—perhaps because so many e-charting vendors show flashy split-window examples of a chart alongside a corresponding radar scan. Realistically, connecting a radar is not that simple. Furthermore, incorporating data from a radar can mean many things, ranging from basic collision avoidance to full-color radar scans overlaid on a chart.

The simplest radar option is ARPA/MARPA radar, which refers only to collision avoidance data. An ARPA-equipped radar works like an AIS receiver, but creates tracks of other vessels using radar contacts rather than radio transmissions. ARPA/MARPA provides collision warning by tracking the course, speed, and closest point of approach of nearby objects. It is not a full image of your radar scan.

The more expensive applications, and several of the mid-priced choices, include ARPA/MARPA radar support.

An Elephant Has the Right of Way: *AIS capability can add visual clarity to a busy seaport.*

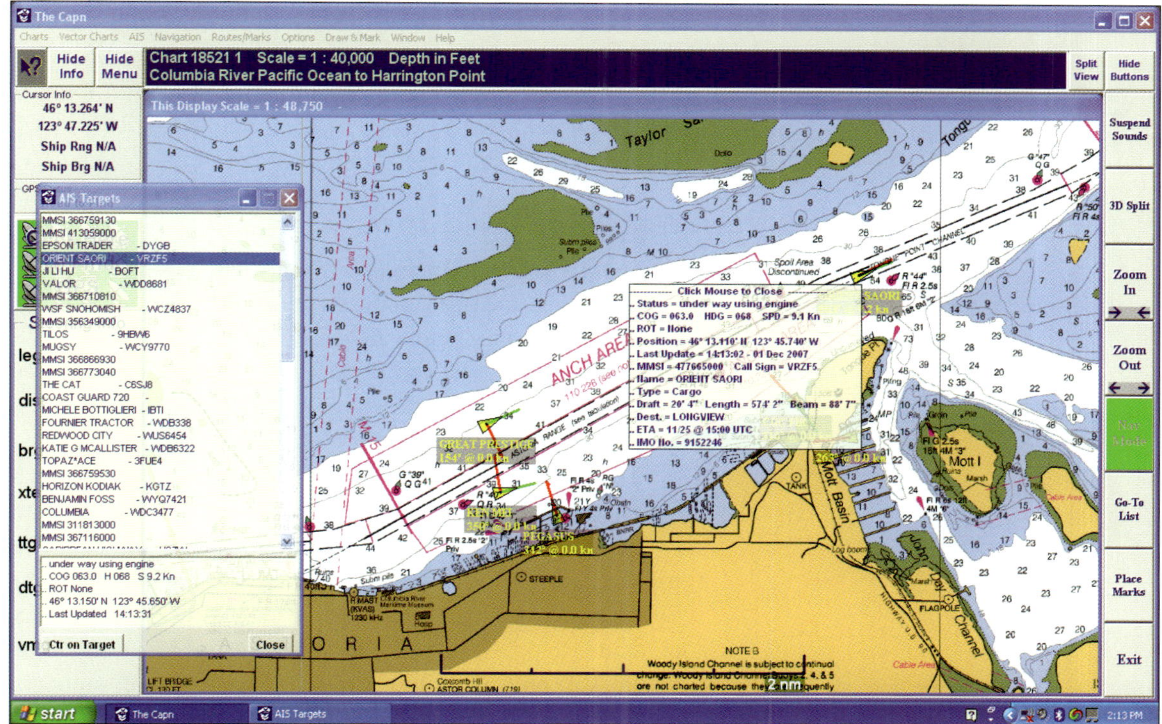

Don't forget to load the appropriate driver for your GPS. For our BU-353 USB GPS sensor, we went to USGlobalSat's website and downloaded drivers for Windows XP and Mac OS X (www.usglobalsat.com). Other GPS manufacturers have similar **support>download** areas where you match drivers for your particular GPS model and computer operating system.

An application is more likely to locate a peripheral if the device is connected to the computer *before* starting the application.

Displaying or overlaying a full image of the radar scan is a separate and more complicated feature. Because radar data is analog, it must be converted to digital in order to display on a computer. Implementing a radar display on a personal computer requires not only e-charting software that supports this feature, but also either a newer-model digitally-enabled radar or a translation device commonly called a *black box*. In addition, rendering a radar overlay requires more granular heading data than a GPS sensor provides. Most configurations also require a heading sensor or electronic compass. All these extras, ranging from digital radars, black boxes, to heading sensors, can easily add up to several thousand dollars.

If you want full radar overlays on your personal computer—useful for boaters in foggy areas such as Maine and the Pacific Northwest—you must use a higher-end software package such as Coastal Explorer, Chart Navigator Pro, RayTech RNS, Max-Sea, or Nobeltec Admiral MAX Pro. The *Software Comparison Tables* summarize which packages support ARPA/MARPA radar versus full radar display overlays.

Wind/Speed/Depth/Temperature

Although most recreational boats are equipped with wind, speed, depth, and sea temperature displays, feeding this information to a computer only makes sense for boaters with specialized needs. This information is typically unnecessary to perform e-charting piloting or navigation functions and is best viewed directly from other instrument sources.

For example, wind data is important if you intend to use a software's sailing route optimization features. In this case, accurate and timely data on wind speed and direction must be transmitted from an anemometer to your personal computer. Some sailors may also want to mirror wind conditions to a computer to provide a belowdecks navigator with wind information for route planning. However, for most small sailboats, the traditional instrument display at the helm is sufficient.

Speed data is now automatically provided by GPS calculations. In fact, many boats no longer use a traditional knot/log instrument that collects data via a thru-hull transducer. The old way punches a hole in your hull, becomes impeded with barnacle growth, and provides speed-through-the-water rather than speed-over-ground data. In contrast, a GPS sensor (connected easily using the computer's USB port) provides accurate speed-over-ground and heading data.

Depth displays are a common feature of nearly all e-charting applications. Depth data is an important component of a software's alarm features. However, e-charting applications that support vector charts also set depth alarms using the combination of chart data and boat position. Most traditional depth sounders include simple, but effective, depth alarms. For most boaters, it will not be worth the cabling hassle to have a depth alarm sound on a computer versus a traditional helm-mounted depth instrument. Before running depth instrument cables, consider moving to vector cartography to supplement in-place depth alarms, or investing in the new wireless instruments.

Water temperature is very important to a select group of boaters. Most, but not all, e-charting applications can display water temperature data when connected to a temperature sensor. This feature is particularly important to offshore fishermen and Gulf Stream racers, who can integrate this data with their e-charting features to plan trolling or racing routes.

Video

Several e-charting applications tout video camera support. BoatCruiser, The Capn, Nobeltec, and RayTech RNS can display a small video window on your laptop while you perform your regular navigational work.

Depending on your situation, video monitoring may make sense. There are many scenarios where an "extra eye" adds an important safety layer. You can monitor an engine room. It can be used as a nursery monitor, keeping an eye and ear on a small child alone in a stateroom. Sportfish captains can steer in response to events out of their vision at an aft deck fighting chair. Video is also an effective theft deterrent, surveying decks against intruders or dinghies against theft.

Video is relatively easy to set up using common consumer electronics components. It's a flashy feature, and useful to some boaters. But do realize that video is a notorious hog of computer memory, bandwidth, and chart display area.

Chartplotters and Beyond

At the highest level of instrument support, a chartplotter, multifunction display, or entire shipboard local area network (LAN) can be connected to your personal computer.

Chartplotter connectivity requires e-charting applications specifically engineered for high-speed, simultaneous networking. Computer/chartplotter data exchange is even more data-intense than computer/radar exchanges. In order for a computer and chartplotter to communicate, many different types of data must be simultaneously sent and received. Typically this data is graphic-intensive, and in some cases involves data streams as large as the chart files.

RayTech RNS and Nobeltec Admiral MAX Pro are the software leaders in networking between a personal computer and chartplotter. RayTech RNS is specifically designed to integrate with its Raymarine E-Series and G-Series multifunction displays. The entire system connects with a simple Ethernet cable between a PC and a Raymarine SeaTalk network switch. Nobeltec's networking feature is called GlassBridge Network. GlassBridge sends data—including charts—across computers, chartplotters, and mirrored monitors.

Networking to a chartplotter or multifunction display is covered in more detail in each application's review (Chapter

Instruments and Sensors[1]	GPS	Autopilot	AIS receiver
SeaClear II	Yes	Yes	Yes
Software-On-Board	Yes	Yes	Yes
NavimaQ	Yes	Yes	Yes
MacENC	Yes	Yes	Yes
TIKI Navigator	Yes	Yes	Yes
Marine ENC	Yes	Yes	Yes
BoatCruiser	Yes	Yes	Yes
Coastal Explorer	Yes	Yes	Yes
The Capn	Yes	Yes	Yes
Nobeltec VNS	Yes	Yes	Yes
Chart Navigator Pro	Yes	Yes	Yes
RayTech RNS	Yes	Yes	No
MaxSea	Yes	Yes	Yes
Nobeltec Admiral	Yes	Yes	Yes

Software Comparison: *The "Big Three" Connections*

Instruments and Sensors[2]	Wind	Depth	Water temperature	Radar	Heading sensor	Video camera
SeaClear II	Yes	Yes	No	Yes	Yes	No
Software-On-Board	Yes	Yes	Yes	Yes	Yes	No
NavimaQ	No	No	No	No	No	No
MacENC	Yes	Yes	Yes	Yes	Yes	No
TIKI Navigator	Yes	Yes	No	No	Yes	No
Marine ENC	No	Yes	No	No	No	No
BoatCruiser	Yes	Yes	Yes	Yes	Yes	Yes
Coastal Explorer	Yes	Yes	Yes	Yes	Yes	No
The Capn	Yes	Yes	Yes	No	Yes	Yes
Nobeltec VNS	Yes	Yes	Yes	Yes	Yes	Yes
Chart Navigator Pro	Yes	Yes	Yes	Yes	Yes	No
RayTech RNS	Yes	Yes	Yes	Yes	Yes	Yes
MaxSea	Yes	Yes	Yes	Yes	Yes	No
Nobeltec Admiral	Yes	Yes	Yes	Yes	Yes	Yes

Software Comparison: *Standard Vessel Electronics; Heading Sensor; Video*

Radar Overlay Capability: *More and more e-charting vendors are interfacing digital radar to provide true overlay capability. Keep abreast of these developments by Googling, "Digital Marine Radar."*

23 for RayTech RNS and Chapter 25 for Nobeltec Admiral MAX Pro) and in Chapter 8, *Networking and Data Exchange*.

TRENDS IN EXTERNAL SENSORS

Trends are often driven by efforts to improve weak areas in existing products. After reading this chapter, you may have noticed two inconveniences in connecting marine instruments to personal computers: cabling and analog data. Fortunately, improvements to both arrive with each season's boat shows.

As evidenced by the popularity of Bluetooth and wireless phones, keyboards, and mice, no one likes to be constrained by a cable. Even worse, cabling on a boat may require drilling through bulkheads or locating inaccessible wires. And once a cable is run, there is always the future question of wiring and port connection compatibility.

The move toward USB and Ethernet standards has greatly improved the plug-and-play aspect of external sensors and e-charting. Even sophisticated networked systems, such as Raymarine's SeaTalk, now use a simple Ethernet connection—as simple as plugging in a phone jack. More and more instruments will be offered using easy USB or Ethernet protocols.

Wireless is also the new wave, and it has reached marine instruments. Relatively simple data, such as wind or speed displays, can now be received wirelessly to your onboard computer. The current industry leader is Tacktick, producing wireless solar-powered marine displays for wind, speed, or depth data (www.tacktick.com).

Similarly, more and more instruments are going digital. We discussed the problem of displaying radar scans, given that most radars currently generate analog data. But your next radar purchase will likely be High Definition (HD) digital. Because they are digital, these radars can scan and display more than one range simultaneously. The latest radar products from Furuno and Raymarine include this new technology, with multiple digital scans each customized with their own individual gain and clutter settings. Even better, this new class of digital radar immediately integrates with e-charting software and multifunction displays.

Chapter 6
Charts and Supplemental Data

In order to view a nautical chart, your computer must have access to a database of charts in electronic format. These charts can be stored on the computer's hard drive or accessed through an external card reader.

Chart files open and display fastest when loaded directly from the computer's hard drive. Although charts may be read from a CD or DVD directly, this approach is not recommended—and in many cases will not work at all. Copy the charts or chart folders you need from the CD or DVD into the appropriate folder on your hard drive.

Some proprietary charts are sold on flash cards and must be read using an external card reader. Think of the card reader as an external drive. Those that use the newer USB 2.0 connection are fast, but still not as fast as charts accessed from the hard drive.

Although most personal computers can store a staggering number of chart files, performance improves if you avoid bloating your hard drive with unnecessary regions. You'll be surprised how sprightly your laptop behaves with 200 rather than 2000 charts!

Develop regular habits of chart management, loading and unloading chart regions as needed. For example, southbound we load Intracoastal Waterway charts as we clear the Potomac River, planning ahead for transiting south of Norfolk, Virginia. As we approach Coinjock, North Carolina, we unload the Chesapeake charts.

TYPES OF ELECTRONIC CHARTS

Despite the recent trend toward international standardization, chart files are still produced in a surprising number of formats. Variations arise for numerous reasons: incompatible chart production technologies, isolated standards of chart-producing countries, and commercial interests of chart distributors. In fact, the choice of a type of cartography is as important as the choice of a software package. No cartography is perfect and there is no single best choice for all boaters.

Although there are less than ten different electronic chart formats, no e-charting application can read them all. Since the fundamental structure of the data files differs, accommodating many file formats is a difficult and time-consuming task for code developers. Many e-charting applications purposely support only one or two of the more popular formats. A small handful of e-charting vendors pride themselves on their utilitarian ability to read a broad matrix of file formats. The *Cartography and Supplemental Data Tables*, later in this chapter, summarize which e-charting applications support which file formats.

At the highest level, a nautical chart can be stored as a raster or vector file. The format of these charts, namely raster or vector, is an important distinction to understand. The digital representation affects what you see on your computer screen, and more importantly, how you use the charts on the water.

Raster Navigational Charts

Raster charts have been described as analog charts disguised as digital charts. A raster chart displayed on a computer screen looks like a traditional paper chart. In fact, most boaters like raster charts for the simple reason that they "look right." A raster chart is produced by scanning a paper chart, creating an image file comprised of millions of color picture elements or *pixels*.

In the United States, Raster Navigational Charts, called by their acronym RNCs, look familiar since they are simply copies of NOAA paper charts. Raster chart files are most commonly BSB/KAP files, the format established by NOAA and Maptech as part of a Cooperative Research and Development Agreement (CRADA). Other raster formats include

Chapter Terms

BSB or BSB3: A digital file format for raster charts co-developed by Maptech and NOAA. Now the standard format for North American raster chart files.

ENC: Electronic Navigational Chart. *NOAA ENC* is the trademark name for its vector-format chart data files.

IENC: Inland Electronic Navigation Charts. Vector-format digital charts of the U.S. Inland Waterway System, created by the U.S. Army Corps of Engineers.

Raster Chart: A digital chart in raster format is a scanned image of a paper chart. The data file is comprised of millions of pixels to reproduce a color-scanned image.

RNC: Raster Navigational Chart. *NOAA RNC* is the trademark name for its raster-format chart data files. The North American standard for raster chart data.

S-57: The updated international format for vector digital charts, developed by the International Hydrographic Organization (IHO).

Vector Chart: A digital chart in vector format is created from a data file of chart features. The data file consists of a collection of geospatially referenced points, lines, polygons, symbols, and areas.

Chapter Questions

- *What are the differences between raster and vector charts?*

- *What are the different raster and vector chart formats?*

- *How do I download free U.S. charts?*

- *Where do I get international electronic charts?*

- *What kinds of supplemental data are included with electronic charts?*

Overscale and Pixelization: *Raster charts are scanned images of paper charts. Rows and columns of colored pixels display a picture of the chart. They are extremely popular because of their familiar appearance. A charted area is comprised of a suite of charted scales (including a main panel, extensions, and insets) showing varying detail—but all subject to overscale and pixelization (above left).*

Chart management is a *process*, not an *event*. Get comfortable loading and unloading chart regions as you migrate. Copying files from a CD or DVD does not erase them from the disc. By keeping the CD or DVD, you still have the charts available to reload later for your return transit.

GEO/NOS files, once sold under the brand SoftChart; ARCS, produced by the U.K. Hydrographic Office; and NV.Digitals produced by NV-Verlag. NOAA's free raster charts are BSB3 format, typically called simply BSB. Maptech also produces "locked" BSB4 and BSB5 raster charts.

BSB4 and BSB5

Although most e-charting applications support NOAA BSB raster format charts, not all support the proprietary Maptech BSB formats called BSB4 and BSB5.

Maptech initially had many reasons to protect its raster charts. In part, it was a defensive move by a private company with an exclusive government contract coming to an end. In addition, any international charts on Maptech chart portfolios require encryption as part of the licensing agreement. However, even Maptech's DVD of "free NOAA charts" are locked, requiring purchasers to register and thus be added to Maptech's marketing database.

Although BSB4 charts are locked, many e-charting applications work with BSB4 chart files. On the other hand, the use of BSB5 charts—which are the locked raster files provided on Maptech's paper ChartKit companion discs—is severely limited. BSB5 files are currently only supported by Maptech's free viewer Offshore Navigator Lite, Maptech's The Capn, and MacENC, an e-charting application for the Mac. In order to read BSB5 charts, GPSNavX, the creator of MacENC, worked with Maptech to integrate a utility that unlocks BSB5s.

Before purchasing BSB4 or BSB5 charts, be sure your e-charting application supports these raster formats.

Vector Charts

Vector charts are fundamentally different from raster charts. Rather than a scanned image, vector charts are a database of charted objects and their attributes, organized in layers, that can be quickly drawn at any scale. They are called *vector* files because the chart is created from a database of points, lines, polygons, symbols, and areas that correspond to features on a chart. Each data point is geospatially located by its latitude and longitude.

In addition, each object on a vector chart is linked to a list of item-relevant attributes. For example, a navigational aid is a vector chart object, comprised of its latitude and longitude, number, lateral significance, height, light color and rhythm, sound characteristics, and so on. A bridge data object is comprised of its latitude and longitude, structure, vertical and horizontal clearances, and other bridge-relevant information.

Vector charts are becoming popular for their uncluttered appearance and efficient storage and display of information. However, unlike raster charts, vector charts only show the primary layer of land forms and some depths. This sparse presentation leads some boaters to mistakenly think the information is missing and something is wrong with their file!

Nothing is wrong with the display; vector charts are displayed in layers. This format reduces chart clutter and lets you select only the information you'd like to view, which is very handy in densely noted areas. Clicking on a vector object typically opens a data window with detailed information on that chart object.

Vector charts come in many file formats, including ENCs (produced by NOAA), IENCs (produced by the U.S. Army Corps of Engineers), DNCs (produced by the U.S. Department of Defense), S-57s (an international vector standard), and S-63s (an international encrypted vector standard).

International S-57 and S-63

Through the work of the International Maritime Organization, international vector charts are produced in S-57 format. This file format has been accepted by the International Standardization Organization (ISO) as the standard for a commercial electronic chart display and information system (ECDIS).

Private companies also tend to produce S-57 charts that comply with commercial ECDIS requirements. For example, C-Map, Navionics, and Transas Marine offer a worldwide portfolio of S-57 vector electronic charts.

While S-57 is a standard format for the structure of data files, it does not extend to a standardized viewing experience. In fact, the same S-57 chart file looks very different across e-charting applications. Vector chart objects may display differently or at different levels of detail. Colors and contour level gradations may vary. And of course, chart-related features and functionality are dependent upon the e-charting software.

Although international vector charts are standardized, only the U.S. government distributes its charts in the unencrypted S-57 format. Other countries typically choose to uniquely lock their vector charts. At best, they wrap them within an internationally-recognized standard encryption, creating S-63 charts. From the developers' standpoint, standard encryption saves time and effort, avoiding the need to program a slew of unlock utilities for different encryptions. From a user's standpoint, charts are sold with an unlock code which then allows the charts to be displayed.

Despite this international standard, only a handful of recreational e-charting applications display S-57s (which include the free U.S. ENC and IENC vector charts). An even smaller subset, most notably MacENC, display international S-63 vector charts. Many software companies limit their chart support to the more traditional and familiar raster file formats.

To the consumer's detriment, the number of e-charting applications that support S-57s has been shrinking in the past couple of years. Both MaxSea and Nobeltec recently removed S-57 support in favor of their proprietary vector chart formats.

Raster and Vector Compared

If you understand the fundamental difference between raster and vector formats—that one is a scanned image from a paper chart and the other is an image created from a database—you may already foresee several important implications.

One obvious implication is file size. A raster chart is an image comprised of millions of saved pixels. Because raster charts are scanned pictures, and every pixel on the chart must be represented in terms of color and intensity, the files are relatively large.

A vector chart is a database of navigational objects linked to a chart position. Since vector charts are a database rather than an image, the information is compressed into much smaller files, taking up less space on your hard drive.

Furthermore, because raster charts are scanned images, you cannot "zoom in" on a raster chart to see more detail. Zooming the display past the accuracy of the digitized image, called *overscale*, results in an unreadable mosaic of colored pixels. (*Underscale*, which applies to both raster and vector charts, occurs when a more detailed chart is available than the one you are using.) Vector charts, because they are created on-the-fly using a database of objects and placement, allow you to zoom in to see more detail.

A similar distinction occurs when you attempt to rotate electronic charts, such as to display the chart course-up or north-up. Since raster charts are large image files, they don't rotate as quickly. Even worse, the text is fixed so it rotates with the image. Now you have a course-up or north-up chart with depth and aid information upside-down or sideways on your screen!

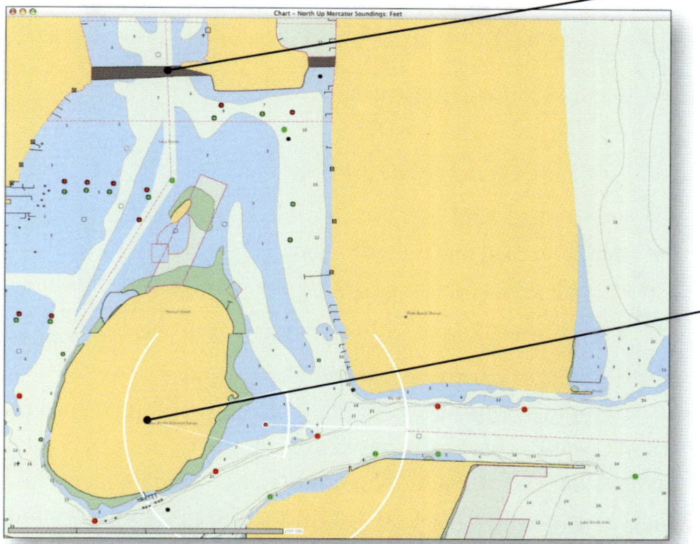

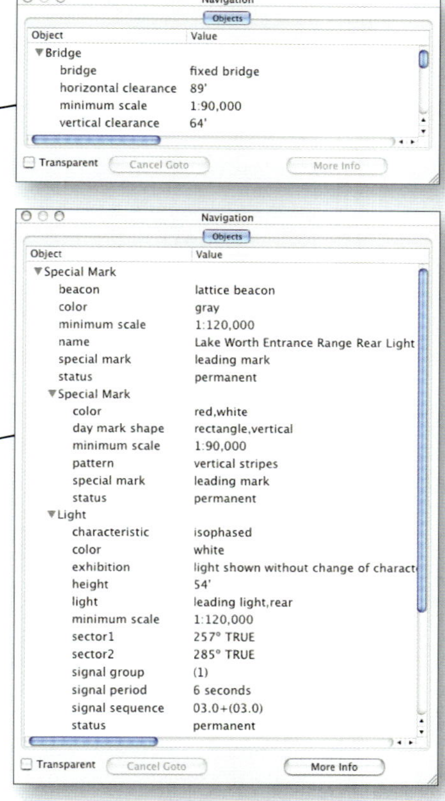

Intelligent Chart Features: *Clicking on a bridge or an aid to navigation on a vector chart displays windows with complete information about the charted object.*

Before international S-57 charts can be displayed, you must register your permit (like a software license or hardware key). Permits are not needed for free NOAA or USACE RNCs, ENCs, or IENCs.

RASTER CHARTS	**VECTOR CHARTS**
✓ Looks like a paper chart	Different appearance; some similarities to a paper chart
Scanned data: rows/columns of pixels	✓ Data as points, lines, and polygons; attributes in database
Standard appearance	✓ User-defined appearance
Large size (MB)	✓ Small size (KB)
Slower display and screen re-draw	✓ Faster display and screen re-draw
✓ Easier and cheaper to create	Difficult and costly to create
Difficult to correct and update	✓ Easy to correct and update
Zoom subject to pixelization	✓ Zoom unrestricted
Non-intelligent chart features	✓ Intelligent chart features and symbology
Cluttered look in densely charted areas	✓ Uncluttered appearance; user-definable
Whole chart rotates	✓ Chart base rotates; text/objects maintain screen orientation

Raster vs. Vector Comparison: *Aside from paper chart appearance and ease of conversion, vector charts win on most metrics, including intelligence and uncluttered appearance.*

Chart scale notation can be confusing. Just remember: the smaller the number, the more detail. For example, a 1:24,000 chart translates as "one inch on this chart equals 24,000 inches (2000 feet or one-third of a nautical mile) on Earth." This chart has more detail than a 1:100,000 chart. By the same logic, a 1:24,000 chart shows a smaller area than a 1:100,000 chart. So a chart with a smaller number is called a *large-scale chart* and has a *large* amount of detail.

In contrast, rotated displays are not a problem for vector charts. Vector chart software has the ability to create a display independent of orientation, with text on a separate layer. Text can be held to an easy-to-read horizontal alignment while chart features rotate.

The most important implication of the differences between raster and vector charts lies in their *intelligence*. A depth contour of "60" on a raster chart is simply a collection of black pixels to create the numeric characters six and zero. A same depth contour on a vector chart is a value in a database linked to its latitudinal and longitudinal positions. Because vector charts are databases of navigation items, they can be intelligently linked to the navigation and charting software (hence why a commercial ECDIS requires vector cartography). Raster charts are only pictures, so the software is even unable to differentiate between a land pixel and a water pixel.

With vector charts, navigational features can be identified and linked to the position of your vessel. The navigation software computes distances and provides warnings based on these calculations. Alarms can be programmed to warn the helmsman if the vessel crosses a depth contour (say 60 feet), enters a traffic separation lane, or is approaching a particular aid to navigation. And, best yet, users can tell the computer to display only the information they need at the time while the software continues to process all available information, whether it is displayed or not!

Although vector charts seemingly have the technological advantage, raster charts still currently win the popularity contest on two counts: availability and familiarity.

Because converting a paper chart to raster format is infinitely easier than creating a vector database, a complete U.S. raster catalog is already available through NOAA for free. But only about 60 percent of NOAA's projected chart cells are available in vector format. Creating a vector database of all chart features is an enormous project and the priority for conversion is driven by commercial ship traffic.

In 2001, NOAA initially identified 40 major ports for ENC completion by 2003. After these were completed, another 100-plus ports were identified. NOAA's priority is to begin with major ports, followed by their seaward connections. For example, the Intracoastal Waterway between the commercial ports of Norfolk and Charleston is not a priority. Instead, the seaward connection between these two ports, including Cape Henry and Cape Hatteras, is a priority. So far NOAA has completed about 600 of its roughly 1000 projected ENCs.

This huge task is hampered by government budgets and further slowed by the demands of simultaneously maintaining and updating an increasingly larger chart library. Alexandra Heliotis, Deputy Chief of NOAA's Marine Chart Division, points out that an average year produces 5,900 individual updates that result in corrections to 16,000 chart products.

Because they look familiar, raster charts are also much easier to use for planning and general chart viewing. Most boaters still have trouble assimilating the sparse display of a vector-format electronic chart. Fortunately, there is an effort underway to standardize paper chart symbols. When complete, this symbology should flow to both raster and vector charts, reducing the visual discrepancy between paper and electronic chart formats.

Encrypting Charts

To prevent unauthorized copying, chart files are often encrypted. Both raster and vector charts can be encrypted. Without some sort of encryption, chart files can be copied across computers or burned onto a CD, letting a single boater pass chart files on to hundreds of their closest boating buddies.

Although most countries, and all private cartographic companies, encrypt their charts, the U.S. government does not. Encryption became unnecessary with the decision to release the chart library for U.S. waters. In fact, as of 2008, only the U.S. considers its chart portfolio to be a free public good for its taxpayers.

Encrypted charts must be unlocked prior to their use. Unlocking an electronic file is accomplished with a hardware key and/or a software key.

Several chart distributors use a hardware key to unlock chart files (see *What's a Dongle* in Chapter 3). For example, both Nobeltec C-Map MAX Pro and MaxSea MapMedia charts unlock with a dongle. The advantage of a dongle is that charts can by used in different computers simply by copying the files and plugging the dongle into another computer's USB port.

Like software applications, charts can also be unlocked by entering an alphanumeric software key. For instance, Maptech uses a chart registration unlock process. Unlock codes may be generated through the chart distributor's website at the time of purchase. Alternately, a registration code may be issued online after you register your physical purchase of charts from a licensed retailer, such as with Canadian Hydrographic Service charts. Once you have your unlock code, you can use the charts in any e-charting application that supports that encrypted format.

Some software companies now include world chart coverage on DVD, but limit your access by unlocking regional purchases. For example, Nobeltec MAX Pro includes the entire world on a set of DVDs. But you can only access a region of charts after purchasing that region's unlock code. The advantage of this approach is you already hold the physical chart assets; you can access any new region of charts at a moment's notice with a phone call or Web login to purchase an unlock code for a new region. A disadvantage is that the chart data is only accurate as of the date of manufacture, not the unlock date!

SOURCES OF ELECTRONIC CHARTS

The latest scuttlebutt among U.S. boaters is "free charts." Can one really download free electronic charts on the Internet? Yes and no—it depends what types of charts and for what geographic regions.

Not every hydrographic office (the government agency in each country responsible for issuing charts) is ready to hand over the keys to its kingdom. Most countries sell their charts, either directly or through licensing agreements with private companies. In addition, a handful of companies also compile their own chart data, selling *proprietary* chart portfolios on DVD or flash card.

Downloading Free NOAA Charts

Free electronic charts are indeed available through the NOAA website—and have been since fall 2005. Both raster and vector charts for the U.S. are distributed using Web-enabled software called *ChartServer*. You can access this download site at http://chartmaker.ncd.noaa.gov/.

With free U.S. charts online, why isn't every American boater carrying the latest up-to-date charts in electronic format? That's NOAA's goal, but a few hurdles remain before the average boater is ready to click-and-chart. When downloading e-charts the first time, users face two challenges: knowing which charts they need, and implementing an over-the-Internet download strategy.

Before proceeding with ChartServer, you have to do a little homework.

First, you'll need to con-

If you prefer to store compact versions of chart catalogs on your laptop, download the individual state catalogs as PDFs at http://nauticalcharts.noaa.gov/mcd/ccatalogs.htm#state.

Solve the Cataloging Problem: *Identify chart numbers using a Catalog of Charts & Publications. Free and updated every few years, choose from five regions: (1) Atlantic Coast, (2) Pacific Coast, (3) Alaska, (4) Great Lakes, and (5) Gulf Coast. If unavailable at your chart agent, you can also interactively view them at http://nauticalcharts.noaa.gov/mcd/ccatalogs.htm.*

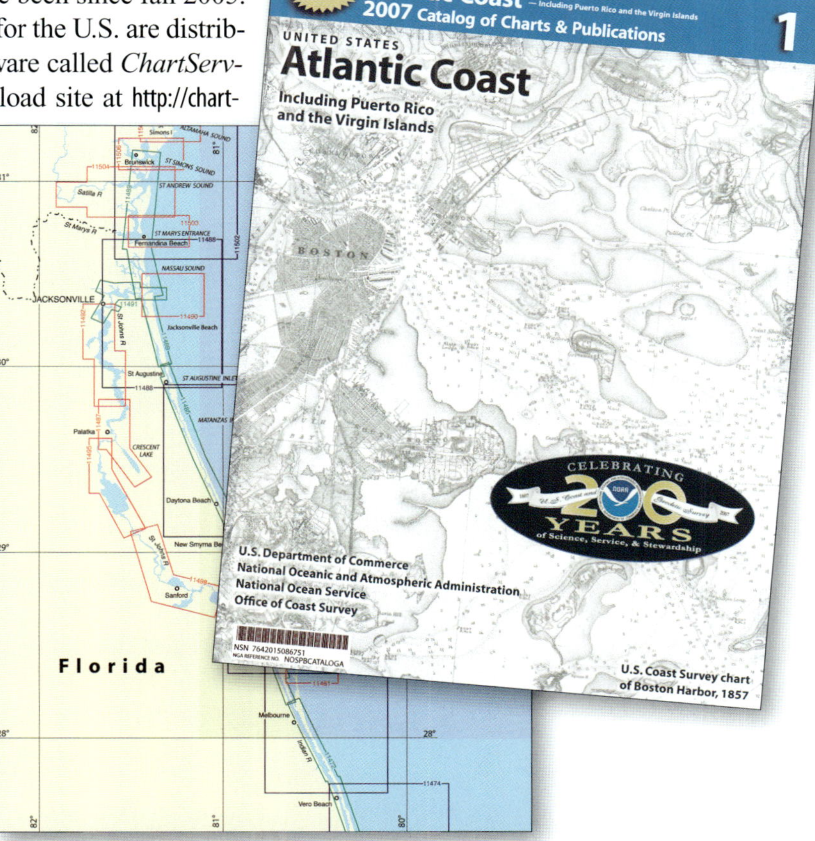

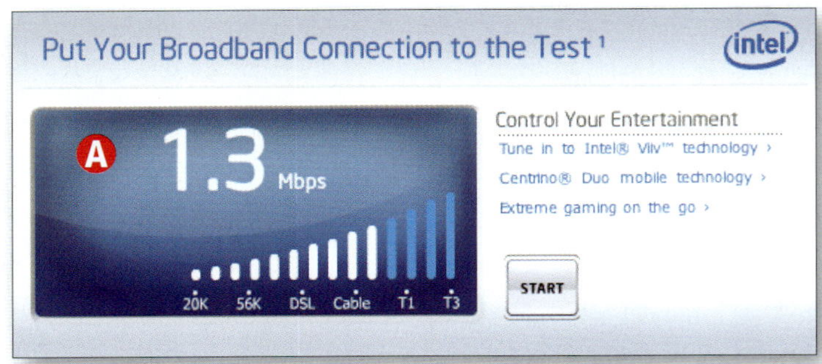

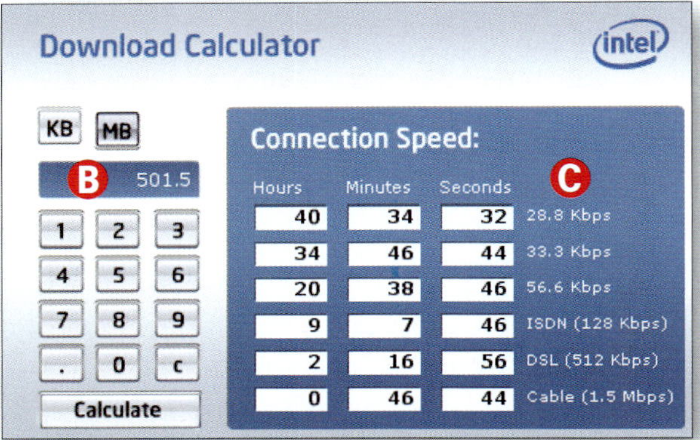

What's Your Speed Limit? A *Measure your Internet connection speed* **B** *Add up your chart files and enter the total download size in MB* **C** *View your estimated results!*

Download Speed Tests: *To find currently available speed test sites, Google "Bandwidth Calculator."*

struct a list, identifying the chart numbers you'd like to download. Knowing which charts to download, what we call the "cataloging problem," is the major reason many boaters continue to purchase charts on DVD instead of downloading them directly.

Unfortunately, ChartServer doesn't have a "New England" button, programmed to intelligently download all the charts you need to cruise New England. For complete chart coverage of New England, you first catalog, then download 107 chart folders, 209 chart panels, and 734 discrete files.

To identify the charts you require, use NOAA's *Catalog of Charts & Publications*. This resource lists all charts visually by their NOAA designator number. For example, the chart covering Martha's Vineyard to Block island is Chart 13218. Cape Cod Canal and Approaches is Chart 13236.

Second, you'll want to calculate your download time before blithely hitting the Order Selections button. If you're on a pay-per-minute dial-up connection, think carefully before you click. The New England example—with 209 charts and 734 files—involves a 501.5 MB (over one-half gigabyte) download. With a common 56 Kbps dial-up connection, the download time would be over 20 hours. Even with a high-speed DSL connection, it's still over two hours!

If you are willing to take the time or only want to download a handful of charts, use a download calculator to plan your download session. Interactive calculators are available on the Web at Intel's site (www.intel.com), at Speakeasy (www.speakeasy.net/speedtest), or at Speedtest.net (www.speedtest.net).

With your chart list in hand and an accurate estimate of your download time, visit ChartServer to download your chart selections. Begin by choosing the chart format: Raster Navigational Charts for raster (BSB) files, or Electronic Navigational Charts for vector files. After agreeing to the terms of use, you enter the ChartServer database. Online instructions guide you to Select, Review, and Order your chart list. Depending on the number of charts you'd like to download, you may have to plan multiple sessions. (ChartServer currently limits downloads to 100 charts per session.) ChartServer bundles, compresses, and downloads a single folder containing all chart selections to your desktop.

Stemming from its past partnership with NOAA, Maptech also maintains a chart download portal (www.freeboatingcharts.com). However, this site requires email registration.

If downloading chart files seems cumbersome, you can purchase electronic charts pre-cataloged on CD or DVD for less than $50. These chart sets are significantly cheaper than proprietary charts, such as the ones purchased on cards for a chartplotter. Chart compilations of NOAA charts are largely a convenience product, saving you the cataloging, organization, and time of downloading hundreds of files.

Example: Raster Download

When you download charts, ChartServer bundles, compresses, and delivers all of the charts in a single folder to your desktop. In the case of raster downloads, the name of this folder is a download ticket number that has nothing to do with the particular charts you have selected.

Let's take a look inside the RNC download folder for the Port Canaveral area (see NOAA Download Examples). In this example, NOAA RNC Chart 11478 Port Canaveral was selected. Two panels were downloaded: main panel 11478_1 Port Canaveral (1:10,000) and extension panel 11478_2 Canaveral Barge Canal Extension (1:40,000).

Each chart download includes several files. RNC 11478 downloads seven files: two KAP files, one TXT file, three PTC files, and one BSB file. Although raster charts are colloquially referred to as BSBs, the charts are actually the larger KAP files. A chart has at least one KAP file. Each KAP file represents a panel of the chart, such as a main panel, extension, or inset.

The TXT file is a small text file of the user agreement detailing legal terms and conditions. PTC files provide some applications the ability to download smaller update files to *patch* charts with the latest corrections. This compression technology was much more important prior to today's high Internet data transfer speeds. The BSB file is a companion parameters file informing the application software about scale, metrics, and geospatial references. Chart notes are also included.

Although not all files are necessary to display the chart, it's better not to meddle. The KAP and BSB files are fundamental. Although you don't technically need the TXT or PTC files, they are very small and it's always safer to keep chart files as a complete package.

RNC Chart Codes

Raster charts share the same naming convention as their paper counterpart. Since raster file names are the chart number, it helps to understand this coding system. A paper chart's first digit refers to a world region. Region 1 covers U.S. and Canadian waters. The second digit, coupled with the first digit, represents a subregion. The Gulf of Mexico to Cape Hatteras is subregion 11. The remaining three digits are assigned counterclockwise within the subregion and typically sub-grouped by chart scale.

Example: Vector Download

ChartServer bundles, compresses, and delivers ENC charts to your desktop in a folder labelled ROOT, containing a CATALOG.031 file and nested chart folders.

Let's take a look inside the nested ENC US5FL82M download folder for the Port Canaveral area (see NOAA Download Examples). Selecting ENC US5FL82M downloads six files: three chart files 000, 001, and 002, and three TXT files. Notice that while the RNC had two chart *panels*, the ENC has three *cells*. ENC cells do not match RNC panels. The S-57 standard used in international ENC production is limited to 5 MB, resulting in a different "quilting and stitching" algorithm.

There are also three TXT files in this case, rather than a single generic user agreement text file. ENCs typically include TXT files for notes and cautions. In addition, some charts also include JPEG or TIFF image files (not shown in this case) for photo or illustration references. For instance, the vector charts produced by the U.S. Army Corps of Engineers include numerous illustrations of bridges and locks with important annotations and photos of recent wrecks or other hazards to navigation.

ENC Chart Codes

NOAA vector charts use a different numbering convention than raster or paper charts. The first two digits indicate the country hydrographic office, in this case the U.S. Office of Coast Survey. The third digit communicates the scale: 1-overview, 2-general, 3-coastal, 4-approach, 5-harbour, and 6-berthing. The fourth and fifth digits are typically state *postal* codes. The next-to-last two digits are the ENC number with a final digit M indicating the use of metric units.

Downloading Free USACE Charts

The U.S. Army Corps of Engineers (USACE) is responsible for U.S. inland waterways, especially the navigable rivers and canals of the Midwest. USACE charting efforts focus on shallow draft inland waters, meaning 14 feet or less at minimum maintained depth.

The Army Corps distributes its charts only in vector format. Inland Electronic Navigation Charts, called IENCs, are also available for free download through the USACE website (www.tec.army.mil/echarts). The download process is similar to NOAA's ChartServer.

In 2008 the Army Corps completed all rivers and waters on the Great Loop, including the Illinois, Mississippi, Black Warrior, Tennessee, Tombigbee, and Mobile Rivers. They have currently issued vector charts for about 90 percent of inland river system usage. Over time, coverage for an additional 2500 miles of Inland Waterway with lower vessel traffic will be added.

IENC Chart Codes

IENCs use a slightly different naming convention with several important clues to their location. Rather than a state postal code, the two-letter designator is a *river* code, such as UM for Upper Mississippi. The last three digits indicate that chart's beginning mileage. For example a chart name ending in UM155 begins at mile 155 of the Upper Mississippi.

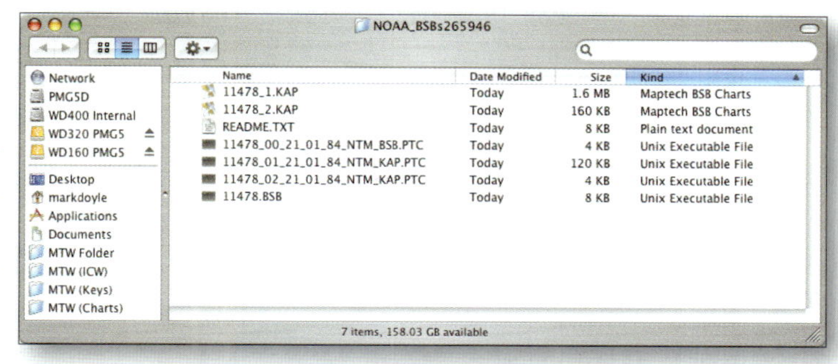

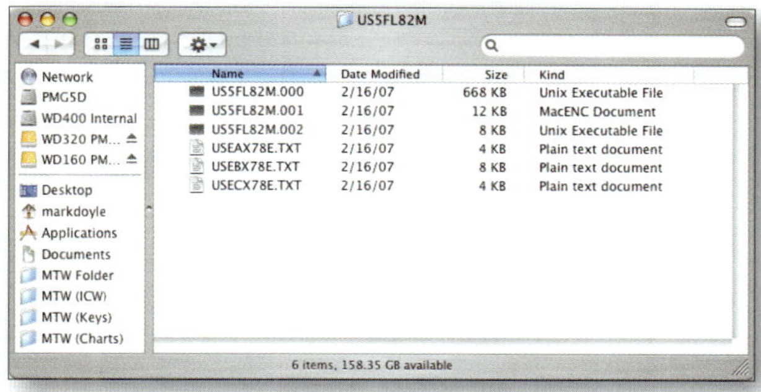

NOAA Download Examples: *Raster chart RNC 11478 has two chart panels (above). Vector chart ENC US5FL82M has three chart cells.*

Unlike raster charts, when loading vector charts, you must select a chart catalog (CATALOG.031) in your Chart Manager, rather than an individual chart cell.

Charts on the Cheap: *To find currently available commercial electronic chart compilations, Google "DVD NOAA Electronic Charts."*

Once you understand chart naming conventions, you can easily search and display charts. For RNCs, simply type in a chart number instead of resorting to a slow scroll-and-redraw process. You can search for ENCs on your hard drive by postal code. If you want to view IENCs of the Upper Mississippi, type UM in the search field. Any IENC of the *Upper Mississippi* River has this two letter designator in its file name.

More is Better

E-charting applications differ widely in their support of various chart formats.

Some programs, such as DigiBOAT Software-On-Board, go with a single chart source (C-Map). Unfortunately this single-sourcing excludes free U.S. charts, popular with North American cruisers.

More typically, applications support multiple formats, and may be unique in supporting a particular format, such as MaxSea and British Admiralty ARCS.

Overall, two companies stand out in the cartography category: both Fugawi Marine ENC and NavSim BoatCruiser support many formats and embrace both free U.S. cartography and one of the high-quality proprietary chart sources (Navionics and C-Map respectively).

Digital Nautical Charts

There is another, less well-known, source of free electronic cartography. The National Geospatial-Intelligence Agency (NGA), the division within the U.S. Department of Defense responsible for geospatial intelligence, also produces non-classified vector charts.

These vector charts, called Digital Nautical Charts (DNCs), supplement NOAA's vector coverage and include international coverage. Eventually the portfolio will include approximately 5,000 nautical charts between 84° North latitude and 81° South latitude. They are unclassified and available for download (www.nga.mil/portal/site/dnc/).

Unfortunately, there is a major glitch for recreational users: no e-charting application currently reads native DNCs.

Coastal Explorer includes 250 DNC files, re-formatted for their software, but does not support DNCs downloaded from the NGA. The NGA offers a DNC viewer, called *EDGE Viewer*, intended for NGA government or commercial customers. Many boaters currently purchase the paper chart counterparts of DNCs for non-U.S. waters (www.naco.faa.gov). DNCs may be supported in the future as this chart resource expands.

ARCS Worldwide Raster Charts

Outside the U.S., electronic cartography is the responsibility of each country's hydrographic office. In 1996, the United Kingdom Hydrographic Office began production of raster navigation charts under the name ARCS (Admiralty Raster Chart Service).

Cartography and Supplemental Data[1]	Free U.S. rasters BSBs (RNCs)	Free U.S. vectors S-57s (ENCs, IENCs)	DNCs
SeaClear II	Yes	No	No
Software-On-Board	No	No	No
NavimaQ	Yes	No	No
MacENC	Yes	Yes	No
TIKI Navigator	Yes	No	No
Marine ENC	Yes	Yes	No
BoatCruiser	Yes	Yes	No
Coastal Explorer	Yes	Yes	Yes
The Capn	Yes	No	No
Nobeltec VNS	Yes	No	No
Chart Navigator Pro	Yes	Yes	No
RayTech RNS	Yes	No	No
MaxSea	Yes	No	No
Nobeltec Admiral	Yes	No	No

Software Comparison: *Free U.S. Government Chart Support*

Cartography and Supplemental Data[2]	International S-57s	International S-63s	British Admiralty ARCS	Seafarer	CHS Canadian rasters	Scan and geo-reference paper charts
SeaClear II	No	No	No	No	No	Yes
Software-On-Board	No	No	No	No	No	No
NavimaQ	No	No	No	No	Yes	No
MacENC	Yes	Yes	No	No	Yes	No
TIKI Navigator	No	No	No	No	Yes	No
Marine ENC	No	No	No	No	Yes	Yes
BoatCruiser	Yes	No	No	No	Yes	No
Coastal Explorer	Yes	No	No	No	Yes	No
The Capn	No	No	No	No	Yes	No
Nobeltec VNS	No	No	No	No	Yes	No
Chart Navigator Pro	No	No	No	No	Yes	No
RayTech RNS	No	No	No	No	Yes	No
MaxSea	No	No	Yes	Yes	Yes	No
Nobeltec Admiral	No	No	No	No	Yes	No

Software Comparison: *International Chart Format Support*

Leveraging one of the world's largest paper portfolios, ARCS has near worldwide coverage, with over 3000 raster charts covering the world's commercial shipping routes and main ports and harbors (www.ukho.gov.uk).

Admiralty charts, although produced by the British government, are licensed and sold through a network of distributors (www.ukho.gov.uk/amd/howtobuy.asp).

Seafarer Charts

The Australian Hydrographic Service also produces raster charts. Seafarer charts cover waters of significance to Australia. Australia is also in the process of producing vector charts to be licensed and incorporated into C-Map and Transas portfolios.

Like ARCS, Seafarer charts are sold through the licensed distributors listed on their website (www.hydro.gov.au). The complete set of Seafarer RNCs, covering regions surrounding Australia, includes about 400 charts and costs about $1,500.

Canadian Hydrographic Service Charts

Until recently, Canadian charts were distributed through a public/private partnership between the Canadian Hydrographic Service (CHS) and Nautical Data International (NDI). In effect, NDI had exclusive licensing of Canadian government charts.

In 2007, the CHS and NDI relationship ended in a split similar to the NOAA/Maptech separation of 2005.

At that time, the Canadian government decided to continue to distribute their charts through licensed dealers, rather than making them free to the public. Canadian charts must be purchased through an authorized CHS dealer, listed on their website (www.charts.gc.ca). The CHS produces both raster (BSB) and vector (S-57) format electronic charts.

ENCs for the World

Every day, government hydrographic offices complete vector chart coverage of new regions. For an up-to-date list of completed ENC regions, visit the IHO website. It lists all completed or in-production ENC cells for the world (http://services.ecc.as/ihocc/public).

Alternatively, the International Centre for ENCs, an association of national hydrographic organizations that coordinates the production and distribution of official electronic charts, has a worldwide ENC catalog (www.ic-enc.org).

Example: The U.S. Great Loop

Longer transits typically cross chart jurisdictions and require sourcing charts from several different government agencies. For example, the popular Great Loop—circling through the U.S. inland river system and U.S. and Canadian canals and Great Lakes—requires assembling a patchwork of U.S coastal, U.S. inland, and Canadian charts.

Many boaters are confused about electronic charting for the Great Loop—and their confusion is warranted. A Great Looper needs to source charts from multiple charting agencies, both within and outside the U.S. Furthermore, some regions are available only in raster or only in vector format. An important rule of thumb is that vector charts are only available for "first-phase" areas including ports and commercial connections. Fortunately, both the U.S. Great Lakes and the major inland river systems are considered commercial transit routes so were an early priority for vector conversion.

To understand different chart sources and formats it's easiest to "armchair sail" the Great Loop. Leaving Mobile, the Gulf of Mexico has free NOAA raster coverage. The Intracoastal Waterway, Chesapeake and Delaware Bays, and Atlantic Seaboard are also covered by NOAA, so free raster charts are available for their entire lengths. However, free vector charts are only available for areas deemed important to commercial traffic, such as major ports and coastal transit areas.

The Hudson River is also charted by NOAA, so the same rule applies: free raster charts for its entire length and free vector charts for the ports of New York and Albany. At Troy and the Waterford Locks, you enter the New York State Canal System, the so-called Erie Canal. Although technically within the jurisdiction of the New York State Canal System, NOAA produces both paper charts and free raster charts for these waterways. Future vector conversion, if deemed necessary, would be the responsibility of New York State.

Approaching Oswego, both raster and vector charts are available through NOAA. Beginning at the port of Oswego,

For any questions about U.S. charts, call NOAA's new toll-free Navigation Inquiry Line at 1-888-990-NOAA. You can also submit your question by Web form at http://ocsdata.ncd.noaa.gov/dr/inquiry.asp.

Flash Cards: *C-Map and Navionics provide charts and supplemental data on flash cards of varying formats.*

the U.S. Great Lakes have complete RNC and ENC coverage. However, NOAA's charts, whether raster or vector, only cover U.S. waters. The only exceptions are small-scale charts that show a large region and may include portions of Canadian waters.

Therefore, the transit from Oswego to Chicago requires a mix of Canadian and NOAA charts. Canadian charts are not free and must be purchased through licensed resellers.

At the port of Chicago, you enter waters surveyed and charted by the U.S. Army Corps of Engineers. From Chicago to the Gulf of Mexico destinations of Mobile or Baton Rouge, vector charts are the only electronic chart option. There are no government raster charts for this stretch. Raster charts may be purchased, but with a caveat. Since these files are not government-produced, they are created by scanning dated paper chart books, making them expensive and of questionable accuracy.

Fortunately, the Canadian Government, NOAA, and the USACE work together using what is called an edge match system. For example, NOAA produces an ENC to the head of Mobile Bay. Their cell for that region, "Mobile Bay Alabama," abuts the USACE IENC running from Mile 0 to Mile 88 of the Tombigbee River, "Mobile Bay to East Bassetts Creek."

The biggest hurdle for Great Loopers and other long-distance cruisers is not the availability or seamless alignment of electronic charts, but gathering them from multiple agencies and sources.

Proprietary Charts

In addition to government hydrographic offices, private companies are also involved in the production and distribution of electronic charts, either through licensing agreements and/or their own data collection.

C-Map and Navionics are best-selling names for world chart coverage. Their proprietary charts are sold on disc, or more commonly, on flash card. A flash card, often called a *chip*, is smaller than a book of matches. If you purchase charts this way, you must use the card reader provided by the respective manufacturer. Both C-Map and Navionics sell USB card readers that unlock and read their charts on either PC or Mac.

C-Map and Navionics are both suitable for e-charting. Garmin emphasizes BlueCharts, designed for Garmin chartplotters rather than personal computers. Although they do offer chart CDs, these only work with Garmin PC viewer software.

Many boaters complain about the high cost of a chart card, typically costing over $200 for a region of the U.S. This price difference is exaggerated by the availability of free NOAA charts or low-cost charts on DVD. The price difference is not in the media—a plastic flash card doesn't cost much more than a plastic disc—but in the source of the charts. Charts sold on cards are typically vector charts, which are not available yet for all regions through NOAA. You are not paying for the card per se, but for the manufacturer's cost of converting the original paper chart information into their proprietary vector format.

You are also paying for the manufacturer's supplemental data mining. All of the facility information, aerial photography, and satellite imagery costs do add up.

However, although proprietary cartography is some of the best in the world, we're always surprised by some of the inaccuracies in *everyone's* chart collections. To be fair, international chart and port coverage is a phenomenal amount of data to manage. But we often notice gross omissions or inaccuracies, such as phone numbers without area codes, inaccurate city names, humorously misplaced ports, marina icons placed well inland, entire regions with omitted navigation aids, and even aerial photographs with the wrong image linked to a location. As a warning, always be alert to possible inaccuracies when viewing a single source of information.

World Coverage

When it comes to electronic charts, "world coverage" should not be taken literally. Although several chart companies have impressive portfolios, gaps in coverage, particularly outside the U.S., Canada, and Western Europe, are common. Carefully evaluate chart coverage for any cartographic manufacturer you choose as a partner. Remember, these cards are married to devices and it will be fairly painful (and expensive) to change chartplotters or ECS software later on.

Several companies are considered worldwide chart distributors, including Garmin, Maptech, NDI, MapMedia, Transas, Nobeltec, C-Map, and Navionics. However, each sells charts in different formats, on different media, and viewable with different software or hardware limitations.

As already mentioned, Garmin BlueCharts are intended for chartplotters rather than personal computers. Even with a card reader, you cannot use a Garmin BlueChart for e-charting on your PC except with Garmin's limited proprietary planner software.

Maptech has aggressively partnered, collecting an extensive world portfolio of raster charts in their locked BSB4 format. Not all e-charting applications that support general BSB charts also support BSB4s, but many do. Be sure to check specifically for BSB4 software compatibility. Maptech charts are sold on disc, which easily load into a personal computer.

NDI is a chart distributor, focusing on Canadian charts (it was the Canadian equivalent of Maptech through the 1990s), but also has a growing worldwide portfolio. It is an official distributor of Primar S-57 format vector charts.

The European marine software company MaxSea has its own chart division, called MapMedia. They have an impressive chart portfolio of nearly 6,000 charts specifically designed for use with MaxSea's e-charting software. In fact, the new MaxSea Time Zero software depends on MapMedia charts in the .mm3d format. These cartographic collections include raster and/or vector charts, bathymetric data, and satellite photos, fueling the seamless display of realistic 3D charts.

Although most Americans are less familiar with MapMedia charts than sources such as C-Map or Navionics, they are well established in Europe. In fact, MapMedia's vector charts are based on Navionics data. Coverage includes the U.S. (and Hawaii), Canada, Mexico, Central America, Europe, Australia, New Zealand, and the Caribbean.

Transas charts, known for its established worldwide coverage, are for commercial vessels. Their native cartography only runs on Transas software. Nobeltec purchased licensing rights from Transas many years ago, incorporating their gold mine of data into the Nobeltec Passport chart portfolio. In the past, recreational boaters gained the benefits of Transas data through the purchase of sets such as Nobeltec Passport World Charts. Today, this Transas data lives on in most proprietary cartographic world base data.

Nobeltec recently acquired C-Map, meaning a shift away from its Passport charts to C-Map MAX Pro cartography.

C-Map charts are sold on flash card or on disc and include NT+, MAX, and MAX Pro chart collections. Each upgraded version adds extra chart features such as tides, currents, perspective views, aerial photos, marina charts, and flashing aids to navigation. C-Map charts are compatible with many e-charting applications, have extensive global coverage, and include excellent supplemental data.

Navionics charts, sold on flash card, are also compatible with computer-based e-charting. Navionics sells charts by region at Silver, Gold, Platinum, and Platinum+ levels. Platinum and Platinum+ are their premium offering, adding 3D bathymetrics, complete Coast Pilots, satellite photos, and aerial photos of marinas and ports.

Regional Coverage

Sometimes small geographic regions fall through the electronic charting cracks. The heavily-boated Bahamas are a perfect example. For years boaters lamented that they couldn't get good charts—much less good digital charts—of the Bahamas. With small companies stepping in to fill these gaps, excellent specialty regional charts are now available.

Explorer Charts began as a personal project of cruisers Sara and Monty Lewis, who created cruising guides and paper charts of the Bahamas (www.explorercharts.com). Their color charts were largely created by hand, supplementing their own surveys with satellite image tracings. Wavey Line charts, also created by a cruising couple, Bob Gascoine and Jane Minty, cover the Bahamas, Turks & Caicos, and Hispaniola (www.waveylinepublishing.com).

Both C-Map and Navionics offer update services: C-Map MAX Pro updates automatically online through an annual membership and Navionics with a trade-in program and free discrepancy updates on its website.

Cartography and Supplemental Data[3]	Navionics cards	C-Map cards	C-Map CD/DVD	SoftCharts
SeaClear II	No	No	No	Yes
Software-On-Board	No	Yes	Yes	No
NavimaQ	No	No	No	Yes
MacENC	No	No	No	Yes
TIKI Navigator	No	No	No	Yes
Marine ENC	Yes	No	No	Yes
BoatCruiser	No	Yes	Yes	Yes
Coastal Explorer	No	No	No	Yes
The Capn	No	No	No	Yes
Nobeltec VNS	No	No	Yes	Yes
Chart Navigator Pro	No	No	No	Yes
RayTech RNS	Yes	No	No	Yes
MaxSea	No	Yes	Yes	Yes
Nobeltec Admiral	No	No	Yes	Yes

Software Comparison: *Proprietary Chart Formats*

Ask the Experts: *To find currently available info on a cruising destination's cartography, Google the specific region and broader topics "charts" and "forum."*

Chart cards (often called *chips*) are now burned on demand at certain marine stores rather than stored as inventory. But just because the card was burned that day, it's only as fresh as the date of the master owned by the store. Confirm you'll receive the latest edition available before purchase.

These smaller companies, often actively engaged in surveying a geographically small but under-charted area, typically don't directly sell their charts as electronic files. Instead their chart assets are licensed to another company, which folds the regional data into its chart library database.

For example, Explorer Charts are incorporated into C-Map NT+, C-Map MAX, Garmin BlueChart 6.0, and Nobeltec Passport Deluxe electronic charts. Wavey Line electronic charts are sold through EarthNC, planning software built upon the Google Earth platform.

A relative newcomer in the chart-making business, NV-Verlag, produces the Caribbean Yachting Charts (CYCs) in raster format and has recently cut deals with all the European hydrographic offices to produce their charts in a new raster format called nv.digital. These files are BSB/EAP pairs, a variant of Maptech's BSB/KAP standard with a different encryption scheme. Early supporters of this new format are Chart Navigator Pro, Coastal Explorer, Fugawi Marine ENC, MacENC, OziExplorer, and TIKI Navigator. NV-Verlag specializes in charts of the Caribbean (www.cyc-charts.com) and Europe (www.nv-verlag.de).

If you want the best charts of a particular area, choose chart collections that license the best cartography and supplemental data for that region. It's widely acknowledged that chart quality varies by region across chart vendors. However, chart collections and licensing arrangements are constantly changing. Research the differences in chart quality for your particular cruising geography by talking to other boaters and searching the forums.

SUPPLEMENTAL DATA

Collections of digital charts now encompass much more than contour lines, depth readings, and navigation aids. With today's personal computing power, digital charting is no longer restricted to flat blue-and-brown images of nautical charts. Most e-charting applications also display tide and current predictions, shoreside overlays, satellite and aerial images, topographic maps, and 3D bathymetric views of the sea floor.

Higher-end proprietary chart portfolios, such as those provided by C-Map, Navionics, Nobeltec, Maptech, or Map-Media, include significant data extras. However, these bonus files must work together with the e-charting software. You not only need the supplemental data files, you must have an e-charting program specifically written to support those data features. The software reviews in Part Four, *Choosing an Application*, summarize which packages support extras such as 3D viewing, aerial photos, or bathymetric charts. Here we focus on the types of supplemental data commonly included when you purchase a chart disc or card from one of the major cartography vendors.

Tide and Current Predictions

Tide and current data comprise a basic feature of more expensive proprietary cartography such as C-Map or Navionics. Any e-charting application that supports one of these proprietary chart formats takes advantage of this data to display tide and current predictions. Vendors who produce both e-charting software and cartography, such as Maptech, Nobeltec, or MaxSea, integrate their own tide and current data.

For obtaining tide and current data without integrated cartography, some e-charting applications turn to a third party utility. For example, Mr. Tides (http://homepage.mac.com/augusth/MrTides) or XTide (www.flaterco.com/xtide) are two popular sources. But, again, the e-charting application must support the viewing of Mr. Tides or XTide data.

Each e-charting application displays tides and currents in a different way. Tide predictions are typically accessed by clicking on a "T" or tidebar icon on a chart, designating a tide station. This, in turn, opens a tidal prediction window. Currents may be displayed as text tables, line graphs, and/or 24-hour polar graphs. E-charting applications utilizing C-Map MAX charts may even support the option of overlaying colored arrows showing direction and intensity of currents.

Shoreside

Supplemental shoreside data typically includes street and road maps, points-of-interest (POIs), and/or marina locations or descriptions.

Although many of the higher-end chart sets include shoreside data, the quality varies. Because Maptech has produced cruising guides for decades, their collection of supplemental data is known to be strong for shoreside marina locations. Navionics and C-Map also provide good shoreside data.

Nobeltec's Passport charts (superseded by C-Map MAX Pro cartography) had poor supplemental data. This was a critical factor in Jeppesen's decision to acquire and integrate C-Map offerings into the Nobeltec software portfolio.

Coast Pilots

Coast Pilots, produced by the U.S. government, are an incredible wealth of information. Published by region, they include supplemental information on ports, anchorages, and marinas.

Although many chart sets include Coast Pilots for their region of coverage, most e-charting applications only view the information as PDF documents using Adobe's Acrobat Reader. Coastal Explorer and Chart Navigator Pro are notable exceptions, integrating geo-referenced Coast Pilot entries directly accessible from the chart location.

Satellite Images

Satellite images are most useful for boaters who transit areas with poor charting and/or shifting channels. They are useful for visualizing a channel or harbor entrance. They can also be used to overview the land features adjacent to an anchorage.

Many chart sets include satellite images, which are supported by an increasing number of e-charting applications. Some applications, most notably EarthNC and Fugawi Marine ENC, also provide a way to access free satellite images through Internet sources such as Google Earth.

Aerial Photos

Aerial photos provide a bird's-eye preview of a harbor entrance, channel, or marina layout. Maptech's aerial photo collection is particularly good, well-supported by Coastal Explorer or Chart Navigator Pro software. C-Map and Navionics also include aerial photography, often called *nav photos*.

However, few applications can access these photos directly by clicking on a chart icon. Simply downloading a folder of Maptech nav photos only results in the proverbial needle in a file folder haystack. Unless the e-charting application is written to geo-reference and display that particular photo database, you will not be aware of—nor be able to directly access

via the chart—an aerial photo of a particular marina. Today, Coastal Explorer, Maptech, and chart card suppliers such as C-Map and Navionics are your aerial photography options.

Topographic Maps

Topo (pronounced toh-po) maps may also be included with marine chart sets. These files work with e-charting applications that support this feature to display elevation contours for land adjacent to coastal waters. Some applications go a step further and transform the contours into rotating 3D perspective views.

About half of the e-charting applications support topographic maps. Boaters in topographically-rich areas such as Puget Sound or Maine benefit most from this data extra. Where the terrain is relatively flat, topographic maps don't contribute much to on-the-water navigation. For these boaters, traditional navigation charts, which show conspicuous topographic contours, should be sufficient.

Bathymetrics

Bathymetric refers to the measurement of undersea depth data. Although bathymetric charts cover harbor and channel entrances, they are intended primarily for offshore fishing. Several e-charting packages support the detailed 3D bathymetric charts included in premium cartography sets.

For the serious offshore sport or commercial fisherman, two of the vendors, Nobeltec and Furuno, provide support for custom bathymetric recording. With this feature, which requires optional add-on software modules and hardware, you record the sea bottom topography to create your own bathymetric charts. Furuno MaxSea even blends your data with its bathymetric cartography—and can display the sea floor in stunning rotating 3D.

Cartography and Supplemental Data[4]	Satellite geo-referenced photos	Aerial informational nav photos	Topographic maps	Customizable Bathymetric data
SeaClear II	No	No	No	No
Software-On-Board	No	No	No	No
NavimaQ	No	No	No	No
MacENC	No	No	No	No
TIKI Navigator	Yes	Yes	Yes	No
Marine ENC	Yes	Yes	Yes	Yes
BoatCruiser	Yes	Yes	Yes	Yes
Coastal Explorer	Yes	Yes	Yes	No
The Capn	Yes	No	No	Yes
Nobeltec VNS	Yes	No	No	Yes
Chart Navigator Pro	Yes	Yes	Yes	Yes
RayTech RNS	Yes	No	Yes	No
MaxSea	Yes	No	No	Yes
Nobeltec Admiral	Yes	No	No	Yes

Software Comparison: *Supplemental Data*

TRENDS IN DIGITAL CARTOGRAPHY

Digital charts are already evolving in three important ways: a shift toward "intelligent" vector charts; the emergence of user-generated chart data; and the prospect for online automatic chart updating.

Although boaters were initially reluctant to embrace vector charts, preferring the traditional look of raster charts, vector charts are rapidly gaining ground. Vector files are smaller, more intelligent, and incorporate frequent updates more easily. Vector charts are needed for important and useful navigational e-charting features such as route checking, alarms, and boundaries.

In fact, NOAA's long-term vision is a reversed cartographic process: printing paper charts from an accurate and up-to-date vector database.

The source of chart data is also changing. The incorporation of publicly-available satellite imagery is a step in this new direction. With more and more free data available online, supplemental data need not come from a boxed chart set. You can now go to the Web for layers of additional marine data, ranging from user-posted photographs to the locations of West Marine stores.

A new planner, EarthNC, is the best example of this trend toward user-generated nautical data sources. Its collection of Google Earth KML layers is analogous to a Wikipedia of e-charting. (EarthNC is reviewed in detail in Chapter 13.)

Finally, the old days of hand-transcribing Notice to Mariner changes on your paper charts are long gone. Today's e-charting packages are just beginning to incorporate plug-ins that automatically update chart files. NavSim BoatCruiser was the first e-charting application to include an auto-chart update option. Coastal Explorer's new 2009 Edition is introducing an auto-chart update Synchronization Tool.

Eventually real-time updating over the Internet for charts will be standard. In fact, C-Map MAX Pro charts, their premium professional collection, are now issued with a one-year license for free downloads of chart corrections. These corrections no longer come in as text, but automatically make individual corrections within each chart file.

Loading and Unloading Electronic Charts

In order for a charting and navigation application to display charts, you must tell it where the files are located. This task is called *loading* the charts and is separate from *copying* the charts to your hard drive. Charting applications typically use one of two metaphors for locating and recognizing charts: (1) a *smart folder*, or (2) a *directory*.

Smart folder applications, such as Nobeltec or GPSNavX, have a special designated folder on your hard drive. Any chart dropped into this folder is identified as a new chart. Each time you start the application, it scans that special folder, says, "Aha, those are the charts I have," and loads those charts as its database.

Directory-based applications, such as Fugawi Marine ENC or MacENC, build a directory of charts and chart folders. You can have multiple charts in multiple folders, but you need to actively tell or "point" your application to those new charts. Every application accomplishes this differently, described in detail in the application's manual or help files. The terminology varies, but the idea is the same: after the first set of charts that came with your software, you must direct it to new chart files.

Smart folder applications are easier to manage, but they aren't a panacea. Because everything is automated, they have a lot less flexibility and can cause problems. If your software uses a smart folder metaphor, do not get creative and rename files and folders!

With smart folders, it's tempting to be sloppy with file hygiene. Becoming a "chart collector," and indiscriminately dropping chart files into this folder, will severely bog down your laptop's performance. Furthermore, since every chart must first be read into the directory, a hefty smart folder will make your application significantly slower to open.

Directory-based applications require some attention to the manufacturer's operating instructions, but they are a precise chart management system since you choose to load and unload charts or regions as you need them. A little bit of time invested in learning your charting application's loading and unloading process will save directory clutter and maintain optimal laptop performance.

Even if your laptop is a screaming demon, only load those charts you need. Selectively pre-load and unload regions as you transit. We have literally thousands of charts on our laptop, but rarely need access to more than a few hundred at a time.

Chapter 7
E-Charting Software

In order to view an electronic chart on your computer, you must have charting and navigation software that can open, display, and manipulate chart files. Just as opening a PDF file requires Adobe Acrobat, or opening a DOC file requires Microsoft Word, chart files are opened by software applications designed to work with particular file formats.

Fortunately there are many e-charting software choices available, and they run the full gamut in price and features. You can download free versions or opt for several-thousand-dollar packages designed for megayachts.

E-charting software is typically divided into three categories. At the simplest, an e-charting *viewer* lets you open and view chart files. The next step, a *planner*, displays charts and performs simple planning operations such as creating waypoints. Finally, a *full-featured application* performs navigation and piloting functions, integrates tide and current predictions, downloads weather forecasts, and integrates with marine electronics.

VIEWER SOFTWARE

Viewer software is e-charting at its simplest: an application that displays a chart on a computer screen. But this simple feature means you can pan and scroll over a nautical chart, "armchair sailing" your computer to plan and preview routes.

Several e-charting vendors provide free, albeit de-featured, viewer versions of their full-featured software. These "starter sets" are intended to entice users to eventually purchase the full-featured option.

For example, Fugawi provides a free viewer, Fugawi View ENC, that displays S-57 format ENCs (www.fugawi.com). It displays a maximum of three charts at one time. A version without this limitation is available for about $20.

Similarly, Global Navigation Software provides a free download of its NavPak Professional Edition, but without the GPS plotting, called NavPak Demo (www.globenav.com). NavPak Demo can display raster and vector charts with an included conversion utility.

Maptech also provides a free application for printing and planning with Maptech BSB charts. Chart Navigator Viewer looks similar to Maptech's Offshore Navigator and Offshore Navigator Lite software, but does not integrate with a GPS. It is often used to demo Maptech charts and is available as a free download (www.freeboatingcharts.com).

NavSim, the makers of BoatCruiser and SailCruiser, provides a free NavSim Viewer with the purchase of two charts from their MapServer site (www.navsim.com). NavSim Viewer includes chart panning and zooming, seamless chart quilting, and reads BSB, SoftChart, NDI, and S-57 charts.

dKart Look and SeeMyDENC are two other options for free chart viewers. dKart Look, provided by HydroSERVICE (www.hydroservice.no), displays S-57 and S-52 charts. SeeMyDENC is distributed by SevenCs and displays SENC, DNC, and S-57 files (www.sevencs.com).

NOAA's On-Line Chart Viewer is currently the only free viewer option that also works on a Macintosh computer (http://ocsdata.ncd.noaa.gov/OnLineViewer). Because it's an online viewer, it works on any computer or operating system. However, you must have an Internet connection and Adobe Flash Player installed (www.adobe.com).

A large digital mapping company, CARIS, provides the free viewer with the most flexibility (www.caris.com). CARIS Easy View opens nearly any kind of graphic file, conveniently including electronic chart files. This free utilitarian chart opener has recently been discovered by boaters.

Chapter Terms

Full-Featured Application: A software application that displays charts; performs planning, navigation, and piloting functions; and integrates with a vessel's marine electronics.

Planner: A software application that displays charts and is capable of creating and storing waypoints and routes for navigation.

Software: The computer code that uses the capabilities of a computer to perform tasks. Also called an application, package, or program.

User Interface: The hardware and operating software by which a user executes commands on a device and the device conveys information to the user.

Viewer: A software application that displays charts on a personal computer.

Chapter Questions

- *What is the difference between a viewer, planner, and full-featured e-charting application?*

- *Where can I get free viewer or planner software?*

- *What basic and advanced features should I expect in a full-featured e-charting package?*

- *What are the differences among the Windows, Macintosh, non-traditional, and chartplotter user interfaces?*

- *What are some new trends in e-charting software?*

Planners: *Both C-Map and Navionics provide card readers and planner software to be used with their chart cards. C-Map's planner is called PC-Planner; Navionics planner is called NavPlanner.*

PLANNER SOFTWARE

Planner software adds an important tool for boaters: the ability to create waypoints and routes. However, these navigation assets must be transferred to a chartplotter for active use underway. Planner software is intended as a way to prepare a route in advance using the convenient mouse-and-keyboard and graphical user interface of a personal computer.

Most planners are provided by companies that sell charts, chartplotters, or charting software. Free planner software is often included with the purchase of a chartplotter or cartography set. In addition, some e-charting software companies offer a low-cost planning version of their full-featured version.

Card-Provided Planners

C-Map and Navionics, two major producers of cartography on flash card, provide planner software compatible with personal computers running Windows. Intended to leverage their flash card format, the planners let you plan routes on a PC for transfer to a chartplotter via a card reader.

C-Map's PC-Planner is bundled with their C-Map card reader, currently at $149 for the card reader and planner software CD. PC-Planner transforms your PC into a virtual chartplotter, letting you view charts and create waypoints, routes, and marks. Most importantly, this work can then be transferred to your chartplotter by saving the data to a flash memory card using the C-Map card reader.

Navionics provides a NavPlanner CD for $129, also bundled with a card reader. This planner software, compatible with Navionics charts, is used in the same fashion as C-Map's PC-Planner, but using a Navionics Multi Card Reader.

The planner software produced by cartography companies takes advantage of extensive supplemental data. For example, C-Map's PC-Planner is specifically designed for C-Map charts, known for their excellent data extras. With PC-Planner, your PC can display these extras, such as tide and current predictions, point-of-interest data, and marina information.

Chartplotter-Provided Planners

Chartplotter manufacturers also provide planner software so their users have the option of creating waypoints and routes at home. Again, data is transferred from computer to chartplotter using a blank flash card, or uploaded using a temporary cable connection.

Garmin and Lowrance produce their own planning software: MapSource for Garmin and GPS Data Manager for Lowrance. MapSource is included on Garmin's CD-ROMs of BlueCharts. GPS Data Manager is available as a free download from Lowrance's website (www.lowrance.com). Both planners transfer waypoint and route information between their brands of GPS or chartplotter units and a PC.

Furuno and Raymarine, producers of both chartplotters and e-charting software, also provide planner software. Because they sell full-featured e-charting software, their planners are covered in the following section on e-charting planners.

E-Charting Planners

Several e-charting software companies provide "lite" planner versions for free or at significantly reduced prices. These planners are based on the same software as their full-featured siblings, but limit features or functionality. For instance, they generally do not connect to a GPS sensor, meaning you cannot display your boat position. The number of waypoints and routes may be restricted. Instrument connectivity and advanced piloting and navigation features are often omitted.

Furuno distributes its MaxSea Planner for free, available on DVD from Furuno dealers. TIKI Navigator sells a light version of its TIKI Navigator Pro software, TIKI Navigator Planner, as a $37 download (www.tiki-navigator.com). Unfortunately, Nobeltec's eChart Planner is no longer available.

Maptech bundles several different planners or viewers with their digital charts or paper chartkit choices (www.maptech.com). (For a full accounting of all Maptech software choices, see Navigating Maptech, in Chapter 22.) Offshore Navigator Lite is included on the Companion CDs that accompany their paper Chartkits and Waterproof Chartbooks. In the past, Offshore Navigator Lite has also been included with other chart CDs, usually for waters outside of the U.S. It is very similar to Offshore Navigator, but does not include features such as tracks, instant and named waypoints, autopilot support, alarm zones, or a ship log.

Raymarine offers its RayTech Planner as a free download through its website or as a disc bundled with a Navionics USB card reader (www.raymarine.com). RayTech Planner is specifically designed to display Navionics charts and transfer waypoints and routes to Raymarine chartplotters and multifunction displays.

A new player on the e-charting scene takes the planner concept one step further. Although it is not yet a full-featured navigation application, EarthNC is more than a planner. EarthNC views charts and exports waypoints and routes, but with Google Earth's satellite imagery as the backdrop.

FULL-FEATURED APPLICATIONS

No one can deny that e-charting has come a long way. It's a different e-charting world than the early days of jury-rigged MacSea prototypes on world-girdling race boats, or bootlegged copies of The Capn passing dinghy-to-dinghy among Caribbean cruisers.

With laptops now comprising two-thirds of all U.S. computer sales, pocket-sized USB GPS sensors costing less than $60, and Internet access as easy as popping in a wireless card, e-charting has become commonplace. The arsenal of advanced tools now available to any boater armed with a laptop and electronic charting software would have seemed downright futuristic just a few years ago.

Want to scan and geo-reference paper charts of Croatia to create your own digital chart files? How about displaying a small window in the corner of your screen to stream video from your engine room? Or create custom bathymetric charts

Many software packages now include manuals as PDF files rather than hardcopy. In order to view and print PDFs you must have a copy of Adobe's Acrobat Reader, available free at www.adobe.com/downloads.

Out of Box Experience	Packaging	Documentation	Software Install	Load Charts	Load Supplemental Data	GPS Hookup	Technical Support
SeaClear II	N/A	2	3	2	N/A	2	2
Software-On-Board	N/A	4	3	3	N/A	4	2
NavimaQ	2	3	3	3	N/A	2	3
MacENC	N/A	2	3	3	N/A	3	3
TIKI Navigator	N/A	2	3	3	3	3	4
Marine ENC	3	3	3	2	2	3	3
BoatCruiser	4	2	3	2	2	3	3
Coastal Explorer	4	3	4	4	4	5	4
The Capn	5	3	3	4	4	4	4
Nobeltec VNS	4	3	3	3	3	4	3
Chart Navigator Pro	3	3	4	4	4	5	4
RayTech RNS	3	4	3	3	3	2	4
MaxSea	4	5	3	3	3	3	4
Nobeltec Admiral	4	3	3	3	3	4	3

(N/A) Not applicable; (1) Poor (3) Average (5) Excellent

Software Comparison: *Out of Box Experience*

User Interface	GUI metaphor	Screen space management	Program responsiveness	Chart display speed	Chart display quality	Customizable GUI	Customizable metrics	Split windows	Tabbed interface
SeaClear II	NT	4	4	4	3	2	1	No	No
Software-On-Board	NT	4	3	3	3	3	4	No	No
NavimaQ	Mac	3	3	2	3	2	3	No	No
MacENC	Mac	3	2	2	5	2	3	No	No
TIKI Navigator	NT	5	4	4	4	2	3	Yes	No
Marine ENC	Win	2	3	3	3	3	3	No	No
BoatCruiser	Win	4	3	4	3	4	4	No	Yes
Coastal Explorer	Win	3	4	3	2	3	3	Yes	No
The Capn	Win	3	3	3	2	4	4	Yes	No
Nobeltec VNS	Win	3	3	3	3	5	4	Yes	No
Chart Navigator Pro	Win	3	4	3	2	3	3	Yes	No
RayTech RNS	CP	2	4	3	3	2	2	Yes	Yes
MaxSea	Win	4	5	4	4	4	4	Yes	Yes
Nobeltec Admiral	Win	3	3	3	3	5	4	Yes	No

(NT) Non-traditional, Macintosh, Windows, (CP) Chartplotter; (1) Poor (3) Average (5) Excellent

Software Comparison: *User Interface*

Full-Featured E-Charting Applications

Chart Navigator Pro (PC) (R and V)
www.maptech.com

Coastal Explorer (PC) (R and V)
www.rosepointnav.com

Deckman (PC) (R and V)
www.sailmath.com

DigiBOAT Software-On-Board (PC) (V)
www.digiboat.com.au

Expedition (PC) (R and V)
www.iexpedition.org

Fugawi Marine ENC (PC) (R and V)
www.fugawi.com

Furuno MaxSea (PC) (R and V)
www.maxsea.com

Global Navigation NavPak (PC) (R and V)
www.globenav.com

GPSNavX (Mac) (R)
www.gpsnavx.com

GPSy (Mac) (R)
www.gpsy.com

Henry Navigation System (PC) (R and V)
www.chersoft.co.uk

MacENC (Mac) (R and V)
www.gpsnavx.com

MacGPS Pro (Mac) (R)
www.macgpspro.com

Maptech Offshore Navigator (PC) (R)
www.maptech.com

NavGator (PC) (R)
www.navgator.com

NavimaQ (Mac) (R)
www.barcosoft.com

NavSim BoatCruiser (PC) (R and V)
www.navsim.com

NavSim SailCruiser (PC) (R and V)
www.navsim.com

Nobeltec Admiral MAX Pro (PC) (R and V)
www.nobeltec.com

with your own vessel's soundings? Would you like to display the current location of a buddy boat?

All of these capabilities–and many more–are possible with the electronic charting packages currently on the shelf at your local marine store. Sharp competition in a crowded software market, and steady advances in the power of the average computer, have combined to produce software with enough functionality to indulge the most tech-savvy boater.

Getting Started

Although most of us look forward to a newly-purchased software package, we dread the nuts-and-bolts of getting started. After being wowed by a colorful box and the prospect of an exciting new nautical toy, often nothing seems to work as it should. Software installation, chart loading, GPS and instrument connections, learning new features, transferring files— it can all be too frustrating to classify as recreation. In this section we look at the specific components that combine to create a trouble-free or even pleasant out-of-box experience.

The good news is that no e-charting application reviewed in Part Four is beyond the abilities of an average computer user, or inappropriate or over-featured for a recreational boater. In fact, despite features such as AIS tracking, bathymetric recording, or polar optimization, they are all relatively easy to use. You don't need to be a computer scientist or marine engineer to use even the most advanced features on any of these applications.

But you should consider your own abilities. If you are uncomfortable with computers or are simply impatient learning new software, then a trouble-free getting-started experience is important. Software should have packaging that makes its features easy to understand. It must have well-written documentation. Software, charts, and supplemental data should install effortlessly. External devices should connect easily. And finally, if you do need help, there should be strong technical support.

Many boaters like to purchase a packaged set, including data discs, documentation, and sometimes even hardware extras such as multimedia card readers. Several companies provide very nice retail packages. Most notable is The Capn's professional-looking zippered portfolio. Nobeltec, Maptech,

and MaxSea also provide upscale out-of-box experiences. Fugawi Marine ENC, Coastal Explorer, and NavSim do a fine job. But SeaClear II, Software-On-Board, MacENC, and TIKI Navigator are available only via download.

Whether a software application is delivered on disc or via Internet download, what really matters is the ease of set-up. Thorough, accurate, and easy-to-read user manuals are important. Although one can muddle through any of these applications without a manual, reading the documentation makes the learning process much smoother.

However, traditional printed manuals are going the way of the abacus. Because of printing costs and frequent software updates, most e-charting packages do not include a printed manual. In fact, Nobeltec charges an additional $30 for its User's Guide. RayTech RNS includes a hefty spiral-bound User's Guide, but it is surprisingly thin on software content. The truth is, although a glossy color manual is impressive at first, we found it is more important that the overall documentation *portfolio* be well-organized, accurate, and current.

To avoid the constraints of printed documentation, most e-charting applications include some combination of integrated help menus or PDF files. For example, TIKI Navigator does not currently have an English manual, instead relying on tutorials and integrated help. Coastal Explorer and Chart Navigator Pro also rely on hot-linked chapters through a help menu. Although help files are useful for troubleshooting particular issues as you use the software, they are more difficult to use to broadly understand the software's features. Going through help menus screen-by-screen is a bit like reading an e-book. You can resort to printing the help files in some cases, but this is a slow and tedious way to put together a manual.

The most practical form of documentation is a PDF file included with the software. This format saves printing costs, is easy to keep up-to-date, saves weight aboard, can be printed section-by-section as needed, and most important, is searchable. Software-On-Board, Fugawi Marine ENC, and MaxSea have the best documentation in PDF format. MaxSea covered all the angles, combining three excellent documents, video tutorials, and an advanced integrated help system.

Getting started with an e-charting application requires

overcoming three main hurdles: loading the charts, loading supplemental data, and connecting instruments. The most pleasant installations include setup wizards, manage the chart installation, and then scan and assign ports for the GPS and other devices. Our most difficult getting-started experience was NavSim BoatCruiser, followed by Fugawi Marine ENC and SeaClear II. Many other packages also had isolated problems, requiring guesswork to locate the GPS port, multiple attempts to load certain chart formats, or calls to tech support to locate missing drivers. In contrast, Coastal Explorer, The Capn, Chart Navigator Pro, and MaxSea easily loaded all our cartography. Coastal Explorer, The Capn, and Chart Navigator Pro also loaded an extensive collection of supplemental data without difficulty. And we found Coastal Explorer and Chart Navigator Pro, both based on the same basic software program, to have the best port detector, making it easy to connect a GPS.

Naturally, the next question is the quality of technical support. Understandably, the quality of service varies considerably with the price. The very small companies must focus on web-based support such as FAQs, user forums, a knowledge base, and email inquiries. However, this doesn't imply the service is substandard—just that the mode of communication is more limited. For example, DigiBOAT, MacENC, and TIKI Navigator all rely on email technical support (with TIKI Navigator returning your email by telephone) and we found them to be very responsive. SeaClear II, a free non-commercial product, cannot reasonably provide any technical support, but it has developed an extensive self-help forum community among its users. Chart Navigator Pro, RayTech RNS, and MaxSea all have top-rate technical support, supplemented by robust forums, knowledge bases, and/or FAQs.

Four Types of User Interface

Although all of the e-charting applications are appropriate for anyone with an average level of computer literacy, some do have a steeper learning curve. We've already mentioned the important aspect of documentation and extras such as tutorials and FAQs to help you get started.

Another aspect that will impact your learning curve is the user interface, in particular whether an application offers a familiar format or a new approach. Although traditional interfaces are easier to learn right out of the box, non-traditional approaches can also be easy to learn. The reason companies redesign their graphical user interface (GUI) is to make it easier to use and better suited to a marine environment. Their success, of course, depends on specifics, such as the design of the menus, icons, and data windows.

E-charting applications currently span four types of graphical user interfaces: non-traditional interfaces, traditional Macintosh interfaces, traditional Windows interfaces, and those styled like a chartplotter.

Non-traditional user interfaces are more common among lower-priced start-up contenders. In all three instances—SeaClear II, Software-On-Board, and TIKI Navigator—the interface is designed afresh as a way to make the laptop more compatible with marine use. For example, SeaClear emphasizes efficient right-click mousing rather than pull-down menus. Software-On-Board has no menus and few actions that require dragging a mouse. TIKI Navigator avoids small-type menus, minimizes the use of the keyboard, and emphasizes mouse-driven large-font contextual menus. Surprisingly, these three pioneering approaches are not that difficult to learn. SeaClear's simple interface can be self-taught despite its lack of documentation. Software-On-Board gets you started with its top-notch documentation, and TIKI's tutorials makes learning painless.

The two Macintosh programs, NavimaQ and MacENC, use a traditional Mac interface. In fact, we have criticized both applications for minor inconsistencies with well-established Mac conventions, such as the absence of pull-out drawers and standard shortcuts.

Understandably, most PC-based e-charting applications choose a traditional Windows interface based on pull-down menus and toolbars filled with icons. Fugawi Marine ENC and The Capn both would benefit from some streamlining and updating of their menus and icons. The best Windows implementation—no surprise given the founder is an ex-Microsoft employee—goes to Coastal Explorer (and by extension, Chart Navigator Pro). These two programs get high marks for their clean and consistent Windows interface, de-

E-Charting Software (Cont'd.)

Nobeltec VNS MAX Pro (PC) (R and V)
www.nobeltec.com

OziExplorer (PC) (R)
www.oziexplorer.com

PassagePlus (Mac) (R)
www.windvector.com

RayTech RNS (PC) (R)
www.raymarine.com

SeaClear II (PC) (R)
www.sping.com/seaclear

SeaTrack (PC) (R and V)
www.seatrack.co.uk

The Capn (PC) (R and V)
www.maptech.com

TIKI Navigator (PC) (R)
www.tiki-navigator.com

Tridentnav (PC) (V)
www.tridentnav.se

(PC) Microsoft Windows; (Mac) Apple Macintosh
(R) Raster BSB support; (V) Vector S-57 support.

signed for seasonal or weekend boaters who don't want to remember shortcuts and menu choices each time they open the software.

One company, Raymarine, has a chartplotter interface designed to turn your PC into an on-screen version of a keypad. Boaters who are more comfortable with chartplotters than computers may prefer this interface. However, if you grew up on PCs, then a chartplotter interface may feel more cumbersome and constrained than a properly-executed Windows interface.

The most important side-effect of the user interface is what we called *screen real estate*, which is the amount of screen space available for viewing the most important window: the nautical chart. Laptop screens are limited, and nautical charts can be notoriously small and hard to read, so any program that maximizes screen real estate by reducing menu, window, and icon clutter scores high points.

RayTech's chartplotter interface, with large soft keys and databoxes, creates an incredibly small chart display area. Similarly, Fugawi Marine ENC's extensive menus, toolbars, and library windows often obscure the chart display. SeaClear II, Software-On-Board, TIKI Navigator, and MaxSea have the cleanest and largest chart display areas.

Because boater needs vary, applications work best when their interface and metrics can be customized. Customization is important for usability (such as larger fonts for far-sighted users), safety (such as twilight or night displays), and convenience (such as statute miles for Intracoastal Waterway travelers). Many customizations are simply nice extras, such as personalizing icons, colors, and toolbars.

RayTech RNS has the most rigid user interface, constrained by their choice of a chartplotter look-and-feel. SeaClear II, NavimaQ, and TIKI Navigator also have interfaces that are essentially pre-set. BoatCruiser, The Capn, and MaxSea allow for quite a bit of customization, but Nobeltec VNS

MAX Pro and Admiral MAX Pro, with their customizable Toolbar and Console, are exceptionally flexible. In fact, a Nobeltec user can even save separate customized work spaces for different activities, such as fishing, cruising, or racing. Not surprisingly, many of the most flexible interfaces are also the most customizable when it comes to metrics, including Software-On-Board, BoatCruiser, The Capn, Nobeltec, and MaxSea.

In order to display multiple charts, instrument windows, and position data, each e-charting application must shovel the proverbial ten pounds of navigation information into a five-pound screen display. Two features help accomplish this feat: split windows and a tabbed interface.

Split windows partition a single display window into multiple panes, often displaying one for the main chart and one for something else, such as a satellite image or depth sounder data. A *tabbed interface* stacks separate displays of information in layers, which can be called to the foreground using tabbed buttons at the bottom of the screen.

About half of the e-charting applications—generally those in the higher price range—use some form of split-window display. RayTech RNS has the most elaborate functionality, with up to four simultaneous split-window panes (plus a tabbed interface). Tabs are less common, although we found them to be a more effective solution because they allow larger single windows to be customized and sequenced, rather than relying on a collection of tiny windows. This Microsoft Excel-like trick is used by BoatCruiser, RayTech RNS, and MaxSea.

Basic Features

Some features are so fundamental that any e-charting application worth its salt must include them. Basic features include rudimentary navigational tools, waypoint creation, route planning, chart management, and the ability to print chart excerpts.

For example, a one-button man overboard (MOB) marker is a basic safety tool. With all this navigational computing power at your fingertips, there should be a darn quick way to mark a position if someone goes into the water. Although every application we reviewed has an MOB feature, the user interface dictates how long it will take you to find it for the first time or to remember where it is hidden. All the PC applications have very intuitive one-button MOB markers. In a break from the PC and Mac stereotypes, the two major Mac applications, NavimaQ and MacENC, did not have a one-button MOB. Perhaps Mac users don't fall overboard?

Basic features include rudimentary navigational tasks such as creating waypoints, stringing waypoints together to form a route, logging a series of positions as a track, dropping a quick "steer to" position ahead, or converting a track into a route. These tasks are the primary reason to combine a GPS sensor and a digital chart, fundamental tools needed by any boater who has bothered to bring a laptop on board.

Of course, every one of the full-featured e-charting appli-cations we reviewed performs the majority of these tasks. But they differ widely in ease-of-use, influenced by little details we noticed only after spending quite a bit of time with a feature. Some applications are more intuitive, have more functionality, or are more flexible.

As an example, basic tasks such as waypoint or route creation vary in the number of mouse actions or keystrokes an application requires to get the job done. The fewer, the better. Route creation is most convenient when you can "rubber-band" instant waypoints. A clear presentation of route data is also important, either in a window or along a route leg. The cleanest waypoint and route interface overall was Coastal Explorer and Chart Navigator Pro. TIKI Navigator, with its non-traditional user interface also did very well.

Coastal Explorer and Chart Navigator Pro have an unusual proprietary exchange option (in addition to standard GPX) that should be familiar to Microsoft users. Waypoints, routes, and tracks can be packaged and saved in a special "user documents file." A simple pulldown menu choice of "Send To" lets you share your navigational information with other users of Coastal Explorer or Chart Navigator Pro as a data-rich seamless import/export package.

Basic Features[1]	One-button MOB	Waypoint creation	Route creation	Steer-to function	Track creation	Convert track to route
SeaClear II	Yes	3	2	No	3	Yes
Software-On-Board	Yes	3	2	No	3	Yes
NavimaQ	No	3	3	No	3	No
MacENC	No	4	3	No	3	No
TIKI Navigator	Yes	4	4	Yes	3	Yes
Marine ENC	Yes	3	4	No	3	Yes
BoatCruiser	Yes	3	4	No	3	Yes
Coastal Explorer	Yes	4	5	Yes	3	Yes
The Capn	Yes	3	3	Yes	3	No
Nobeltec VNS	Yes	3	4	Yes	3	Yes
Chart Navigator Pro	Yes	4	5	Yes	3	Yes
RayTech RNS	Yes	3	3	Yes	3	No
MaxSea	Yes	3	4	Yes	3	Yes
Nobeltec Admiral	Yes	3	3	Yes	4	Yes

(1) Poor (3) Average (5) Excellent

Software Comparison: *MOB; Waypoint, Route & Track Functions*

Basic Features[2]	Waypoint and route management	Chart management	Search waypoint and route	Range/bearing tool	Chart annotation	Chart printing
SeaClear II	2	4	3	3	No	1
Software-On-Board	3	N/A	3	4	No	1
NavimaQ	2	2	2	3	No	3
MacENC	3	4	4	3	2	4
TIKI Navigator	2	3	2	5	No	No
Marine ENC	4	4	2	3	2	4
BoatCruiser	3	3	4	4	4	2
Coastal Explorer	4	3	5	4	4	3
The Capn	3	3	4	4	No	4
Nobeltec VNS	3	4	2	3	4	3
Chart Navigator Pro	4	3	5	4	4	3
RayTech RNS	4	2	2	4	No	No
MaxSea	5	3	4	4	4	4
Nobeltec Admiral	3	4	2	3	4	3

(N/A) Not applicable; (1) Poor (3) Average (5) Excellent

Software Comparison: *Asset Management; Charting Tools*

If you are *bandwidth-impaired*, Saildocs is an email-based service that delivers Internet documents as text (www.saildocs.com). For example, Saildocs can send text weather forecasts by request or by subscription. Saildocs also provides custom GRIB weather data, downloaded from NOAA, and delivered to your email account as simple text attachments for use with your e-charting software. Send a blank email to info@saildocs.com or gribinfo@saildocs.com to receive start-up information.

For many boaters, route creation is less important than a simple *SteerTo* function—namely the ability to create an instant route and steer to a single waypoint ahead in order to minimize cross track error. A surprising number of applications do not integrate a convenient SteerTo function, including SeaClear II, Software-On-Board, NavimaQ, MacENC, Fugawi Marine ENC, and NavSim BoatCruiser. Of the lower-priced choices, only TIKI Navigator includes this tool. The higher-priced choices, from Coastal Explorer up to Nobeltec integrate a SteerTo function.

All the applications can store a vessel's track, but a handful cannot convert that track to a saved route for future use. NavimaQ, The Capn, and RayTech RNS do not convert tracks to routes.

For boaters who store only a handful of waypoints or routes for their local waters, these basic features are sufficient. But once you start amassing collections of waypoints, routes, or charts, you start to care about the management of these assets. Waypoint, route, and chart management means there is a system for the organized storage and retrieval of these elements. Scrolling through a menu of 100 waypoints, routes, or chart numbers is completely infeasible. These types of management tools will be particularly important in the future when boaters will likely carry larger and larger collections of electronic chart files, waypoints, routes, and markers.

Fugawi Marine ENC, RayTech RNS, and MaxSea have well-developed folder-based waypoint and route libraries. With these more advanced systems, you can create subfolders to organize your waypoints. In many cases, waypoints can be moved between folders simply by clicking-and-dragging.

Some applications, such as SeaClear II, store all waypoints in one single "bucket." A single-bucket approach is feasible if the search function is exceptionally strong, in which case you can at least retrieve what you need quickly. Unfortunately, most of the applications have substandard search tools, particularly when compared to Apple's *Spotlight* or *Instant Search* in Windows Vista. Coastal Explorer and Chart Navigator Pro outshine the other applications with their search capability, letting you type in a word or phrase and pull up any asset—including chart content—that contains that word or phrase.

For some boaters, pre-planning with waypoints and routes is too rigid. Instead, they prefer to use electronic navigation interactively, with cursor-based range and bearing lines. With a range/bearing tool you can instantly measure to a navigational landmark, take bearings for fixes, or maintain a distance off a headland. Although all the applications include a range and bearing tool, some are easier to use; some allow both points (beginning point and destination) to be altered by dragging-and-dropping; and some are better at displaying the information. TIKI Navigator has the most flexible and sophisticated range/bearing tool. Several other applications are also strong, including Software-On-Board, BoatCruiser, Coastal Explorer, The Capn, Chart Navigator Pro, RayTech RNS, and MaxSea.

Similarly, boaters who grew up using paper charts typically like to annotate their charts with personal notes or comments. BoatCruiser, Coastal Explorer, Chart Navigator Pro, Nobeltec, and MaxSea have excellent chart annotation features. Some of the applications take the annotation concept a step further, linking your text, sound, or image files. Personally we're not sold on the concept of using your e-charting application as a way to geographically organize your photo library, but it is available in TIKI Navigator, BoatCruiser, and Fugawi Marine ENC.

One of the most common questions we're asked regarding

e-charting is, "Can I print the charts?" It seems obvious that the answer is—or should be—yes. And, in fact, most packages claim high-resolution printing as a feature. Only two companies state upfront that their software does not print charts: TIKI Navigator and RayTech RNS (NavimaQ prints charts, but without your annotations). Yet despite the claims, we found printing to be the single most bug-ridden feature tested. For example, SeaClear II, Software-On-Board, Fugawi Marine ENC, and BoatCruiser all exhibited some form of printing glitch or artifact in the printout. The highest quality prints came from Fugawi Marine ENC (despite one minor bug), The Capn, and MaxSea. In all cases, a screen capture utility such as SnagIt for the PC or Snapz Pro X for the Mac provides a feasible low-resolution workaround.

Advanced Features

E-charting packages offer much more than chart displays, boat position, waypoints, and routes. They now include incredibly advanced navigational tools and extensive resources, such as integrated supplemental data, situational awareness aids, calculators, route planners, and weather downloads. Yet by labeling these features as "advanced," we do not mean to imply that they are difficult to use. Quite the opposite: navigation software has made these powerful tools simple and accessible to the average recreational boater. The task again is deciding which features are relevant for your boating needs. Do you need a fuel calculator? Do you need to optimize your sailing route based on polar diagrams? Do you need offshore weather? Do you need a celestial calculator? For some boaters, certain advanced features may add important convenience and safety. For others, they will be unused add-ons that simply ratchet up the cost or complexity. The answers lie in the type of boat you own and what you plan to do with it.

A perfect example is tide and current data. If you boat on the coast of Maine or Puget Sound, integrated tide and current predictions are an important part of your daily boating. Sea-Clear II and TIKI Navigator do not include tide and current data. MacENC obtains tide and current predictions through a free third-party application. Software-On-Board, BoatCruiser, and Fugawi Marine ENC display tide and current predictions, but obtain the data through C-Map or Navionics cartography.

(In other words, unless you use this optional premium cartography, the application will not have tide and current predictions.) The remaining applications have integrated tides and currents. Coastal Explorer and Chart Navigator Pro have the best presentation, combining the most data in a format that is easy to read.

E-charting has also grown to display much more than simple navigational aids and landmarks. Most packages now incorporate supplemental data such as Coast Pilots, points-of-interest (POIs), street maps—and even Google Earth capability. Again, the applications handle these materials differently. SeaClear II, NavimaQ, MacENC, TIKI Navigator, and RayTech RNS are less strong in this area, focusing more exclusively on marine data. The Capn does a good job including

Street Maps: *Supplemental data, such as street overlays, makes going ashore less stressful. Pre-plan dinghy landings, taxi pick-ups, shopping trips, or fun excursions with familiar navigation tools.*

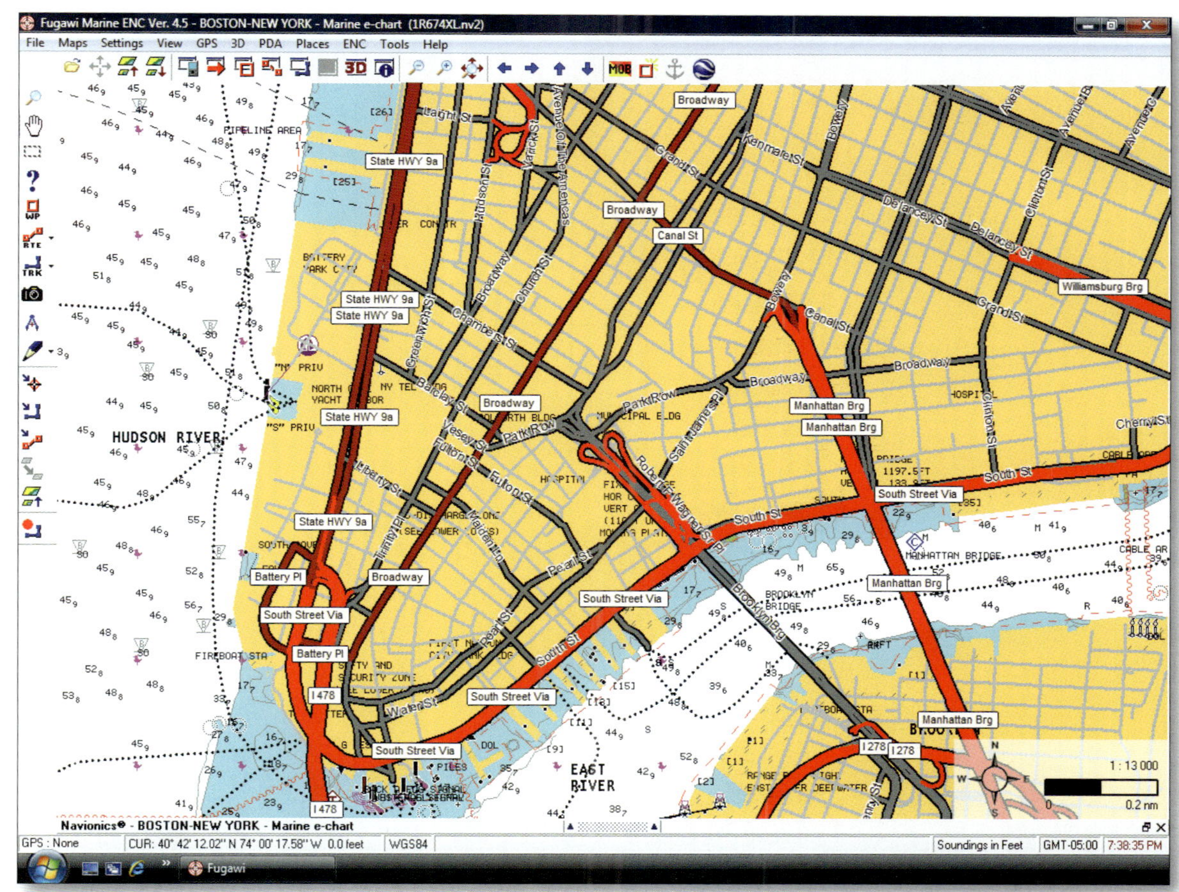

Location Tagger: You can add latitude and longitude data to your digital camera pictures by adding a GPS geo-tagger such as the Sony GPS-CS1.

Coast Pilots, but does not include point-of-interest data, street maps, or Google Earth. Software-On-Board and BoatCruiser piggyback on C-Map cartography, which includes street maps, to provide supplemental data. MacENC and Fugawi Marine ENC work with Google Earth, which can show street and land data in addition to its trademark satellite imagery.

If you want integrated information on ports, anchorages, marinas, and points-of-interest, Coastal Explorer's (and by extension Chart Navigator Pro's) supplemental data is the best. Maptech's extensive disc set of cartography comes bundled with Chart Navigator Pro, or you can add it to Coastal Explorer. Either way, the integrated and searchable Gazetteer in Coastal Explorer and Chart Navigator Pro is impressive. Click on any port to pull up a window with information and photos. These two applications also have the best advanced search feature, letting you quickly and easily access these extensive data extras.

That said, some boaters can benefit from advanced collision avoidance systems. Boaters in areas heavily used by commercial traffic, whether cargo ships, ferries, or commercial fishing, are most in need of these features. This includes areas such as the Chesapeake Bay or Puget Sound. These features may also provide some peace of mind for offshore cruisers in heavy traffic areas, such as those who make nighttime Gulf Stream crossings. But a weekend boater or an Intracoastal Waterway traveler? Think twice.

If you decide AIS, ARPA/MARPA radar, or buddy boat tracking are important to you, realize that even the lowest priced applications now include at least one of these features. AIS has become such a popular feature that it is included in all the applications except RayTech RNS. Two low-cost choices, SeaClear II and TIKI Navigator, both have surprisingly robust AIS functionality. Only Software-On-Board and TIKI Navigator include buddy boat tracking, and both had a nice, but very different, implementation of this feature.

All the higher-priced choices, and several of the mid-priced choices, include the option of displaying ARPA/MARPA radar data. Note that these features typically only display collision avoidance data, not a full image of your radar scan. Many boaters mistakenly assume a display of radar data is a standard option in an e-charting package, perhaps because so many e-charting promotions show a fancy split window with a chart alongside the corresponding radar scan. Although some e-charting applications support radar overlays, the implementation is tricky. Because radar data is analog, it must be converted to digital in order to display on a PC. This may require a digitally-enabled or compatible radar and/or a translation device, often referred to as a *black box*. If you do decide you want full radar overlays on your PC—useful for boaters in high fog areas such as Maine and the Pacific Northwest—you will need a high-end package such as Coastal Explorer, RayTech RNS, Nobeltec, or MaxSea.

One of the advantages of having a computer onboard is it can double as a tool for complicated calculations. Custom cal-

Advanced Features[1]	Tides	Currents	Coast Pilot	Points-of-Interest (POIs)	Streets	Google Earth support	Advanced search
SeaClear II	No	No	No	No	No	No	No
Software-On-Board	C	C	No	C	C	No	No
NavimaQ	3	No	No	No	No	No	No
MacENC	MT	MT	No	No	No	Yes	4
TIKI Navigator	No	No	No	No	No	No	No
Marine ENC	N	N	No	Yes	Yes	Yes	Yes
BoatCruiser	C	C	No	C	C	No	4
Coastal Explorer	5	5	5	5	No	No	5
The Capn	4	4	4	No	No	No	4
Nobeltec VNS	4	4	2	2	2	No	3
Chart Navigator Pro	5	5	5	5	No	No	5
RayTech RNS	4	4	No	No	No	No	No
MaxSea	4	4	2	2	2	No	4
Nobeltec Admiral	4	4	2	2	2	No	3

(C) C-Map (MT) Mr. Tides (N) Navionics; (1) Poor (3) Average (5) Excellent

Software Comparison: *Supplemental Data; Searching*

culators for fuel consumption, transit time, or celestial sight reductions are relatively simple programming extras. Many of the packages include one or more of these options.

If you are a power boater on a trawler or cabin cruiser, you'll appreciate the fuel calculators offered by BoatCruiser, Nobeltec, and MaxSea. For all boaters, a transit calculator is handy, letting you easily compute the time to an anchorage, port, or restricted bridge. In general, applications with a fuel calculator also include some type of transit calculator. The Capn only includes a transit calculator, but it is a very nice implementation. For hobbyists or transoceanic passagemakers, The Capn, Nobeltec, and MaxSea also include celestial calculators and/or Nautical Almanacs.

Route planners are more complicated than calculators. They integrate information about your vessel and prevailing sea conditions such as wind and current to calculate an optimal route. Again, although planning features are incredibly impressive, consider whether your boating habits require optimized route planning. Passagemakers and offshore racers are most likely to use these advanced features.

Coastal Explorer, Chart Navigator Pro, Nobeltec, and Max-Sea include auto route planners. Software-On-Board, Fugawi Marine ENC, Nobeltec, and MaxSea had excellent Great Circle Route tools. For sailors who make long passages—such as to Europe, Bermuda, the Caribbean, Mexico, or Hawaii—several applications include optimization routing for sailboats, or offer it as an option. MaxSea's Sailing Performance Module and Weather Routing Module are the benchmark.

Although many of the e-charting packages that include sail optimization features provide a folder of common polars, you can substitute the polar diagram for your particular vessel. U.S. Sailing sells a Performance Package that includes polar diagrams and target boat speeds for hundreds of sailboats (http://store.ussailing.org).

Advanced Features[2]	AIS	Buddy Boat	Radar (ARPA/MARPA)	Radar (display overlay)	Fuel calculator	Transit calculator	Celestial calculator and/or Nautical Almanac
SeaClear II	Yes	No	No	No	No	No	No
Software-On-Board	Yes	Yes	Yes	No	No	No	No
NavimaQ	Yes	No	No	No	No	No	No
MacENC	Yes	No	Yes	No	No	No	No
TIKI Navigator	Yes	Yes	No	No	No	No	No
Marine ENC	Yes	No	No	No	No	No	No
BoatCruiser	Yes	No	Yes	No	Yes	Yes	No
Coastal Explorer	Yes	No	Yes	Yes	No	No	No
The Capn	Yes	No	No	No	No	Yes	Yes
Nobeltec VNS	Yes	No	No	No	Yes	Yes	Yes
Chart Navigator Pro	Yes	No	Yes	Yes	No	No	No
RayTech RNS	No	No	Yes	Yes	No	No	No
MaxSea	Yes	No	Yes	Yes	Yes	Yes	No
Nobeltec Admiral	Yes	No	Yes	Yes	Yes	Yes	Yes

Software Comparison: *Advanced Instruments; Calculators*

Advanced Features[3]	Auto route planning	Great Circle Route planning	Sail performance	GRIB weather integration	Weather options	Customizable bathymetric recorder
SeaClear II	No	No	No	No	No	No
Software-On-Board	No	Yes	2	3	No	No
NavimaQ	No	No	No	No	No	No
MacENC	No	No	3	3	No	No
TIKI Navigator	No	No	No	No	No	No
Marine ENC	No	Yes	No	3	No	No
BoatCruiser	No	No	SC	No	No	No
Coastal Explorer	Yes	No	No	5	No	No
The Capn	No	No	No	No	No	No
Nobeltec VNS	Yes	Yes	PP	2	PP	PP
Chart Navigator Pro	Yes	No	No	5	No	No
RayTech RNS	No	No	4	4	No	No
MaxSea	Yes	Yes	5	4	4	Yes
Nobeltec Admiral	Yes	Yes	4	2	4	Yes

(SC) SailCruiser (PP) Plus Packs; (1) Poor (3) Average (5) Excellent

Software Comparison: *Planning; Racing; Weather; Bathy Recording*

Real-Time Response: With all of the available data, e-charting vendors are working overtime to provide instantaneous display response. Whether you request a satellite image overlaid on a nautical chart in perspective view (below) or want to "fly-through" a 3D representation of a fishing canyon, these compound requests put a large strain on the hardware, software, and memory capabilities of your system.

example, MacENC achieves superior image quality by using Mac OS X's built-in Quartz technology. Nobeltec makes an effort with a technology they call *CrystalView*, which improves, but doesn't make stunning, their chart displays. TIKI Navigator uses some programming tricks to improve image quality without either Quartz or CrystalView technology.

Fortunately, the trend toward more graphic-intensive personal computers is spilling over into e-charting. Some companies, most notably Furuno, are just beginning to take advantage of new processor and gaming technologies to create increasingly vivid and responsive chart displays. Furuno's MaxSea software uses DirectX graphics acceleration to create incredibly fast chart rendering. Furuno recently took display speed one step further with their Time Zero NavNet 3D. Their full-time 3D chart rendering lets you pan and zoom a chart at any angle and at any scale, instantly and seamlessly shifting from 2D to 3D. (See What's New: MaxSea Time Zero in Chapter 24.) To further accelerate its NavNet 3D technology, Furuno sells a several-thousand-dollar hardware processor unit.

Internet connections are also getting faster. This trend is influencing e-charting with the possibility of an "always-on" system. Geographically-available high-speed Internet may drive e-charting of the future toward live streaming data. This trend is already evident in new online e-charting applications such as the Google Earth-based EarthNC.

TRENDS IN E-CHARTING SOFTWARE

With today's powerful laptops, e-charting applications are relatively quick to open, pan, and zoom over charts. They also have "maxed out" on instrument support, with even the free and low-cost choices including advanced instrument connectivity for new devices such as AIS.

However, there still is room for improvement on chart display quality and speed. Even when displaying the same chart files, such as NOAA's free raster charts, different e-charting applications perform at varying levels.

Unfortunately, the average score isn't overly impressive. Even some high-end packages still struggle with less-than-sharp chart displays, pixelization in the form of broken type, and display delays. All have had to address these issues. For

Part Three

E-Charting in Practice

After you understand the history and individual components of an e-charting system, you're ready to compare your vessel, intended cruising ground, and personal lifestyle with others who have put e-charting in practice.

Chapter 8, *Networking and Data Exchange*, deals with the e-charting topic that is on every boater's mind: how do I get my waypoints and routes from one device to another? This chapter covers the details of data exchange formats and explains how data is actually transferred between devices.

Chapter 9, *Extending the Digital Metaphor*, underscores the point that you haven't used your laptop to its full potential if you simply set up an electronic charting system. Scores of free marine resources are available in the form of government publications, reference texts, nautical calculators, weather observations and predictions, online maps, boating forums, and user-generated (open content) websites.

Chapter 10, *Putting it All Together*, lays out seven real-life e-charting scenarios, ranging from kayakers creating custom chartkits to professionals captaining mega-yachts. Use this chapter to glean ideas on what will work for your vessel and your personal cruising plans.

Chapter 8
Networking and Data Exchange

If a single aspect of e-charting concerns boaters most, it's the exchange of data. Can I create waypoints and routes on my laptop and transfer them to my chartplotter? Can I move my waypoint collection to my new charting software? Can I use the same cartography on my computer as my chartplotter? In theory, the answer to all these questions is "yes." But not all applications can handle all types of data exchange—and they vary considerably in their ease of implementation.

The most important contribution of an e-charting computer is the creation and sharing of navigational assets. A computer adds value to the navigation toolkit as an easier way to create customized collections of waypoints, routes, anchorages, marina locations, fishing holes, or any other chart annotation.

If you intend to use your computer to prepare navigation data for other instruments such as a chartplotter or for other boaters such as flotilla members, then data exchange is an important factor in your choice of an e-charting application. On the other hand, if you envision your computer primarily for previewing charts, or for performing only select dedicated tasks such as fuel calculation, then data exchange is less important. Likewise, bulk data exchange is less of a concern if you only have a handful of waypoints and routes already stored in your chartplotter.

There are many scenarios in which navigation data needs to be extracted from an e-charting application or computer.

Switching to new software or purchasing a new computer should not entail losing all your waypoints and routes. Flotilla or club cruise captains should be able to share float plans of waypoints, routes, marinas, and attractions, distributing trip information as an email attachment file or on CD. Boaters may want to share their dive sites or fishing holes with other boaters, swapping local knowledge of areas they know well. Waypoints and routes should be transferable to a handheld GPS for outings. Even boaters who do not navigate with a laptop need to upload their planning work to a chartplotter or multifunction display.

THE DEVIL'S IN THE DETAILS

Unfortunately, the exchange of data across marine instruments, across computers, and across applications remains the weak link in the e-charting chain.

Some vendors place a high priority on the sharing and exchange of data. They freely publish and share their data format protocol, encouraging other vendors to co-develop integrated products. Other companies choose a strategy of "circling the wagons," protecting their format to encourage consumers to stay within their family of products. They keep their data format secret, forcing companies that want to exchange data to *reverse engineer* their proprietary communication formats in order to integrate products.

The *Data Exchange and Networking Tables* summarize which applications support which types of data exchange. However, be warned: the details and logistics of import/export are not as straightforward as their marketing (and our *Software Comparison Tables*) might suggest. Many applications only export certain navigational objects, such as waypoints but not chart marks or routes. Some exports lose icons or comments. Some applications are robust at exporting data but are limited on importing. Others import and export only to certain devices, or to devices made by certain vendors (such as Garmin or Humminbird).

With so many devices, formats, and frankly, programming quirks, it's impossible to detail all the subtleties of e-charting data transfer. If a particular type of data exchange is important for your needs, download the e-charting application's demoware and rigorously exercise its exchange functions with *your* test files before purchase.

DATA TRANSFER FORMATS

Data exchange is a *handshake*: it takes the cooperation of both sender and receiver to successfully transfer a data file. For instance, simply saving a data file of waypoints and routes onto a flash card has little value if the file cannot be read by your chartplotter or your buddy's e-charting application.

In order to move a file between devices, the sender must package the data in a format that can be interpreted by the receiver. It must be stored in a way other devices or applications can "translate" it back into their native format. For example, in order to move waypoints and routes from your personal computer to your chartplotter, the first step is for the e-charting application to package the file for export.

Although all full-featured e-charting applications are designed with some form of import/export capability, they vary in the data formats they support. At one end of the continuum are manufacturer-specific proprietary formats, which restrict transfer to within that manufacturer's products. At the other end, a new exchange standard, GPX, promises universal *exchange* across all compliant devices.

Proprietary Formats

The most constrained data transfer only works within a company's own software and instruments. This within-vendor communication is more appropriately termed data *interchange*.

For example, Jeppesen takes a stand-alone approach supporting its Open Navigation Format (ONF). Its two software products use ONF and can only, with very few exceptions, exchange data between devices running Nobeltec.

Similarly, RayTech RNS has extremely convenient data transfer to its own devices. It is designed to export waypoints and routes seamlessly to Raymarine products, including its chartplotters and multifunction displays. Data is transferred automatically using their SeaTalk protocol over an Ethernet cable, which plugs into your PC's Ethernet port. However, RayTech RNS can only accomplish rudimentary data exchange with non-Raymarine instruments, and can only do so with a helper application such as GPSBabel (see GPS Helper Applications).

Setting up a navigational system whose software and hardware remains exclusively intra-company—such as RayTech RNS with Raymarine E- or G-Series or Furuno MaxSea with Furuno NavNet 3D—should be smooth sailing. However, realize that the price of this convenience in data exchange is the commitment to stay within-company for all software and hardware choices.

Integrated GPS Transfer

Some e-charting applications do the data exchange formatting work for you. For example, a function known as integrated GPS transfer means the application provides menu choices that send the data to a GPS automatically via a cable connection.

NavimaQ, MacENC, Fugawi Marine ENC, Coastal Explorer, Nobeltec VNS Max Pro, Chart Navigator Pro, Nobeltec Admiral MAX Pro, and MaxSea all provide some form of integrated GPS transfer.

This feature makes it very easy to move your assets to and from a GPS. For example, simply pulling down a series of menu and submenu choices, such as Waypoints>Transfer to>Garmin, moves your data to a connected Garmin GPS. However, realize that an application may only support particular GPS models. Garmin is the most commonly supported brand. Fugawi Marine ENC supports one of the widest selection of GPS brands. Furthermore, this convenient menu approach typically does not extend to more complicated transfers, such as between computers, e-charting applications, or non-GPS devices.

The *Data Exchange and Networking Tables* summarize which software packages support this streamlined process.

CSV or Tab-Delimited Files

Navigational data can also be saved and transferred in generic computer text formats such as comma-separated value (CSV) or tab-delimited files.

In this case, the data (consisting of waypoint names, latitudes, longitudes, and any other fields for transfer) is saved as lines of text in a file. These files are simple text files where data fields are separated by commas (in the case of CSV) or tabs (in the case of tab-delimited files). Sometimes these files are referred to as ASCII files (pronounced ask-ee), referring to their simple text character format. Microsoft Excel files are often grouped into this category, since Excel files are a form of tab-delimited data (albeit more sophisticated).

CSV and tab-delimited files are universal and unconstrained since they simply store the data as plain text separated by commas or tabs. Unfortunately, flexibility can create trouble when transferring waypoint and route data. CSV or tab-delimited data transfers typically require some handwork for a successful import or export. For example, should the "N" in the latitude be placed before or after the numerals? Should there be a space between characters?

Since CSV and tab-delimited files do not dictate the specific data format—only that each data field is separated by a comma or a tab—there is the potential for data transfer errors. The most severe glitches occur when waypoints are omitted or their positions are shifted (sometimes slightly, sometimes grossly). We've experienced waypoint data transfers that resulted in Intracoastal Waterway waypoints appearing about five miles offshore!

Unfortunately, although CSV and tab-delimited files are very generic, most GPS and chartplotters cannot receive them without some help from a translation utility (see GPS Helper Applications).

GPX

In 2002, GPS Exchange Format (GPX), a license-free data exchange protocol for transferring GPS data based on the XML language, was released. It has since become the standard for any software or hardware relating to GPS data, including mapping, marine navigation, and even geocaching.

Latitude and longitude are the minimum properties of a data entry in a GPX file. Other variables are optional, but typically include fields such as waypoint names, sequencing (such as for routes, which are sequences of waypoints), or time-date stamps (for tracks).

For importing or exporting GPS data, nothing compares to GPX. Because GPX constrains the format of the data, it guarantees a direct and accurate transfer. With GPX, literally thousands of waypoints can be easily and accurately transferred between e-charting applications and between computers (even with different operating systems).

Overall, any application that supports GPX has a greater chance of successfully importing and exporting a variety of navigational objects to a variety of other applications and devices. Because of their full GPX support, MacENC, Fugawi Marine ENC, Coastal Explorer, Chart Navigator Pro, and MaxSea score well in data exchange.

KML Files

Keyhole Markup Language (KML) is the equivalent GPX standard for the Google Earth platform. This file format is relevant for boaters who use Google Earth or use e-charting applications based on a Google Earth platform, such as EarthNC.

KML was specifically developed for use with Google Earth, which was originally named Keyhole Earth Viewer. (The name Keyhole pays homage to the original Keyhole eye-in-the-sky reconnaissance satellites of the 1970s.)

Like GPX, a KML file always specifies a latitude and longitude. Then it adds features such as placemarks, images, polygons, or text notes. It is specifically designed to express geographic images and text on Web-based two-dimensional maps

If you need to extract data from an e-charting application that only supports its own proprietary format, you may have to use a GPS unit as an *envoy*. This workaround involves downloading your waypoints and routes to a compatible GPS. First, *download* the assets to the GPS. Then *upload* the assets to the target application. See why GPX is a better mousetrap?

Data Exchange and Networking[1]	GPX support for waypoints	GPX support for routes	Waypoint exchange formats	Route exchange formats	Integrated GPS transfer
SeaClear II	No	No	None	None	No
Software-On-Board	No	No	CSV	CSV	No
NavimaQ	No	No	None	None	Yes
MacENC	Yes	Yes	GPX CSV KML	GPX CSV KML	Yes
TIKI Navigator	No	No	None	None	No
Marine ENC	Yes	Yes	GPX CSV KML	GPX CSV KML	Yes
BoatCruiser	No	No	CSV	CSV	No
Coastal Explorer	Yes	Yes	GPX KML	GPX KML	Yes
The Capn	No	Yes	None	GPX	No
Nobeltec VNS	No	No	None	None	Yes
Chart Navigator Pro	Yes	Yes	GPX	GPX	Yes
RayTech RNS	No	No	Excel CSV	Excel CSV	No
MaxSea	Yes	Yes	GPX	GPX	Yes
Nobeltec Admiral	No	No	None	None	Yes

(CSV) Comma-Separated Value (GPX) GPS Exchange Format (KML) Keyhole Markup Language; Microsoft Excel

Software Comparison: *GPX Support; Data Exchange Formats*

Data Exchange and Networking[2]	Multiple monitor support	Network application sharing
SeaClear II	No	No
Software-On-Board	No	No
NavimaQ	No	No
MacENC	No	No
TIKI Navigator	No	No
Marine ENC	No	No
BoatCruiser	No	No
Coastal Explorer	No	No
The Capn	No	No
Nobeltec VNS	No	No
Chart Navigator Pro	No	No
RayTech RNS	Yes	Yes
MaxSea	Yes	Yes
Nobeltec Admiral	Yes	Yes

Software Comparison: *Networking Support*

and three-dimensional Earth browsers such as Google Earth.

Waypoints or chart annotations saved as KML files can be imported and displayed as a layer on a Google Earth satellite image. MacENC, Fugawi Marine ENC, and Coastal Explorer integrate export of your waypoints to Google Earth. Conversely, EarthNC, a planner-plus software based on Google Earth, can export Google Earth *pins* and *paths* to a GPS receiver or chartplotter. EarthNC is covered in detail in Chapter 13.

Data Transfer Mechanisms

When a file of navigation data is created in a standard exchange format, it still must be transferred to the receiving device. Think of the exchange file as a letter to mail. Once the letter is written and addressed with a standard format address, it still needs to be mailed. The first step is preparing the file; the second step is sending the file.

In general, data files transfer in one of two ways. They can be *physically* transferred using a storage device or *electronically* transferred over a network.

Physical Transfer

Called *sneaker-net* in jest, physically walking a data file between devices is an easy and low-tech way to transfer files.

For example, a file of waypoint and route data can be moved from one personal computer to another via a CD-ROM or thumb drive. Simply prepare the file for export using your e-charting application, save the file to a CD or thumb drive, and insert the disc or drive into the receiving computer. The new computer's e-charting software will then import the data file using the appropriate menu choices for file import.

Navigational data files also can be moved to a chartplotter using a Secure Digital (SD), CompactFlash (CF), or Multi-Media (MM) card. Most chartplotters do not accept CD-ROMs, although some new models now include USB ports for easy thumb drive transfers. Chartplotters most typically rely on flash cards.

In order to save data from your personal computer onto a flash card, you must have a card reader, a small external device that connects to a computer's serial or USB port. Some e-charting packages, such as Fugawi Marine ENC or RayTech RNS, bundle their software with a card reader, realizing its value for reading card-based cartography and for exporting data to a chartplotter.

Network Transfer

Navigational data files can also be transferred over a network, either via cables or wirelessly.

For example, some chartplotters can be connected by cable to a personal computer. Simply connect the manufacturer's cable from your chartplotter to your computer, using an adapter if necessary.

Transfers between networked computers typically use a router and an Ethernet connection, achieved with a simple plug into the computer's Ethernet port (the connection that looks like a phone jack).

The simplest way to transfer a file between computers without any cabling hassle is simply as an email attachment. Once exported and saved in an appropriate exchange format, a waypoint and route file can be emailed to a buddy boater or to members of a flotilla. Ask them what file formats their e-charting software can import or check the *Software Comparison Tables*. Transferring files by email takes advantage of the wireless nature of the Internet.

A handful of e-charting applications integrate their own advanced networking features. For example, larger vessels, in the 50-foot or greater class, may have multiple displays or multiple networked computers on board. With multiple monitor support, you can mirror your PC and its navigation data to different displays at the navigation station, helm, and flybridge. RayTech RNS, Nobeltec Admiral, and Furuno Max-Sea support advanced networking. Each of these applications is covered in detail in Chapters 23 through 25.

Chapter 9
Extending the Digital Metaphor

If you already own a chartplotter, you may be wondering, "Why add laptop e-charting?" Your chartplotter displays digital charts, shows vessel position, stores waypoints and routes, and integrates with marine instruments. And a chartplotter is marinized and designed with a screen bright enough to read in direct sunlight.

True, a traditional laptop (versus a ruggedized model) does not have the sunlight readability or water-resistance of a cockpit-mounted chartplotter. But don't sell the concept short. A laptop may not replace your chartplotter, but it's definitely more than a back-up system or a way to check email. Computer-based e-charting not only adds convenience, it provides access to a staggering array of marine-related resources.

A chartplotter can store and display charts, but it can't compete with the treasure trove of information available through an e-charting package. In addition to basic navigational charts, most e-charting applications include supplemental data such as Coast Pilots, point-of-interest (POI) overlays, street maps, satellite and aerial photos, topographic maps, and 3D bathymetric data.

A computer also easily stores folders upon folders of nautical publications. Most of this information is provided by the government for free, ranging from worldwide sailing directions to quick reference guides on chart symbols. Multi-volume sets, such as celestial sight reduction tables or *Coast Pilots*, are now available as free downloads. In fact, even the complete text of *American Practical Navigator* (Bowditch) can be downloaded for free.

These digital files are easy to add and can be updated regularly. And digital files, no matter how unwieldy, are searchable down to the word level. Onboard paper manuals and reference guides are notorious for their weight and for becoming out of date and mildewed. Selections from digital documents can be printed as needed. Since digital files weigh nothing and your personal computer effectively has unlimited storage space for text files, you can carry reference material that would be infeasible in paper format. Hundreds of publications easily fit on a single DVD weighing less than an ounce.

E-charting also goes hand-in-hand with the Internet and World Wide Web. Although e-charting began as a way to display charts on a computer, the Internet is now an integral part of e-charting's added value. With high-speed wireless Internet available in many regions, most e-charting applications leverage this phenomenal source of information, integrating features such as weather downloads or live satellite imagery. As of this writing, there were over 150 million active unique websites (not web pages!) on the World Wide Web. Literally terabytes of nautical resources are accessible through your personal computer.

Government Publications

A staggering number of government publications are available as free downloads. Knowing what to look for—and then locating them through convoluted government printing office websites—is the only obstacle to creating your own nautical digital library.

The U.S. is unique in its dissemination of free government assets (including its electronic charts). Many other hydrographic offices still exclusively sell their documents through chart agents. Fortunately, with its worldwide maritime presence, many U.S. publications include international coverage.

Although government publications extend to specialized topics, such as *Sight Reduction Tables* for celestial navigation (Publication 229) or the *International Code of Signals* (Publication 102), important basics are also covered. Look for practical topics such as chart symbols, rules of the road,

Chapter Terms

American Practical Navigator: A classic reference text describing the principles of navigation, including piloting, navigation, safety, oceanography, and meteorology. (Also called "Bowditch," after Nathaniel Bowditch, the author of the first edition in the 1800s.)

Chart No. 1: Publication by NOAA with descriptions of the symbols, abbreviations, and terms that appear on U.S. nautical charts.

Gazetteer: Printed or digital documents of place names including information such as geographic coordinates, elevation, and features.

List of Lights: Publication issued by a government tabulating its navigational lights and their characteristics.

Navigation Rules: The official rule book, published by the U.S. Coast Guard, required on all vessels over 12 meters. Contains the International Regulations for Preventing Collisions at Sea (COLREGS) and the Inland Navigation Rules.

Notice to Mariners (NtM): A periodic notice issued by a government regarding changes in aids to navigation and other updates to issued charts.

Open Content Website: An Internet site where the content is created by users collaboratively contributing and editing posted information.

Chapter Questions

- *What other marine resources can I access with my personal computer?*

- *What are my choices for computer-based weather information?*

- *Which nautical publications are available in digital format?*

- *What are some new websites with up-to-date nautical information?*

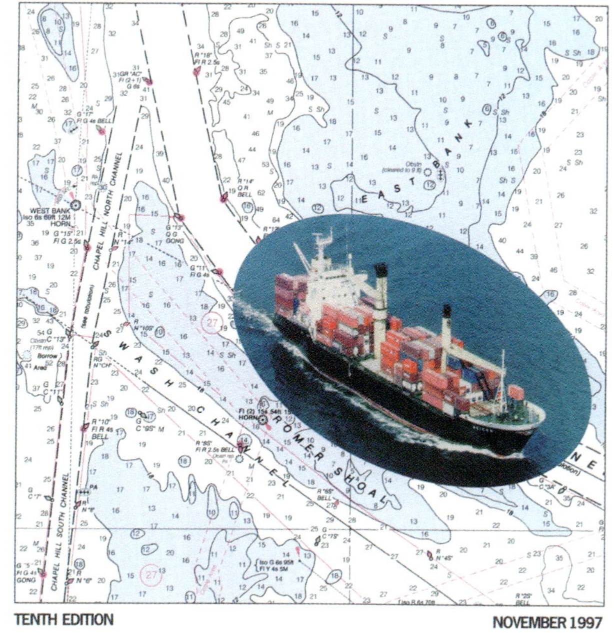

CHART NO. 1

UNITED STATES OF AMERICA

NAUTICAL CHART

Symbols Abbreviations and Terms

TENTH EDITION   NOVEMBER 1997

The Top Three: *Every boater has different needs, depending on their experience and cruising geography. But three free publications stand out as obvious first-choice downloads: Chart No. 1 (above), the Nav Rules, and Bowditch.*

NOAA Chart No. 1
http://nauticalcharts.noaa.gov/mcd/chartno1.htm

USCG Navigation Rules
www.navcen.uscg.gov/mwv/navrules/download.htm

American Practical Navigator (Bowditch)
www.nga.mil/portal/site/maritime

VHF and weatherfax frequencies, and hurricane preparedness.

If you have U.S. charts onboard, you should have a copy of *Chart No. 1*, a NOAA publication that details the symbols, abbreviations, and terms used on NOAA charts. You never know when you might need to look up a particular chart symbol or abbreviation. This document sells for $9.95 in printed form, or is available for free on the Internet.

Similarly, there is no excuse for not having a PDF copy of the U.S. Coast Guard's *International and Inland Navigation Rules*, required on all vessels over 12 meters in length. This $14 softcover booklet is also available as a free PDF download. Unfortunately, the federal carriage requirements are not explicit regarding the format (paper or digital), simply stating that "the operator of each self-propelled vessel 12 meters or more in length shall carry on board and maintain for ready reference a copy of the Inland Navigation Rules."

Even if your vessel is less than 12 meters, there are many circumstances when it's handy to have a reference source on navigational rules. A quick review will remind you of the significance of three black balls or which side to hug when a commercial vessel announces "two whistles." You also can confirm that your own navigation signals comply, such as the rhythm to sound on your horn or bell during fog.

To locate government publications useful for your cruising plans, skim the list of nearly 200 online resources listed in Appendix A: *Nautical Reference Library*. Coastal cruisers will be interested in publications such as the *Coast Pilots* (with supplemental information on ports, marinas, channel descriptions, anchorages, bridges, and currents) or *Light Lists* (detailing lights, sound signals, buoys, daybeacons, and other aids to navigation).

Long-distance cruisers can replace many of those expensive and weighty planning manuals. Why pay $20 to $48 per volume of *Sailing Directions Planning Guides* or $20 each for the *Sight Reduction Tables*? These volumes are all available as free downloads. The invaluable *Atlas of Pilot Charts* (Publications 105 to 109) depicts averages in prevailing winds, currents, and weather conditions for planning the fastest and safest passages. The *Sailing Directions Enroute* include information about coastal weather, currents, ice, dangers, features, and ports for each geographic area. The *Sight Reduction Tables* (Publication 229) consist of six volumes with the tables necessary for celestial navigation. The *NGA List of Lights* (a seven-volume set costing nearly $350 in print but available free online) includes information on lights and other aids to navigation for the world by region.

In addition to charts and atlases, which are collections of charts or maps, gazetteers provide information on places. A small subset of e-charting applications, most notably Coastal Explorer and Chart Navigator Pro, integrate gazetteers with their software, letting you click on a chart location to query for more information. Other gazetteers are available online. The U.S. government maintains two free online gazetteer services. The *GNIS* (Geographic Names Information System) covers nearly two million physical and cultural geographic features in the U.S. (http://geonames.usgs.gov). The *GNS* (GEONet Name Server) contains over 3.5 million features and over 5 million place names for locations outside the U.S. (http://earth-info.nima.mil/gns/html/index.html). Used by the military, the GNS database is updated every week.

REFERENCE SOURCES

With 4.7 GB of words and pictures fitting on a lightweight DVD disc—or unlimited terabytes via an Internet connection—entire tomes of reference sources can be accessed with an onboard computer. Your imagination is the limit: navigational reference books, medical manuals, encyclopedias, field guides, magazines, and directories are all rich sources of information while planning or underway.

For mariners who enjoy honing their craft, some classic nautical reference texts are available for PDF download.

Sometimes the resources are available through odd places, but they are out there. For example, the complete text of *American Practical Navigator* is a free download at the National Geospatial-Intelligence Agency's website. This detailed reference text covers the principles of navigation, including piloting, electronic navigation, celestial navigation, mathematics, safety, oceanography, and meteorology. The *Nautical Chart User's Manual* is out of print but still available online, and is an excellent resource detailing marine cartography.

Don't neglect general reference texts, such as atlases, encyclopedias, or medical manuals. You can access *Encyclopedia Britannica* either online or by CD-ROM to satisfy any of your onboard curiosities (www.britannica.com). Wikipedia and Google are never-ending sources for any question you may have about something you see or a place you are going (www.wikipedia.com) (www.google.com). The *Merck Manual* and *Merck Veterinary Manual* are good references for health and pet questions (www.merck.com/mmpe). All these heavyweight reference texts are now available online for free or on CD-ROM at a fraction of their paper cost.

For natural history buffs, reduce the menagerie of onboard field guides by turning to online guides. For instance, eNature maintains online field guides covering more than 5,500 plants and animals, with color photographs, advanced search, and even a zip code filter to let you know what species are likely in your area (www.enature.com/fieldguides).

Magazines, always a great source for timely information, are also turning to online formats. Overview an entire issue or a single article, reading it online or printing articles of interest. An even more enticing development is the option of free online subscriptions to expensive print magazines. For example, *Ocean Navigator* provides free online subscriptions to members of Seven Seas Cruising Association (SSCA). *PassageMaker* also makes full digital versions of some issues available to SSCA members. Check all your membership benefits for free online magazine subscriptions.

If you are a member of a cruising club, boating association, or yacht club, be sure to add their resources to your onboard computer. Many yacht clubs and cruising clubs maintain databases of anchorages, moorings, marinas, or yacht clubs with reciprocity. Boating organizations, such as BoatU.S. or Discover Boating, maintain marine service directories. Advertising-based directories, such as the MPC network (www.mpcnetwork.com), let you search for marine businesses online.

Nautical Calculators

Boating always seems to involve too much math—especially given that for most people it's supposed to be a recreational activity. There is always something to convert or calculate: speeds, distances, times, tides, weather metrics, or celestial observations.

A personal computer is perfect for these mundane computing tasks. Scores of nautical calculators are available free online, most designed to supplement the calculations used in *American Practical Navigator*. For example, the NGA collection of calculators includes categories of celestial, distance, log and trig, sailing, time zones, and weather data. Each calculator is a self-contained Java program that can be downloaded and used offline. Designed as user-friendly pop-up windows, you simply enter your query into a web-like field, hit return, and view the answer.

There are also some free resources for geographical conversions, such as if you need to convert chart datums used in foreign countries. GeoTrans is a free Windows and Unix program, developed by the U.S. Department of Defense, that converts coordinates such as latitude and longitude, UTM, and map datums (http://earth-info.nga.mil/GandG/geotrans).

Weather Resources

The Internet is becoming an increasingly important source for accurate and timely weather information. In addition to VHF

Distance of an Object by Two Bearings

First Bearings to Object:		(degrees)
		(minutes)
Second Bearings to Object:		(degrees)
		(minutes)
Distance between Bearings:		

Calculate   Reset

Distance to Object at Bearing 2:	
Distance to Object When Abeam:	

Geographic Range

Given the Height of the Light Above Sea Level and the Height of the Eye of the Observer above Sea Level, Compute the Geographic Range

Height of the Light above Sea Level (specify units):		● feet ○ meters
Height of the eye of the Observer above Sea Level (specify units):		● feet ○ meters

Calculate   Reset

Geographic Range:		(Nautical Miles)

Nautical Calculators: *Many navigational tasks involving math can be solved more quickly and accurately using an HTML calculator. Here, the distance off a headland light can be calculated in two ways: taking two bearings to the object (above), or measuring height of object and eye (below).*

The NGA provides scores of general and specialized publications and nautical calculators (www.nga.mil/portal/site/maritime).

Keep a Weather Eye: *Displaying weather data from XM WX Satellite Weather, WxWorx on Water shows radar imagery and buoy data (left) and wave height and direction (right).*

and SSB broadcasts, marine weather is now available via National Weather Service email or FTP, GRIB server downloads, GRIB email, Internet buoy reports, satellite subscription services, and dedicated websites.

Before you fret over the source of your online weather, realize that nearly all weather data comes from a limited number of primary sources. Government agencies, such as the U.S. National Weather Service, are the most important sources of raw weather data. A small group of weather companies, most notably Baron Services and WSI Corporation (the former owner of The Weather Channel), collect this raw weather data from governments, disseminating it to weather distributors who present it through their media outlets.

In order to display GRIB weather files, you must have an e-charting application that supports GRIB weather. GRIB weather is received either through an Internet connection to a server (arranged by the e-charting software) or by the request and receipt of an email attachment. Your e-charting application then displays the downloaded GRIB weather information in its own user-friendly format for interpretation.

If you want to display GRIB weather data without a full-featured e-charting application, you can use a standalone GRIB viewer software such as Ugrib or OCENS GRIB Explorer. Ugrib is a freeware download (www.grib.us). OCENS GRIB Explorer is sold separately or as part of the OCENS WeatherNet package. Both viewers download and view GRIB weather data on a Windows PC. GRIB weather is covered in more detail in Chapter 7, *E-Charting Software*.

For up-to-date coastal weather, more and more buoys are being outfitted to broadcast live weather data. For example, NOAA and the National Weather Service continue to expand the National Data Buoy Center (www.ndbc.noaa.gov). Their website shows maps of weather buoys by region—clicking on a buoy icon brings up the marine data for that station.

Weather is also available by satellite radio transmission. Satellite weather requires a receiver, a subscription plan, and a way to display the received weather data. XM WX Satellite Weather (with data provided by WxWorx On Water) and

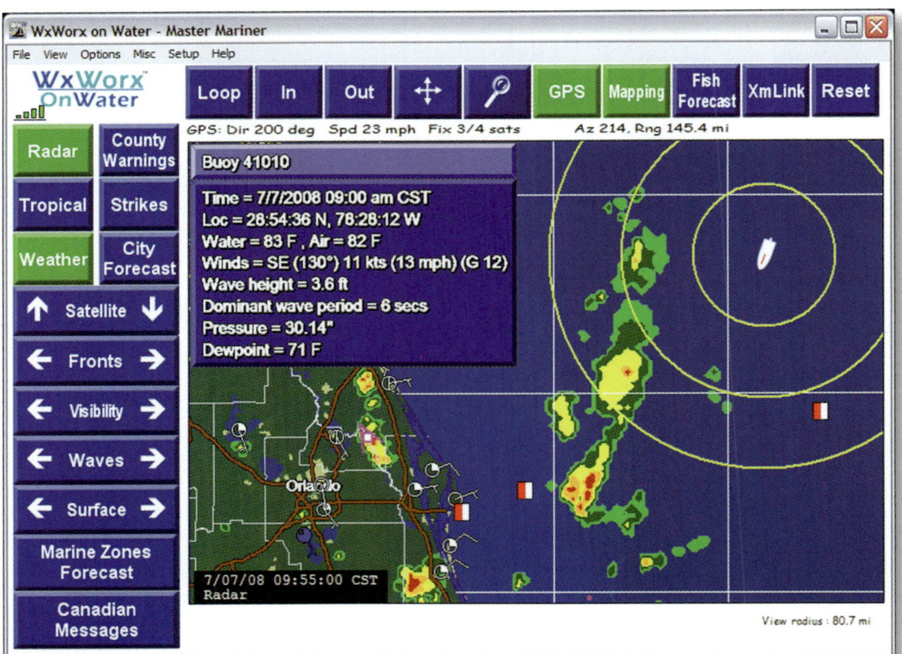

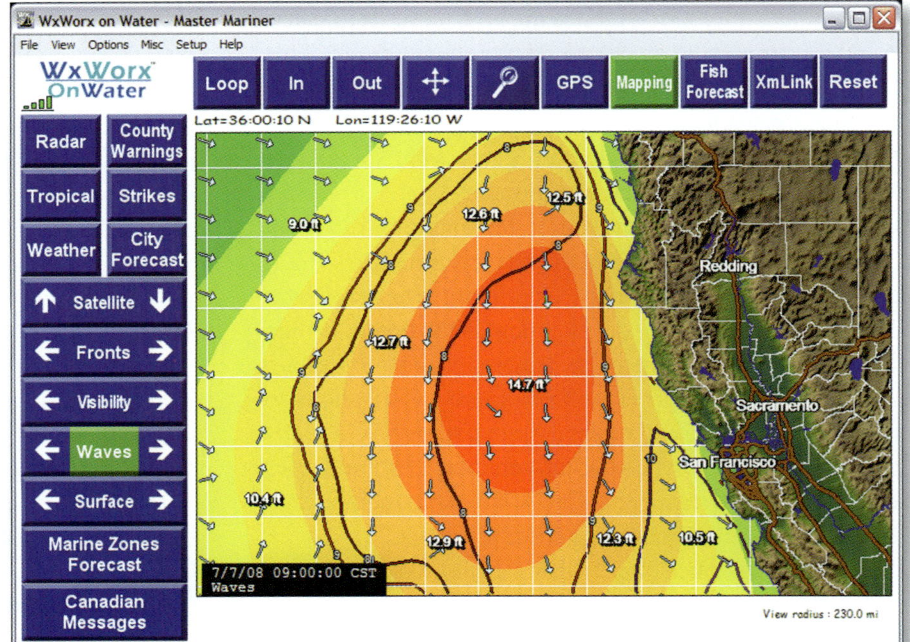

Sirius Marine Weather are two popular subscription services (www.xmwxweather.com) (www.sirius.com/marineweather). OCENS WeatherNet is also well-known, providing optional Internet extras such as email or blogging services (www.ocens.com).

Satellite weather can be displayed on your computer, either with an e-charting application that supports that satellite weather service, or with the subscription service's software. For example, Nobeltec e-charting software supports XM or Sirius Weather through a Plus Pack option or OCENS WeatherNet through a subscription. Alternatively, OCENS weather can be displayed on a personal computer using their own OCENS GRIB Explorer and MetMapper software.

The newer weather receivers are designed to connect directly to a personal computer. For instance, the WxWorx Satellite Weather Receiver connects to (and can be powered from) your personal computer's USB or Ethernet port (www.wxworx.com). It's also Bluetooth-enabled for local wireless use.

Actually, what is commonly called "satellite weather" is in fact accessible in many ways, including broadband Internet, WiFi, cellular phone, or satellite phone. For example, SkyMate provides weather reports as part of its suite of onboard information services (www.skymate.com). Xaxero SkyEye, software for PCs and Macs, includes real-time satellite weather information without user fees (www.xaxero.com). ClearPoint High Definition Weather, another Internet weather subscription, includes specialized weather data such as chlorophyll imagery for fishing, high resolution wind data for racers, and sea ice conditions for extreme northern boaters (www.clearpointweather.com). Roffer's Ocean Fishing Forecasting Service emails offshore fishing weather information (www.roffs.com).

Despite all the specialized Internet weather options, don't neglect traditional weather websites as a fast and simple source for hour-by-hour forecasts, several-day outlooks, and Doppler radar. Weather Underground (www.wunderground.com) and AccuWeather (www.accuweather.com) are two popular sites. Crown Weather (www.crownweather.com) is less known. Since much of this data is simply re-packaged National Weather Service data, you can also obtain it directly through the NWS site (www.weather.gov).

For even more convenience, many of these sites allow for

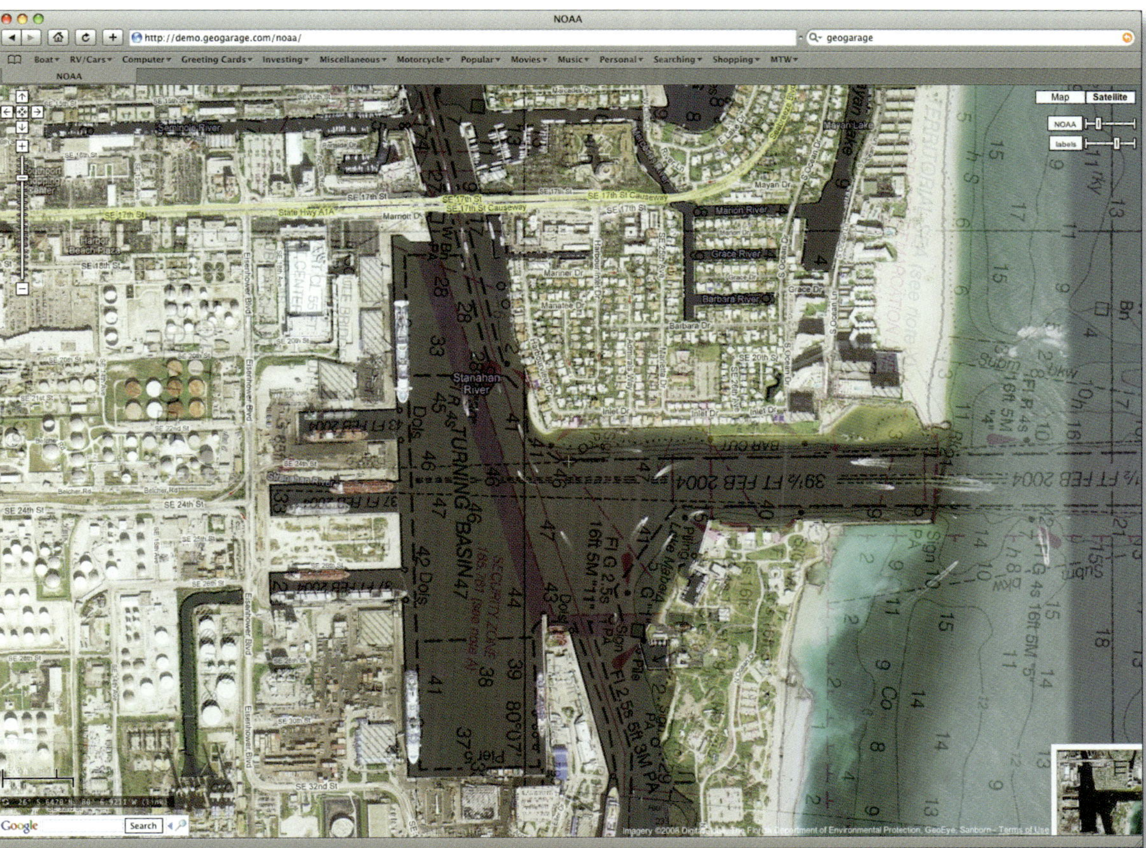

Land and Sea: *GeoGarage is a good example of a Google Earth Mashup, blending satellite views and NOAA charts for free.*

pre-set options. You can set a default forecast location, save weather location favorites, receive forecasts and weather alerts by email, view weather information on your cell phone or iPod, or create a weather widget for your desktop.

Free specialized weather data is also available through government or marine research websites. The Satellite Services Division of the National Weather Service (www.ssd.noaa.gov) posts live satellite images of water temperature, useful for offshore fisherman trying to pinpoint the warm eddies spinning off the Gulf Stream. Sea surface temperature and bathymetric charts are available online through HotSpots Charts (www.sstcharts.com). Many research universities with marine programs maintain online databases of marine and weather data. For example, Rutgers University posts sea surface temperature and chlorophyll satellite data (http://marine.rutgers.edu/

NOAA's National Weather Service website offers many weather-related publications as free downloads. Surf for free weather-related reference publications, handbooks, charts, logs, and pamphlets at www.weather.gov/os/marine/pub.htm.

mrs/sat_data/?nothumbs=0).

Since most weather forecast data is generated by the government, their publications are an excellent source for weather reference material. In the U.S., NOAA and the National Weather Service publish the *Federal Meteorological Handbook,* a twelve-chapter reference text on weather, now available as a free PDF download. NOAA also provides information on hurricane planning, such as *The Mariner's Guide for Hurricane Awareness* and *NOAA Hurricane Basics.* Download the free *NWS Observing Handbook,* a field guide format to recording your own weather observations. The *Marine Weather Service Charts* are also now posted online, listing the frequencies, schedules, and locations of stations disseminating U.S. National Weather Service information. Three of the five *Atlas of Pilot Charts,* which depict worldwide averages of prevailing winds and currents, air and sea temperatures, and wave heights for route planning, are also available as free downloads through NGA's website.

Online Maps

You may be on the water, but don't neglect the excellent online land maps. Satellite images, street maps, direction finders, and service location directories are all available for free using a personal computer and an Internet connection.

When most people think of online satellite images, Google Earth comes to mind (http://earth.google.com). A wildly popular site, Google Earth lets you fly anywhere on Earth to view satellite images, maps, terrain, and even 3D buildings. It now also includes celestial views. Google Earth is free for the basic version, which is sufficient for most people. An upgrade to Google Earth Plus for $20 adds GPS support and higher-resolution printing.

Marine software companies are just beginning to leverage the powerful Google Earth platform. MacENC, Fugawi Marine ENC, and Coastal Explorer are three e-charting applications that let you export and view your navigational waypoints and routes on Google Earth. EarthNC, a new charting application reviewed in detail in Chapter 13, is specifically designed to overlay navigational data on Google Earth.

Several independent websites are also merging Google Earth with free NOAA charts, sometimes called *Google Earth Mashups.* GeoGarage posted demo software that lets you move a slider bar to transition from realistic Google Earth imagery to NOAA chart views, or a blend of the two formats. A demo is available at http://demo.geogarage.com/noaa. Another company, geowake Marina, sees a niche for on-the-water business locators (http://marina.geowake.com). Their software allows businesses such as marinas, boatyards, or dockside restaurants to create websites showing nautical navigation aids with Google Maps.

Traditional map and direction websites are also very useful for mariners. Familiar sites such as Mapquest and Google Maps (www.mapquest.com) (http://maps.google.com) provide more than automobile driving directions. Use these sites to preview marina facilities, to overview a neighborhood with detailed street or aerial views, or to plan walking routes to stores or parks. Before walking out from a new marina, take a quick bird's-eye-view in Google Maps or Mapquest to plan a pleasant walking loop or to locate the closest park or playground for your canine or kid crew members. Google Maps now includes a *Street View* feature for some locations, showing a street-level panoramic photograph of an address, and a *Walk* option to optimize the route for walking rather than driving.

Map websites also let you locate services ashore using their "Find Businesses" features. A quick online search locates the name, address, and telephone number of services— whether it is your crew's hankering for Italian food or your need to find a shop that sells Yanmar impellers. Features such as Search Nearby let you see what services or attractions are near your marina dockage.

If you need to launch a boat, or to come ashore in your dinghy, use Mapquest or Google Maps zoomed in to street mode to locate boat ramps or waterfront dead-end streets. In the U.S., small streets that dead-end at the water often provide public access in an overdeveloped private waterfront.

Like Google Earth, online map websites let you create and save maps, locations, or recent searches. Google Map's *My Maps* lets you drop markers to remember locations or routes of favorite places ashore.

For planning directions and routes on the water, the free

website NavQuest can be thought of as a marine version of Mapquest (www.navquest.com). Although its interface is not as slick as Mapquest (no relation), the site provides free marine navigation planning. Entering your navigational departure and arrival destinations produces a marine trip ticket similar to a AAA *TripTik*, showing the calculated trip time, distance, and fuel consumption based on your vessel.

BOATING FORUMS

Online forums often are the fastest and most informative way to access specialized information. Users post questions and share experiences on diverse and detailed topics.

Most vendors now maintain forums on their website. While Frequently Asked Questions (FAQs) cover a small set of common questions, the forums dig into the nitty-gritty. Look to forums for questions on troubleshooting, cabling, compatible devices, and other users' experiences.

For more general community discussions on boating and cruising, go to one of the many marine forums. Each forum caters to a specific niche, distinguishing itself by vessel or cruising type. For example, there are forums specifically for trawler owners, power catamaran owners, and for most brands of boats. There are also forums geared to boating lifestyles, such as living aboard, cruising, or weekend boating. Some forums focus on a specific geography, such as The Salty Southeast Cruiser's Net (www.cruisersnet.net) covering the coastal southeast United States. Most forums are free, simply asking for registration with an email address.

Don't overlook the excellent forums available through membership-based organizations such as Seven Seas Cruising Association (www.ssca.org) or Marine Trawler Owners Association (www.mtoa.net). Member associations typically archive their members' contributions, sometimes even compiling them as collections for download or purchase on CD-ROM. These electronic logs, journals, and bulletins are an

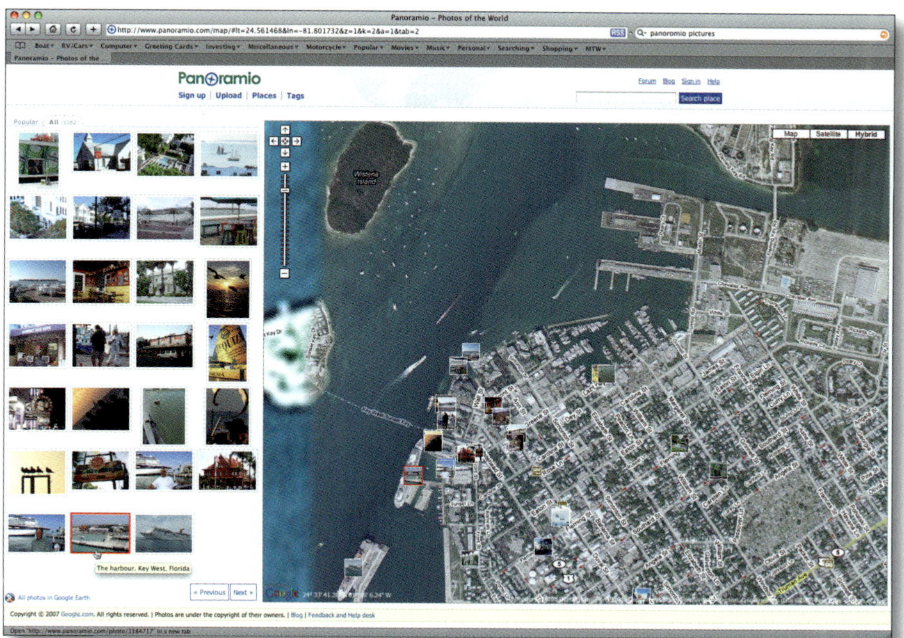

Panoramio: *Still in its infancy, the most popular user-generated photo site is struggling with content quality. Unfortunately, there are at least 100 sunset or building photos for every bridge, marina, or inlet. But expect the value to increase as commercial operations realize the positive marketing impact of a "self-post."*

invaluable source of information for long-distance cruisers. For example, the Seven Seas Cruising Association publishes their archived Commodores' Bulletin on searchable CD. Searching "Belize" pulls up all cruisers' submissions on Belize marinas, customs and immigration, port services, and sightseeing.

Boaters can also maintain their own Web presence with sites like MySpace or Facebook—but specifically geared for boaters. TheBoaters.com is a free online boating community where you create a captain's profile, add a photo of your boat, share photos, and join groups by interest or boat type. TheBoaters.com also has its own forum, covering topics ranging from the debate over power versus sail to servicing fire extinguishers.

USER-GENERATED WEBSITES

When Wikipedia began in 2001, the idea was downright bizarre. Let anyone post information about any topic to create a

You can get information on anything you need, from engine repair tips to fish identification, by using a powerful Internet search engine such as Google. Carefully worded (and punctuated) search commands make it even easier to quickly access online information.

Google: weather Cuttyhunk Island
Access quick local weather by typing "weather" and a city, place, or zip code.

Google: map Friday Harbor
Pull up a map by typing the name of a location and the word "map." Click on the map for a larger version on Google Maps.

Google: marine supplies 33050
Search for a local business by entering the business type and the zip code.

Google: 150 GBP in USD
Instantly convert currencies by entering the conversion in a Google search box.

Google: "Harbour Town Yacht Basin"
Add quotation marks to keep a phrase of common words together as a search string, avoiding a search of each individual word.

Google: ~USCG
Search for a topic and its synonyms (if you are not sure if you are using the correct term) by placing a tilde sign (~) in front of your search term.

Google: bass -music
If your search term has more than one meaning, focus your search by putting a minus sign (-) in front of words related to the meaning you want to avoid, such as if you want information on bass as in fish rather than bass as in music.

Google: Managing the Waterway
I'm Feeling Lucky... For straightforward searches, instead of clicking the Search button after you've entered your search terms, click the I'm Feeling Lucky button. Google will take you directly to the most relevant website without having to wade through the search results page.

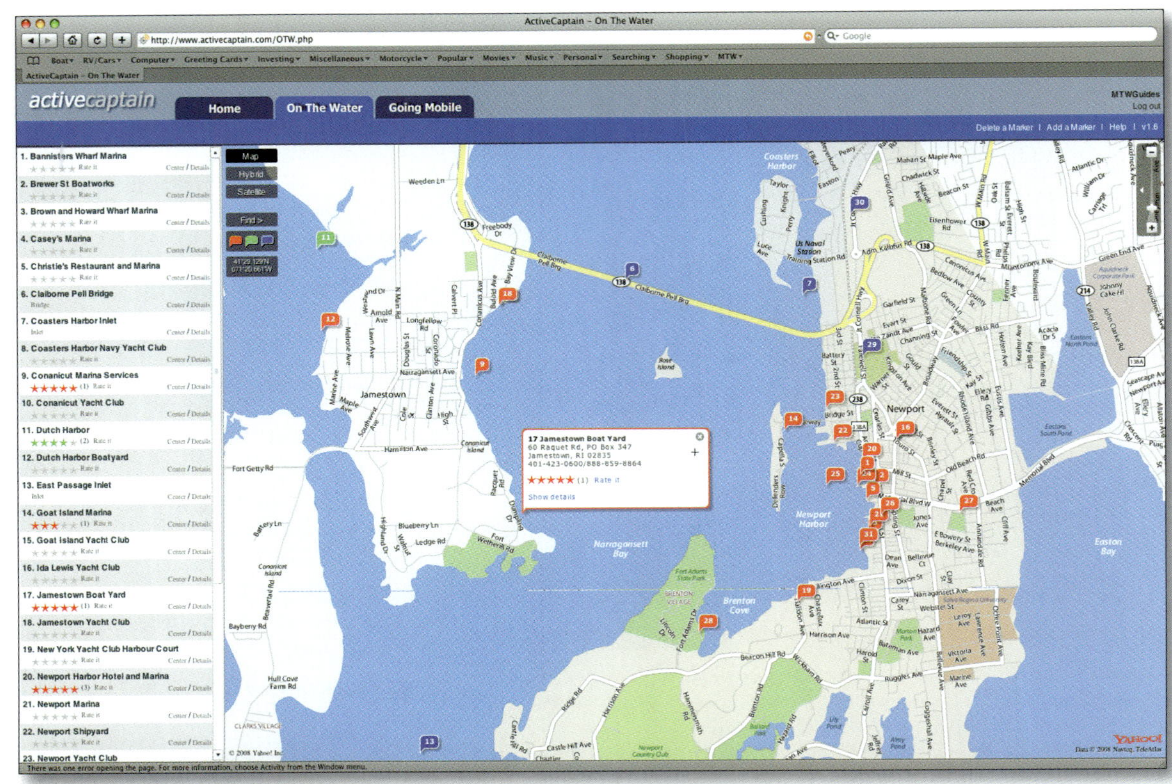

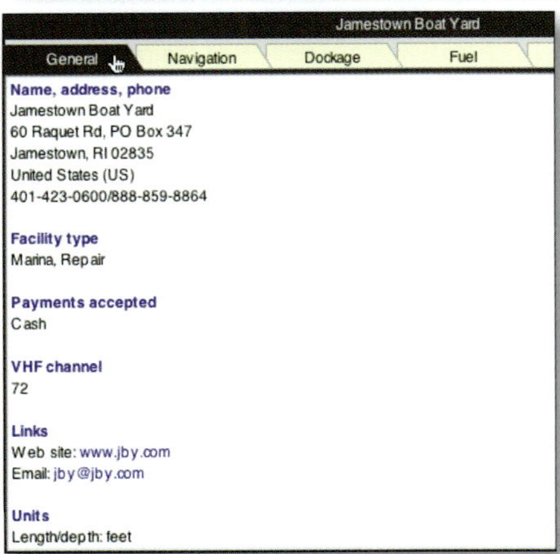

Share and Share Alike: *Many boaters like to share their experiences by posting to a nautically-themed online community such as ActiveCaptain. In less than two years, over 10,000 marinas and 2,000 anchorages have been added.*

dynamic online encyclopedia? This wouldn't be an encyclopedia of knowledge, it would be anarchy.

But today Wikipedia is the largest, fastest-growing, and most popular general reference work currently available on the Internet (www.wikipedia.org). As of April 2008, Wikipedia was well past the two-million article mark, hosting nearly 700 million visitors a year. Vandalism is minimal and submissions are edited by consensus with references.

Wikipedia was one of the first of what are called *open content* websites, where users collaboratively contribute and edit posted information. Other "wiki" sites (referring to the technology that creates collaborative websites) continue to sprout on the Web, including reference sites, social networking sites, blogs, photo libraries, and video sharing.

Panoramio, now owned by Google, is one of the most popular user-generated photo libraries (www.panoramio.com). In fact, the e-charting software EarthNC includes geographically-linked icons to Panoramio photos. If you're curious about a new destination, one of Panoramio's millions of geo-referenced photos likely includes that area. You may not find the precise photo you'd like—such as the view up the marina channel approach—but the selection is impressive. For example, a search for St. Augustine, Florida included a photo of the anchorage by the Bridge of Lions. Each photo's location is marked by a pin on a corresponding Google Earth map, including the photo's latitude and longitude. Google's image library is also a great source for photo references, searching Internet-wide for photos on your topic (http://images.google.com).

For the video equivalent, search YouTube, where users collectively post videos about literally anything (www.youtube.com). Search any location or topic of interest to view the choices for user-submitted footage. Try locations such as "Mt. Desert, Maine," attractions such as "Florida Keys," or topics such as "orca whales." Your computer becomes an amateur Discovery Channel of short clips on that topic.

Waypoint repositories are another new trend in user-generated sites. GPS repository sites collect recreation-oriented waypoints from all over the world. Check out Wayhoo Geographic Coordinates (http://wayhoo.com), TravelByGPS (www.travelbygps.com), or GPSWaypoint (http://gpswaypoint.org).

ActiveCaptain is a new user-generated site for sharing maritime information (www.activecaptain.com). A "wiki" of local boating knowledge, members share their expertise on ports and harbors, including warnings, reviews, things to do, satellite maps, and photos. As of May 2008, ActiveCaptain had a database of over 10,000 marinas, 15,000 local knowledge markers, 4,000 reviews, and 2,000 anchorages. That's a lot of local knowledge available on your onboard laptop.

Chapter 10
Putting It All Together

Ask ten boaters how they use a personal computer for e-charting and you will get twelve different answers. This chapter outlines seven typical e-charting scenarios based on boaters we met at e-charting seminars, boat shows, on the water, and at the docks. Although no one scenario will perfectly match your boating style, existing equipment, or vessel, each scenario illustrates how the pieces of an onboard electronic charting system can come together in very different ways.

Scenario 1: Creating Custom Chartkits

Jack is a dedicated kayaker, preparing to paddle the entire length of the inland river system from Chicago to Mobile, a 1200-mile trip from the Great Lakes to the Gulf of Mexico.

Obviously Jack can't realistically pack even the smallest laptop or subnotebook for his journey. But he can pack a smartphone PDA—prepared with his navigational homework—and he can use a personal computer to prepare for his trip electronically and on paper.

Jack doesn't even own a computer at home, but he has access to a PC desktop at work. He loaded this computer with Fugawi Marine ENC, a worthwhile investment at $280 for its PDA compatibility and smooth GPS exchange.

Because he is transiting the inland river system, Jack needs U.S. Army Corps of Engineer IENC charts. Marine ENC supports this vector chart format, and in fact includes all IENCs

with its U.S. software Data Pack. Jack also loads the street and landmark overlays as supplemental data, a helpful asset as he paddles town-by-town through the Midwest.

Jack had heard that the Army Corps recently released several new IENCs, including missing coverage for sections of the Tennessee and Tombigbee Rivers. He compared chart edition dates on his Fugawi Bonus Pack DVD at the Army Corps chart download site and found several charts that could be updated. These he downloaded for free through the USACE ChartServer, replacing the older chart cells in his Fugawi chart folder with the newer versions.

During his lunch hour at work, Jack prepares for his adventure by creating waypoints, routes, and chartmarks. Marine ENC has the most extensive icon library of all the e-charting applications, letting Jack mark an assortment of locations that are uniquely important to him, such as boat ramps, campsites, and support stops where his wife will meet him along the way.

Most important, Jack carefully creates traditional waypoints and routes on his desktop with Marine ENC. When completed, all these data points will be uploaded to his handheld Garmin GPS, which he will carry with him. Marine ENC has excellent GPS transfer, letting Jack simply connect his handheld GPS to his computer's USB port and select a menu pull-down to initiate the transfer.

Never one to trust electronics alone, Jack is also creating his own custom laminated chartkit. He previews each 10-mile length of the trip on his computer, supplemented with his waypoints and annotations. For each section, he chooses the menu option to print, saving the image as a PDF file rather than sending it to the office printer.

When he has created all his chart files, he burns the folder of files onto a blank DVD. At his local copy center he has the pages color-printed and laminated on 8.5 x 11" paper. Grouping the laminated pages in order, he now has a waterproof chartkit customized for his entire route.

His wife insists he carry a good cell phone on the trip, so he opted for a Windows Mobile smartphone for its PDA features. He's still a bit nervous about bringing this piece of what he calls "electronic jewelry" onboard, so he stores it in a small

Take a screenshot of your track and email it to family and friends. Or consider printing the image for your guests as a great take-home memento of their vacation.

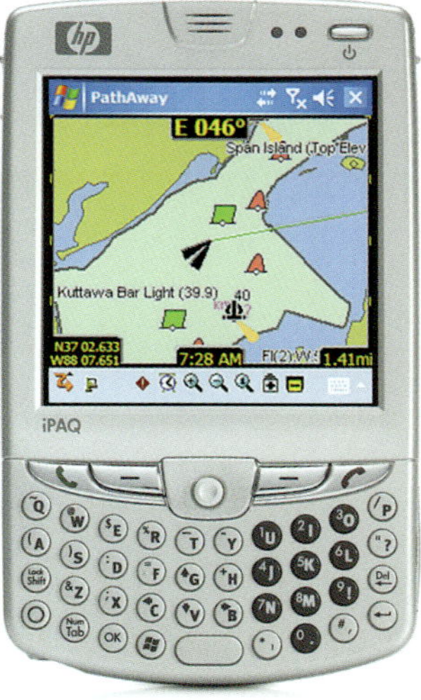

Smartphone Arsenal: *Jack loads up his smartphone with an array of useful applications including Fugawi's PathAway 4 GPS Software for Windows Mobile devices.*

buoyant watertight OtterBox.

This tiny four-ounce smartphone lets him make and receive calls, send text messages, and run a miniature version of the e-charting work he prepared on his desktop computer. He can also use it to check online weather with Weather Underground's mobile version (http://m.wund.com), or by text message from AccuWeather.com (http://wireless.accuweather.com).

When connected to his portable GPS unit, his smartphone also serves as a portable mini-chartplotter, able to store and display the route information he created with Fugawi Marine ENC on his office computer. Marine ENC includes a trial version of Fugawi's PathAway 4 GPS Software for Windows Mobile devices. With this $60 add-on, Jack can view charts,

maintain a backup GPS position, store waypoints, and plan routes on his smartphone. He can even set proximity detection alerts to give him a heads-up as he approaches the many locks and dams on his route.

Because his GPS is Bluetooth-enabled, Jack's PDA can be working while safely stowed in his pocket or OtterBox. With the GPS on, position data is wirelessly sent to the PDA, keeping it up-to-date with position and track information while it saves battery life. Fugawi designed PathAway with a Pocket-Mode feature, turning off the screen and locking the keys while continuing to record GPS data to the track log. When Jack stops to check his charts and route, waking up the PDA updates and displays all his position and track information.

SCENARIO 2: PLANNING ASHORE VIA FLASH CARDS

Mike and Jay use their 30-foot Grady-White Express 305 on the Chesapeake Bay to fish, swim, and sometimes anchor overnight. They usually head out for a day of fishing or with family for a weekend.

They don't bring a laptop aboard when they're on the water, preferring shipboard electronics. Their boat is their escape from the demands of work, so neither wants to be distracted by an onboard laptop. They want the kids to participate in fishing, not be lured by the possibility of onboard computer games. And although the Express 305 has a nice cabin, space is tight with the kids aboard. Anything that can be left at home saves dockcart loads as they pack and unpack each weekend.

Mike treasures his Raymarine E-Series chartplotter with a 12-inch color display. But instead of using RayTech's full-featured e-charting application, RayTech RNS, he uses Ray-Tech Planner to supplement his chartplotter. He already has Navionics Platinum+ cartography for his chartplotter, so he purchased RayTech Planner bundled with the RayTech Navionics Multi Card Reader. The card reader connects directly to a personal computer and provides an interface to display Navionics charts.

Mike and Jay have RayTech Planner loaded on their PC desktop at home, which lets them view the charts to plan fishing and anchoring itineraries. In the evening or during the

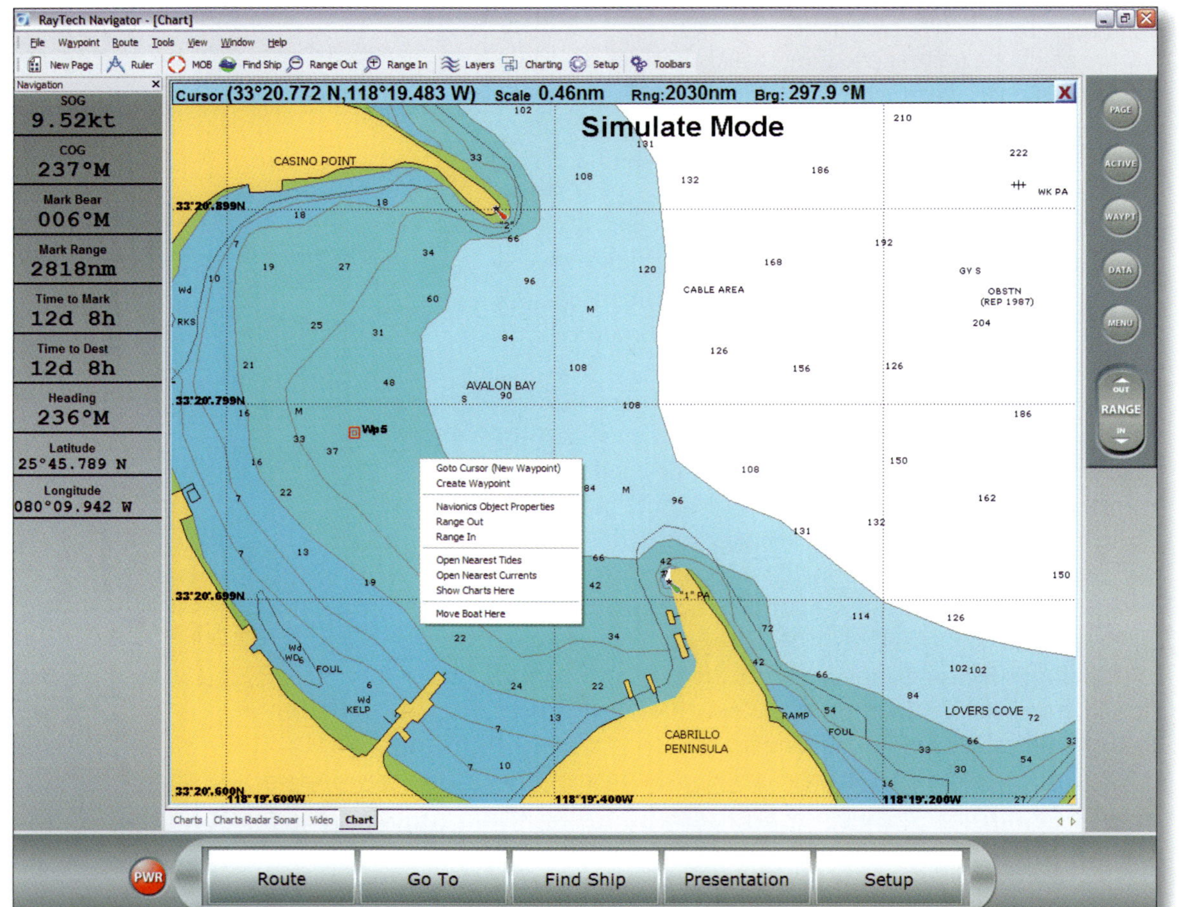

winter, Mike drops waypoints and creates routes for upcoming outings. Before each departure, he plugs the card reader into his computer's USB port, inserts a blank memory card in the reader, and downloads his waypoints and routes.

When it's time to pack for their day out on the water, Mike slips the card into his pocket. Before they cast off, he inserts it into his chartplotter, uploads the waypoints and routes, and is ready to go, navigating with the chartplotter to the waypoints and routes he created at home.

If they are visiting an unfamiliar marina or anchorage, they print detailed chart close-ups of those areas. Since RayTech Planner does not have an integrated print feature, they print using a screen capture utility. They downloaded SnagIt for $39.95 (www.techsmith.com), which lets them print whatever they see on their computer screen.

Mike and Jay decide they don't want Internet access distracting them when they're out on the water. They opt to simply pack their cell phones for emergencies and to stay minimally in touch. Their cell phones also provide a supplemental source for live Chesapeake Bay weather and information. Although they listen to NOAA VHF weather forecasts, they also take advantage of the new live cell phone buoy reports by calling a toll-free number (877-buoy-bay). Created as part of the Captain John Smith Chesapeake National Historic Water Trail—the nation's newest national park—the recordings also tell the story of the Chesapeake (www.buoybay.com). The kids like to listen and learn about Capt. Smith's voyages on their home waters. The new talking buoys, one of several regional networks in the U.S., translate data to voice, providing the latest buoy-by-buoy weather information.

Scenario 3: Piloting with a Standalone Laptop

Jon and Elma cruise the Eastern Seaboard and Bahamas on their Manta 42 catamaran. They've traveled the Intracoastal Waterway a dozen times, usually wintering on the West Coast of Florida, or in some years, crossing over to the Bahamas.

They have two laptops onboard: an older 12-inch Macintosh PowerBook and a new 15-inch MacBook Pro with the Intel processor. Longtime Mac users, Jon and Elma use MacENC as their charting and navigation software. They were

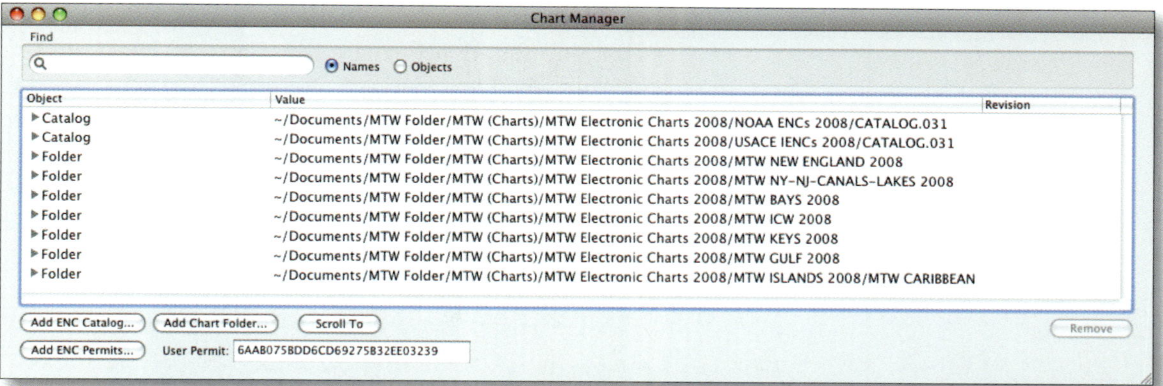

Chart Organization: *Jon and Elma keep their charts organized by cruising geography. This way, they can easily update charts seasonally as they migrate. MacENC's directory-based chart manager also allows them to only load the charts they need, keeping their laptop at peak performance.*

early adopters of MacENC's predecessor, GPSNavX. They are very pleased with MacENC's ability to now view both raster and vector charts.

Their navigation laptop, the newer MacBook Pro, sits at the navigation station in their main saloon, just inside the bridge companionway door. Because of their catamaran layout, the helmsman can easily step into the companionway to review laptop information. While Elma maintains a watch, Jon steps inside to pan large-area charts and make planning decisions. With the catamaran's open layout, he can also look out the forward ports to safely monitor their progress.

Since they cruise primarily in the U.S., they rely on free NOAA raster and vector charts. Although they prefer vector charts for their integrated piloting features, they must rely on raster charts for some regions of the Intracoastal Waterway. Until NOAA completes vector coverage for all U.S. waters, they must use vector and raster charts as available. Fortunately, MacENC displays both formats, can quickly toggle between displays, and even allows for raster-vector overlays.

With free cartography, Jon figures there is no excuse for out-of-date charts. As long-distance coastal cruisers, they update their chart files once a year, purchasing a DVD of all U.S. raster and vector charts. They also maintain a library folder of marine publications, such as *Chart No. 1*, *International and Inland Navigation Rules*, *Light Lists*, state chart catalogs, and hurricane information. These PDF files can be viewed on-screen, or the relevant sections can be searched or printed using Adobe Acrobat Reader.

For Bahamas coverage they purchase Maptech's Region

Be careful using any e-charting application's simulation mode. Because a GPS is not connected and DR is only a hypothetical indication of your course, you should never rely on this display to indicate heading, speed, or position.

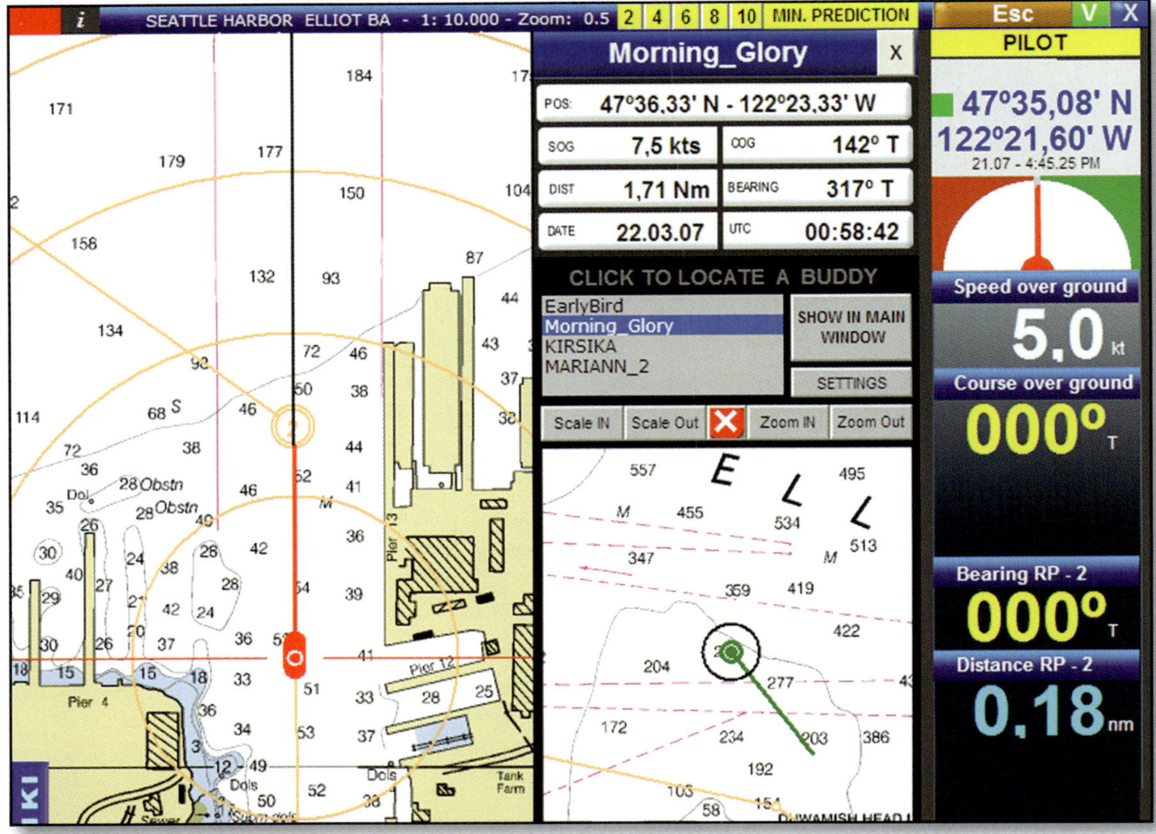

TIKI on a Mac: Using Parallels Desktop, Jon can load the TIKI Navigator trial and see if he likes the program's ease-of-use as well as test the buddy boat functionality.

Take inventory of each of your computer's connections before purchasing a WiFi or air card. For example, if you have two computers, one with an Express slot and one with a PCMCIA slot, then only a USB card will work for both.

9: The Bahamas, which includes BSB raster charts on CD. When traveling in the Bahamas, they take advantage of Mac-ENC's Google Earth integration, exporting their waypoints as a KML file to Google Earth. This lets them view actual satellite images of the shallow Bahamian channels and reefs.

They also purchase regional Maptech Chartkits, which provide paper chart backups. Given the miles they cover each year, printing charts from their laptop isn't worth the time, ink costs, and hassle.

Although MacENC supports a wide variety of marine instruments, Jon and Elma use their laptop primarily as a stand-alone planning and piloting supplement. They decided they didn't need to network the laptop to their other marine devices, which would require running cables inconveniently through their new boat. In the future, Jon may transfer his waypoints and routes from his laptop to his chartplotter using

GPX with a temporary cable connection.

Although the laptop is always on and frequently checked while underway, their helm-mounted Garmin chartplotter is the primary waypoint-to-waypoint marine display, connected to their autopilot and GPS. They don't yet have AIS and are not sure if they need it for their cruising geography.

Instead, the laptop at the inside navigation station is connected to its own GPS receiver, a Garmin GPS 18 that plugs into the laptop's USB port. With a GPS connected, which also serves as a backup to the chartplotter GPS, the charts scroll automatically to show position and route.

Although the laptop isn't networked with other instruments, it is an important planning and logbook tool. For example, they often anchor and are comfortable exploring new anchorages and gunkholes. Jon records a *bread crumb* trail on the laptop as he tries new routes, using this recording not only to backtrack, but as a memory aid when he plans itineraries.

Jon and Elma need to be moderately connected while underway, so opted for a Verizon cellular network card for Internet access. This gives them unlimited Internet access throughout the U.S. for email or Web surfing. They have found cell signals with near broadband rates everywhere they travel.

Now that they have a new Intel Mac, Jon is intrigued with the possibility of running Windows software. He loaded Parallels Desktop for Mac (www.parallels.com) and then downloaded the TIKI Navigator trial version. He's heard TIKI Navigator is the most Mac-like of the PC e-charting packages. Since they've accumulated literally hundreds of waypoints during their travels, they need good waypoint and route management features. They also often cruise in company and are curious about TIKI's new buddy boating feature, which would let them view and exchange position data with their fellow snowbirds.

SCENARIO 4: INTEGRATING SELECT INSTRUMENTS

Bob and Nancy cruise the Pacific Northwest on their Krogen 44 trawler, with plans to someday travel the Inside Passage to Alaska.

They move their Lenovo ThinkPad laptop between home and boat, *docking* at the desk at home or in the pilothouse at their trawler's inside steering station. For e-charting, they use Coastal Explorer, although they realize they could have purchased Chart Navigator Pro and obtained the equivalent software. However Chart Navigator Pro is a bit more expensive and they decided they didn't need the supplemental data DVDs for their cruising geography.

With the luxury of an enclosed pilothouse, Bob and Nancy can pilot actively from a laptop. The laptop sits near the helm, protected from the elements and harsh sunlight. In addition, since it sits near their other marine instruments, cabling the docking station was relatively easy. This connects the laptop to a network of shipboard electronics, including AIS and radar for Puget Sound's busy shipping lanes and notorious fog.

They use free NOAA raster and vector charts. Originally their vector charts were included on a CD-ROM with their purchase of Coastal Explorer. Since then they've downloaded updated chart cells for their region directly from NOAA's ChartServer. For raster coverage, they initially downloaded a few Puget Sound charts from NOAA. Now, for convenience, they purchase a DVD of all available NOAA raster and vector charts. Holding a complete chart library lets them preview some of their future Alaska routes. For cruising into British Columbia, they purchase Canadian Hydrographic Service charts through one of the many CHS retailers in the Pacific Northwest.

Because Bob and Nancy plug their laptop into the boat network, they don't need to transfer navigational data when they go onboard. They prepare this work at home during drizzly Pacific Northwest winters, bringing their waypoint, route, and annotation homework with them on the laptop.

As a backup, however, they GPX transfer their work to their Furuno chartplotter using a blank flash card. In addition to paper charts, the chartplotter/laptop system provides redundancy for both charts and navigational assets.

While underway, the laptop is the main navigational tool, providing route data to the autopilot and receiving GPS position data. Bob and Nancy view charts, port information, Coast Pilots, and tide and current predictions on their laptop. Bob sets alarms for cross-track error and to alert him of shipping lane boundaries. The AIS receiver displays informa-

tion on nearby vessels and projected trajectories for collision avoidance.

Although they currently receive most of their weather from NOAA via their VHF radio, which is the most reliable source for their local weather, they enjoy the GRIB weather feature of Coastal Explorer. Occasionally they download GRIB weather files to stay fluent on interpreting weather data, planning to someday do more offshore boating.

Bob enjoys collecting information on anchorages so he often takes panoramic digital photos of a new gunkhole, georeferencing them to their chart location with the Sony GPS-CS1 (www.sony.com).

Nancy must stay connected for her work, so they opted for Shakespeare CruiseNet for Internet access. The wireless

Penny wise and pound foolish: if your vessel is over 12 meters, either print the Nav Rules or purchase a duplicate paperback copy for $14. This guarantees that your vessel satisfies the carriage requirements of having this document "for ready reference."

Join the Community: *Coastal Explorer's Ships Log stores photos and text which can be shared with other CE users or over the Internet.*

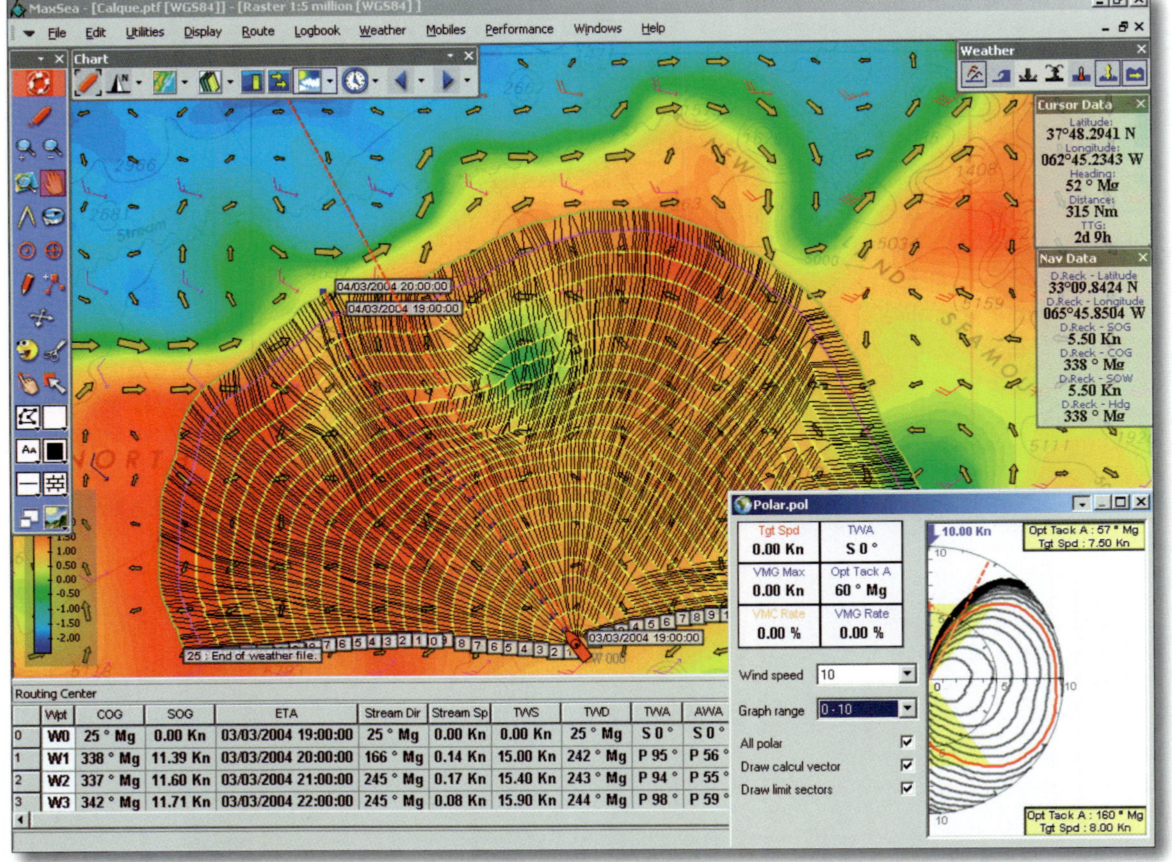

	Wpt	COG	SOG	ETA	Stream Dir	Stream Sp	TWS	TWD	TWA	AWA
0	W0	25 ° Mg	0.00 Kn	03/03/2004 19:00:00	25 ° Mg	0.00 Kn	0.00 Kn	25 ° Mg	S 0 °	S 0 °
1	W1	338 ° Mg	11.39 Kn	03/03/2004 20:00:00	166 ° Mg	0.14 Kn	15.00 Kn	242 ° Mg	P 95 °	P 56 °
2	W2	337 ° Mg	11.60 Kn	03/03/2004 21:00:00	245 ° Mg	0.17 Kn	15.40 Kn	243 ° Mg	P 94 °	P 55 °
3	W3	342 ° Mg	11.71 Kn	03/03/2004 22:00:00	245 ° Mg	0.08 Kn	15.90 Kn	244 ° Mg	P 98 °	P 59 °

Weather Routing: *Brad uses MaxSea's Sailing Performance and Weather Routing Modules for longer passages and ocean races.*

cellular router gives them secure Internet for email or surfing anywhere onboard.

To prepare for their eventual cruise to Alaska, they are gleaning navigation assets from fellow boaters. Several folks in their cruising club have completed the Inside Passage and have collected waypoints, routes, and anchorage locations. Bob and Nancy use Coastal Explorer's user documents file to swap this information with other club members using Coastal Explorer or Chart Navigator Pro software.

SCENARIO 5: PLANNING EXTENSIVE ROUTES

Brad owns a Little Harbor 48, which he sails to Maine every summer and occasionally campaigns in local regattas. When he was younger he was a passionate racer, participating in the Newport Bermuda race. Now that he owns his own sailboat, he sails Block Island Race Week, the Newport races, and his yacht club's local events.

Brad brings aboard an ARMOR X10 Rugged Tablet, deciding the extra cost of a ruggedized computer makes sense given his relatively exposed cockpit, the dampness of Maine, and the roughness of racing conditions. The X10 includes an integrated GPS and has a screen bright enough for outdoor viewing even in sunlight.

His tablet is installed with MaxSea Explorer, Furuno's premium product for recreational boaters. Brad began with MaxSea Navigator+ many years ago and has since upgraded to Explorer, a more full-featured bundled package.

Brad's computer moves around on his spacious sailboat, depending on his cruising or racing circumstances. Sometimes it sits in the cockpit, sometimes down below at the navigation station, and sometimes even in the stateroom.

When he is cruising alone or with a couple of guests, his tablet can sit at the helm, safely under the dodger. Although his sailboat doesn't have the extensive flat area of a trawler or powerboat, his cockpit is spacious and he has custom-fitted a docking station. When at the helm, the tablet cables to the autopilot, GPS, and ship's instruments.

When racing, the ARMOR is down below with his tactician, who calls out headings and updates to the captain, helmsman, and crew on deck.

Sometimes, the tablet sits in his stateroom, used for planning, analyzing weather, Web surfing, and checking email. Because his recent boating has been local he opted for a cellular network Internet plan.

For charts, Brad combines NOAA free raster charts with C-Map MAX Pro vector charts. His original set of NOAA rasters were included with his MaxSea purchase. Now he updates these files by purchasing a DVD of NOAA charts for convenience. MaxSea does not support free U.S. vector charts (ENCs or IENCs).

For vector cartography, his primary chart format, he uses C-Map charts, heeding Furuno's advice to load the charts onto his hard drive rather than using a USB reader. His C-Map cartography also provides all his supplemental data, such as marina information, street maps, tide and current predictions, and aerial and satellite images.

He also owns a C-Map USB Multimedia Reader, which he connects to his laptop only when he needs to transfer waypoints and routes. Moving the data by flash card, his Furuno NavNet 3D System serves as an active redundant system for navigation. He also keeps a USB GPS sensor onboard, a small and inexpensive bit of insurance in case the ARMOR loses its integrated GPS signal.

Brad's MaxSea version includes both the Sailing Performance Module and the Weather Routing Module. He uses these tools to plan sail routes for his longer races and for his overnight deliveries to Maine each season.

Since his computer also contains personal information, Brad has recently enabled the MaxShell Windows Desktop Manager that comes bundled with MaxSea Explorer. This module lets him separate MaxSea from his computer's other functions. The tablet can then be used by race crew members while keeping his other files and applications private.

Although Brad may not need all the features of MaxSea Explorer now, he feels it will serve him well for his planned voyages when he retires fully. He intends to take his boat to the Mediterranean someday, where the best cartographic choices include British Admiralty Raster Charts (ARCS) and MapMedia charts. MaxSea supports both these extensive and popular formats for European cruising.

Scenario 6: Circumnavigating on a Tight Budget

Piet and Inger are living their dream, circumnavigating aboard their Shannon 38. They travel on a very modest budget, working along the way to replenish their cruising kitty.

They have two laptops onboard, originally their respective personal computers, now delegated for different tasks. The newer, more powerful laptop is the primary navigation laptop. The older laptop is for domestic tasks such as email, and is the only laptop taken ashore.

They don't have onboard email, instead relying on free WiFi at Internet cafes. Onshore, they check their Yahoo email accounts, surf the Web, and make telephone calls using Skype Voice over Internet Protocol (VoIP). Before they left home, they downloaded the free Skype software on their el-

derly parents' computers and reminded their grown children to sign up, letting them all take advantage of free Skype-to-Skype calling.

Both laptops have SeaClear II installed, a full-featured charting and navigation software application available as a free download. The older laptop serves as a backup for the software and charts, and lets both Piet and Inger simultaneously browse charts and routes. In fact, they each downloaded SeaClear's manual in their respective native languages, French for Piet and Dutch for Inger.

Given their cruising geography and onboard energy limitations, they still rely primarily on paper charts. They collect charts through chart exchanges or nautical flea markets, storing the stacks of paper charts in custom-made drawers under their V-berth.

But they are also collecting a patchwork of electronic cartography. Before they left to go cruising, Piet had access to a large-format scanner at his office. As they collected paper charts, Piet scanned each sheet, creating an impressive collection of TIFF files. They realized many of their charts were old and potentially inaccurate, but figured they were better than nothing for some remote, poorly charted destinations.

They particularly like SeaClear since it is one of the few charting applications that can geo-reference these scanned chart files. SeaClear is able to convert JPEG, BMP, and TIFF files, letting Piet and Inger create their own supplemental collection of electronic chart files on a shoestring budget. Piet and Inger also trade scanned chart files in WCI format with other SeaClear II circumnavigators. To complete their electronic patchwork, they have a collection of older SoftChart and Maptech unencrypted raster charts on CD-ROM.

SeaClear II does not display vector charts, which rules out

Hundreds of nautical reference publications are available as free downloads on the Internet. Many are expensive-in-print piloting volumes covering the world by region, such as the *Pilot Charts*, *Sailing Directions*, *Sight Reduction Tables*, and *World Port Index*. For a description of nearly 200 free downloadable publications, see the Nautical Reference Library in Appendix A.

Compact and Rugged: *Brad's ARMOR X10 Rugged Tablet is equipped with a built-in GPS and can handle the wet, cold, and rough treatment "cockpit living" dishes out.*

international S-57 files, and does not support other raster formats such as Seafarer or British Admiralty charts. But these limitations do not bother them.

With a wind generator and some solar, Piet and Inger need to watch their onboard energy use. They did invest in a small Furuno chartplotter and purchase C-Map cartography on cards as needed for new regions. They also have an autopilot onboard, but neither the chartplotter nor the autopilot is networked to the laptop.

To limit their power draw while underway on long passages, the chartplotter and the laptop are only turned on occasionally. The laptop sits safely below deck, connected to a USB GPS to show position and as a GPS backup. In SeaClear II, they create their waypoints and routes, enjoying its clean and responsive interface. But they are not heavy waypoint users, instead creating simple routes which they plan and double-check on their laptop, then enter into their chartplotter.

Since they don't have onboard Internet access, they don't mind SeaClear's lack of GRIB weather—they would not be able to access the server download or Saildocs email anyway. Instead, Inger and Piet both have ham radio licenses, using single sideband radio (SSB) to pick up weather information and local cruiser networks.

To keep family and friends back home updated on their adventures, they maintain a weblog (*blog*). They collect material along the way, including digital photos, journal entries, and miniature charts of their ports-of-call. They use SeaClear II to create these thumbnail charts, displaying a close-up of their location on their laptop screen and capturing the image as a JPEG using SnagIt screenshot software. Inger creates the blog pages onboard, uploading them from the traveling laptop when she is shoreside with a WiFi connection.

Scenario 7: Captaining a Large Yacht

Capt. Kent is the captain of a Hatteras 72 Motor Yacht. Kent is responsible for planning itineraries, delivering the vessel, and managing day-to-day operations.

The vessel is equipped with two dedicated *desktop* com-

Converting Paper Charts: *Inger converts the paper chart Piet scanned as a TIFF file to a WCI file. It's an easy process, first selecting* Convert>Single file to WCI *from the* Tools *menu (left). She then chooses her source TIFF file (center) and a target WCI file name and location (right). After geo-referencing, the WCI file is ready for e-charting in SeaClear II.*

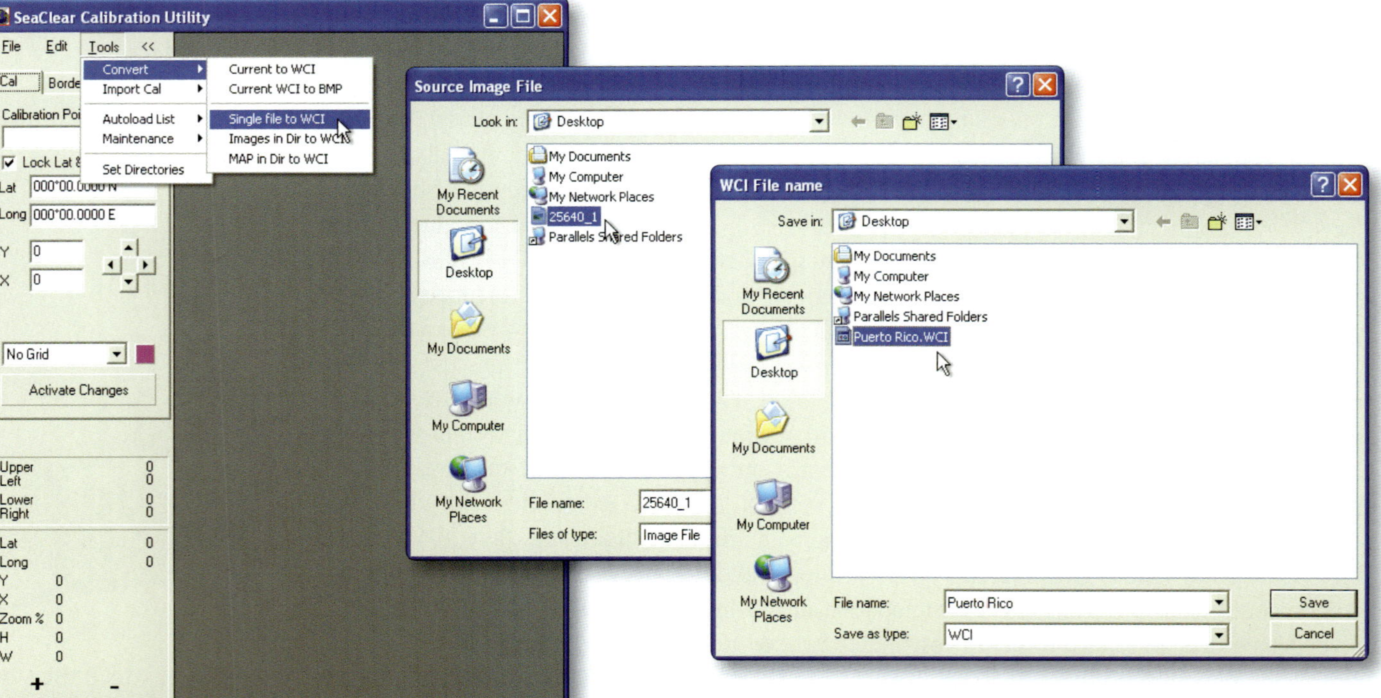

puters. Kent also has his own personal laptop onboard, which he brings ashore for meetings with port services and vendors, crew briefings, and to review itineraries with the owners.

All three computers, the two shipboard desktops and Kent's laptop, are loaded with Nobeltec Admiral MAX Pro. With each computer part of an onboard network, Kent or the owner can view all of Admiral's displays and features from any monitoring station. While underway, the desktop computers and Kent's laptop actively show all navigation information, sharing charts, routes, AIS collision risk updates, radar overlays, satellite imagery, and tide and current predictions.

Kent's laptop becomes part of the Nobeltec GlassBridge Network from either the bridge or his captain's quarters. By plugging in to any onboard Ethernet jack, his laptop becomes a third station in the network. An Ethernet jack in his captain's quarters lets him monitor shipboard activity and plan itineraries at a quiet below-deck navigation station. When "off-duty" in his quarters, he uses Admiral's video monitoring and tender tracking features to keep an eye on security.

The vessel is equipped with C-Map MAX Pro cartography, providing near-worldwide vector charts. These chart portfolios also include supplemental data Kent needs, including marina and port information, aerial and satellite images, and tide and current predictions.

Since all onboard computers are networked, Kent is not burdened with transferring waypoint and route data. In fact, by not supporting any of the three standard formats for export, Admiral assumes a within-network exchange. Data exchange is irrelevant for Kent: he can drop waypoints, create routes, annotate charts, and set alarms on his laptop, sharing this data automatically with the other onboard computers running Admiral.

All marine electronics, including GPS, autopilot, radar, AIS, and engine sensors, are connected to the Admiral GlassBridge Network. Each computer station displays all instrument data independently and shares the same C-Map cartography. In addition, the vessel is equipped with a heading sensor and an InSight Radar 2 BlackBox in order to display radar scans on each networked computer.

Rather than use Admiral's limited GRIB weather features,

Kent suggested using the pre-installed OCENS WeatherNet. Although the software is included with Admiral, the owner had to sign up for the annual subscription and pays usage fees for each weather download. Internet access, as well as voice/data transmission, is available worldwide using a sophisticated Sperry NAVICOM F77 Global Marine Satellite Communication system.

In addition to Admiral's route planning and network features, Kent regularly uses its trip planning features. The Route Wizard provides routes appropriate for the large beam and draft of his vessel. Kent then runs the ETA calculator on each itinerary, providing the owner with preferred departure times and estimates of travel times and fuel consumption.

Automated Routes: *Kent often uses Nobeltec's Route Wizard to find the shortest safe route. After the route is accepted, he then uses the ETA calculator to find the fastest and most fuel-efficient departure time.*

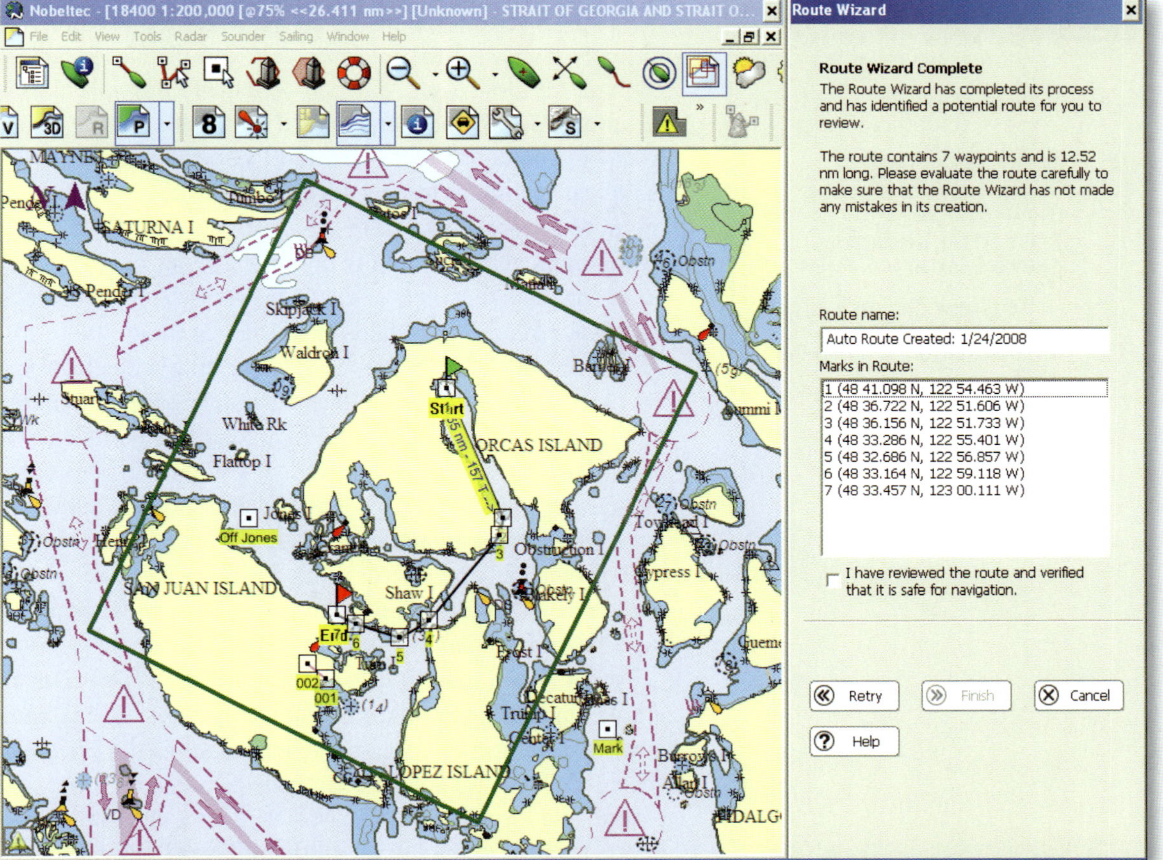

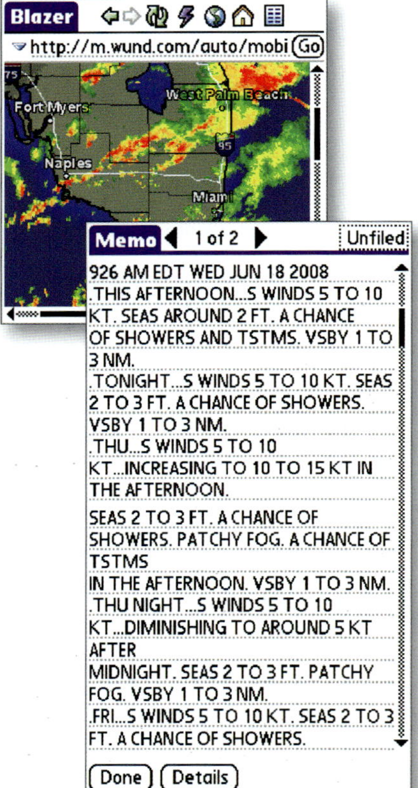

Going Mobile: *Today's smartphones offer remarkable functionality, including email, music, alarm clocks, digital photography, calculators, database logging, and Web surfing. Boaters will especially appreciate mobile sites such as Weather Underground's animated weather radar (top) and NOAA's text weather broadcasts (bottom).*

HOW WE E-CHART

Navigation is as much art as science, so every navigator has his or her own unique style. Here are a few practical e-charting thoughts from our own personal experiences using a laptop onboard our PDQ catamaran.

Rather than choose between raster or vector charts, we've found it's best to use both. Many e-charting applications support both formats, allowing you to use each to its strengths. With both formats available you can choose the chart that works for your current task, whether overview planning, scanning supplemental data, or setting intelligent alarms. For easiest viewing, set up a single keystroke command that toggles between formats.

We're more comfortable viewing raster charts, so we use these for overviews or planning. Vector charts are more informative while underway, since they can display vessel-up, zoom in for more detail, and can be set with multiple alarms. We always use vector charts to read information about an aid to navigation or a bridge. That level of detailed information is typically difficult to read on a raster chart, particularly in high-density areas.

To minimize the pitfall of "porthole driving," namely navigating only the tiny view of the world you see on your screen, we always keep paper charts helmside (in large plastic folios) for the larger bird's-eye-view. Navigating with electronic charts can slip into driving the computer, screen by screen, rather than looking ahead to the big picture. This tendency is exacerbated by screen redraw time: if your laptop bogs down and takes a while to redraw a new chart, your eye and brain have trouble assimilating the total navigational landscape. Since you should never trust your ability to navigate to your hard drive or laptop battery, back-up paper charts should *always* be onboard anyway.

Typically, we have split screen viewing enabled. For example, transiting Norfolk is considerably less stressful when a raster chart accompanies a satellite image of the Elizabeth River's Eastern and Southern Branches. The same can be said for cruising Florida's West Coast and entering the Little Shark River. There is nothing like a little bird's eye assistance in a winding mangrove channel!

As cruising guide authors, we are admitted waypoint junkies, entering literally hundreds of waypoints for a region while underway. We start by entering data points in their approximate locations ahead of time (such as during a Minnesota winter!), then verify them on the water. We typically enter waypoints for all marinas and suspect anchorages, and build safe routes for major passages.

While transiting, we sprinkle additional waypoints liberally—creating tracks using the *bread crumb* analogy. We convert waypoint tracks from our dinghy to routes. Using a handheld GPS, we carefully investigate shallow anchorages for our guides, then safely bring in the mother ship. As we enter an unfamiliar anchorage or marine facility, we turn on tracking so we can follow the trail of waypoints out in the morning if it's dark or foggy.

We diligently load and unload charts as we transit regions. For instance, southbound, we load Intracoastal Waterway charts as we clear the Potomac River. As we approach Coinjock, North Carolina, we unload Chesapeake charts. Our laptop performs much better with fewer charts loaded.

We've also extended the digital metaphor beyond simple charts. In the past, mildewed out-of-date *Light Lists*, *Coast Pilots*, *Chart No. 1*, and other bulky publications came along for onboard reference. Who could predict a night entry into St. Mary's River? Or what a particular arcane chart symbol meant? We now have an up-to-date folder of searchable Adobe PDF publications a mouse-click away (see Chapter 9, *Extending the Digital Metaphor*).

Over the last decade, we've migrated from a BlackBerry to a Treo to an iPhone. What we do with these little demons could easily fill another book. If you're a boater and don't already use a smartphone, buy one today.

Lastly, don't forget the print function. Using the application's print command, or by taking a screenshot, you can print snapshots of annotated charts and more. We use screenshot software regularly, printing chart excerpts for planning or note-taking. Some boaters print excerpts to clip into a log. You can have lots of fun printing and sharing color highlights of your travels, something you could never do with paper charts.

Part Four

Choosing an Application

You've done the homework. The history of global position finding, hardware, instruments, electronic charts, supplemental data, and data exchange are behind you. You're champing at the bit to put reference publications and weather resources on your laptop. You can even imagine your blend of the Seven Scenarios. You're finally ready to choose an application partner, but which one should you choose? Or maybe you've been e-charting for years and after reading some of these chapters think there may be a more suitable match for your boat, geographic intentions, and cruising style. This section helps you sort through the wide range of e-charting choices.

Chapter 11, *How to Choose an E-Charting Application*, summarizes the software players and the state of e-charting software today. Read this chapter first to clarify any confusion about e-charting start-ups, features you need, and using charts you already own.

Chapters 12 through **26** contain in-depth reviews of 16 e-charting software packages. We begin with a simple viewer, then turn to a new planner-plus that uses the Google Earth platform. The next 12 chapters review the major PC software applications, ordered by approximate base price from free to several-thousand-dollar modular packages. The final chapter wraps up with e-charting software for the Mac. Use these reviews to add focus and help narrow your choices.

Chapter 11

How to Choose an E-Charting Application

It's a crowded and confused marketplace for today's boater trying to decide which e-charting package to purchase. Unlike buying a car, where choices and brands are well known, the e-charting market has not yet shaken out with a well-established consumer map of options.

But a crowded marketplace is good news for boating consumers, who benefit from a slew of companies chasing each other's latest feature. This "keep up with the Jones'" product cycle has resulted in recreational software with incredibly advanced features. Previously only available to commercial vessels, weather downloads, AIS support, annotated charts with linked image files, and route optimization planners are becoming standard fare for recreational boaters.

A competitive marketplace is also good for pricing. Imagine the price of Nobeltec VNS MAX Pro ($490) if Fugawi Marine ENC ($279.95) wasn't also on the market. Besides a truly free option (SeaClear II), there are e-charting packages ranging from under $100 (Software-On-Board) to several thousand dollars with add-on modules and subscription services (Admiral MAX Pro and MaxSea Explorer). Even Macintosh-toting boaters are no longer excluded, with choices such as GPSNavX, GPSy, MacENC, MacGPS Pro, NavimaQ, and PassagePlus.

With so many e-charting choices, there's no shortage of knobs and dials to turn or twinkling lights to gaze upon. The competition has combined with recent advances in computing power to bring mind-boggling functionality to recreational e-charting. Want to scan and geo-reference paper charts of Croatia to create your own digital chart files? How about displaying a small window in the corner of your screen to stream video from your young child's stateroom? Or create custom bathymetric charts with your own vessel's soundings? Need to display your buddy boat's current location on your chart display? No problem with today's e-charting choices.

THE E-CHARTING PLAYERS

An amazingly diverse collection of companies created the 14 full-featured e-charting packages reviewed in the following chapters. Six of the applications were developed by programming teams with less than three people. Several were one-person, lone programmer efforts. On the other hand, others were produced by subsidiaries of huge corporations, such as Jeppesen Marine (Nobeltec) which is owned by Boeing.

E-charting software development is also a global project, with companies in the U.S., France, U.K., Canada, Australia, New Zealand, Norway, and Sweden. Yet your software choice should not be constrained by citizenship. For example, many U.S. boaters use Fugawi Marine ENC (from Canada), or RayTech RNS (British), or MaxSea (French). Boaters around the world use Nobeltec (American) or SeaClear (Swedish). Many small start-ups are just beginning to enter the U.S. market, including DigiBOAT Software-On-Board (Australia), TIKI Navigator (Norway), and NavSim BoatCruiser (Canada).

You don't need to choose one of the large-company or high-priced options for a professional, full-featured product. All the e-charting applications reviewed in the following chapters are incredibly robust. For example, even SeaClear II, the freeware option, supports GPS, autopilot, AIS receiver, wind, depth, and water temperature instruments.

In fact, the small companies, such as DigiBOAT, TIKI Navigator, GPSNavX, or NavSim, are typically on a more rapid development cycle than the bigger companies. As an example, within a three-month period, GPSNavX upgraded MacENC to include customizable units for distance, depth, and speed; a Base World map; extended GPX and Google Earth support; and improved chart display speed, workspace

---

E-Charting Application Reviews

Sixteen popular e-charting packages are evaluated in this book. We examine a *viewer*, a *planner-plus,* and the 14 leading PC and Macintosh *full-featured* e-charting applications:

Viewer
CARIS Easy View

Planner-Plus
EarthNC

Full-Featured
SeaClear II
DigiBOAT Software-On-Board
TIKI Navigator Pro
Fugawi Marine ENC
NavSim BoatCruiser
Coastal Explorer
The Capn
Nobeltec VNS MAX Pro
Chart Navigator Pro
RayTech RNS
Furuno MaxSea Explorer
Nobeltec Admiral MAX Pro
NavimaQ
MacENC

Chapter Questions

- *How do I choose an e-charting package?*
- *What are the important feature and price differences?*
- *Is it better to buy from a large marine vendor?*
- *Can I use my existing cartography?*

saving, and chart printing. TIKI and NavSim have both issued significant feature updates in the last two years. Small companies are able to focus on code development rather than attending trade shows, turning out an impressive array of updated functionality.

Small software companies also produce competitive products, taking advantage of their late market entry to learn from others' mistakes. With low overhead costs, they can provide a boatload of functionality at a reasonable price. Their applications are clean and fast, engineered by first-rate programmers rather than fettered by committees struggling with legacyware. Unlike the past, small software company offerings are safe to choose and should be given consideration.

The Marine Software Gap

Unfortunately, marine software has not caught up with the general software industry in terms of pricing, packaging, and user experience.

As with everything else on your boat, expect a marine surcharge. Unlike general consumer software, which often costs $29.95 or $49.95, e-charting packages start at $53 and quickly run into the hundreds of dollars. Before you compare prices between marine and general software, realize that a $200 e-charting package is relatively inexpensive.

Conspiracy theories aside, the up-charge is due to low volume sales, not because you're a boater with a surplus of disposable income. Unlike a $29.95 software application that will be purchased by hundreds of thousands of users, e-charting vendors cater to a small and fractured audience.

Marine software is also immature in areas such as packaging, documentation, automated updates, and stability. For example, most e-charting packages are still purchased as boxed sets of physical media, often leading to stale cartography. Typically, lower-priced packages such as SeaClear, Software-On-Board, MacENC, and TIKI Navigator rely on the faster and more efficient digital delivery option.

Other standard software services, such as integrated help files or auto-checking for updates, also tend to be lacking in e-charting applications. Only two of the applications, Fugawi Marine ENC and MaxSea, have video tutorials, now a standard marketing and getting-started tool in the general computing market. And, perhaps most telling of a software industry still in its infancy, nearly half the applications we tested occasionally crashed or locked up.

Which Features Matter to You?

Many boaters mistakenly buy an e-charting package based on a word-of-mouth recommendation.

Before you go out and buy that software, consider the source of the recommendation. What if your trawler buddy—who never leaves his home port—recommends an e-charting application, but you're taking a sailboat to the Bahamas? Is it still a good choice for you? Or you read a magazine article in which the technical editor (who spends his free time fiddling with e-charting) really likes the software? Is it appropriate for your level of computer comfort and expertise?

Not all boaters need MARPA, AIS, or GRIB data on their laptop electronic charting system. The navigation software package that works best for you depends on how you use your boat. Are you a weekend sailor, a coastal cruiser, or an offshore passagemaker? A weekend sailor wants to plan routes at home and then bring a chartplotter-compatible memory card aboard. The software is primarily used to create waypoints and routes and to locate marinas and anchorages. In contrast, an ocean-going trawler has different software needs, requiring collision avoidance, weather routing, and international chart support.

Because many e-charting packages try to be all things to all boaters, the average boater may only use a small subset of an application's features. In the general software industry, we're all familiar with this "bloatware" trend, generating complaints about software that has become too complicated with too many features. Even popular applications such as Microsoft Excel or Word generate the saying that, "95% of users use 5% of its features."

The same phenomenon has emerged in e-charting. For the average boater, the most expensive e-charting package is not always the best choice—even if price isn't an issue. When choosing an e-charting application, think about what makes sense on your boat and what you can reasonably manage with an onboard computer. For example, we work from our boat—spending many months out at a time—but mainly use basic waypoint, route, annotation, and track features.

If you're going to use only five percent of an application's features, then it should be exceptional on *those* features. Evaluate your computer use, your vessel, and your cruising geography to identify "bread-and-butter" features. Then evaluate the e-charting packages on those features, not on extras you'll never use.

Every boater values different features in various ways. For some, cost is a drop-dead component of the purchasing decision. For those who prefer boating to computing, ease-of-use may be most important. Do you value waypoint management, buddy boat tracking, or a fuel calculator? Can you live without an engine room video monitor, sea temperature display, or Great Circle route planning?

When considering your feature preferences, don't discount your e-charting baggage. Hardly any boater starts with a clean electronic slate. At a minimum you probably already own a laptop—which dictates either PC or Mac-compatible choices. Most boaters own a used boat, which means pre-installed electronics of a particular brand and possibly even hand-me-down cartography or software. Inheriting a copy of The Capn, or a recently-purchased set of C-Map cartography, or a Furuno chartplotter may affect your choice more than whether you own a sailboat or a powerboat. Even new boats come with e-charting constraints. For example, if your newly commissioned yacht comes bundled with a Raymarine E-Series chartplotter, then RayTech RNS may be the right choice for you, regardless of your cruising geography.

Once you decide what matters to you, review our *Software Comparison Tables*, which are presented in the feature and software chapters. Highlight which of the 81 features are important to you, then compare the e-charting applications on those metrics.

What About My Charts?

Although you may already own electronic cartography for your chartplotter, don't let this previous purchase pre-determine your e-charting choice. As your boating life changes,

this will prove to be a "tail wagging the dog" decision.

Except in rare instances, most notably RayTech RNS's unique networking arrangement, chartplotters and personal computers typically don't share cartography interactively. The files are too large and the connection is too slow. So you really can't cut cost by double-dipping. Although some e-charting applications are intended to read Navionics or C-Map charts via a USB multimedia card reader (see the *Cartography and Supplemental Data Tables* in Chapter 6), you can't use the same card in your computer and chartplotter simultaneously.

Instead, when weighing the cartography choices, think about where you expect to cruise long term. It's widely acknowledged that chart quality varies by region. Research the differences in cartography for your particular cruising geography by talking to other boaters and searching the forums.

If you plan to cruise outside the U.S., consider e-charting applications that read a variety of international formats. Although international vector charts are standardized, only the U.S. government distributes its cartography in the unencrypted S-57 format. Other countries encrypt their vector charts, requiring an unlock utility, many using an evolving standard called S-63. A surprising number of e-charting applications do not read international S-57 or S-63 charts, as summarized in Chapter 6, *Charts and Supplemental Data*.

International raster charts also come in country-specific formats. For example, British Admiralty charts (ARCS) do a good job covering British territories, including the British Caribbean, British Isles, and Mediterranean. Yet very few applications, most notably MaxSea and PassagePlus, read ARCS. Canadian Hydrographic Service charts cover Canadian waters. Seafarer charts specialize in Australia and New Zealand. If you intend to travel to remote areas with weak electronic chart coverage, you may want an e-charting application that lets you scan and geo-reference paper charts. Only SeaClear and Fugawi Marine ENC currently have this feature.

In addition, seven of the fourteen full-featured applications support proprietary vector charts on card and/or DVD format. DigiBOAT, NavSim, Nobeltec, and MaxSea support C-Map data. Fugawi and Raymarine team with Navionics. Although proprietary charts are more expensive, they are generally of

- *How easy is this application to set up?*

- *What else do I need to download or purchase?*

- *What chart formats does this application support?*

- *How extensive is the customer support?*

- *How does this application "look and feel" on the screen?*

- *How well does this application open, zoom, and pan charts?*

- *How easy is it to create and manage waypoints and routes?*

- *What other marine instruments are supported by this application?*

- *What additional planning and piloting features are included?*

Make sure your Windows firewall and any virus protection software is turned off prior to installing e-charting software.

high quality and include extensive supplemental data.

Distant destinations aside, the majority of U.S. boaters confine their cruising to U.S. waters, which means they should care about support for free RNCs, ENCs, and IENCs of U.S. coastal and inland waters. Any e-charting application used by U.S. boaters should read free U.S. charts. The U.S. government is unique in providing free, frequently updated digital cartography. Yet a surprising number of e-charting applications do not support free U.S. *vector* files (ENCs or IENCs). For example, SeaClear II, NavimaQ, TIKI Navigator, The Capn, Nobeltec VNS, RayTech RNS, Nobeltec Admiral, and MaxSea only read the *raster* format files (RNCs). Software-On-Board supports neither RNCs nor ENCs.

Some e-charting vendors take a "the more formats supported the better" approach. For example, Fugawi Marine ENC, BoatCruiser, and Coastal Explorer support an impressive assortment of chart formats. Some of the applications even let you view multiple chart formats simultaneously. For example, MacENC and The Capn can display a traditional-looking raster chart while harnessing the intelligence of an underlying vector chart. The Capn can even blend the transparency level of the raster chart over the vector chart.

Other vendors intentionally restrict their chart compatibility. Some do so for philosophical reasons, such as Digi-BOAT's exclusionary bias to vector charts, believing them superior to raster charts. DigiBOAT also condemns user-scanned charts, considering the digital conversion of paper charts to be a bad navigational practice. Other vendors, such as Nobeltec or MaxSea, restrict chart support for commercial reasons, presumably because they are also in the vector chart-selling business.

Finally, realize that your choice of an e-charting application has an economic impact on your long-term investment in cartography. Digital charts sell for very different "nautical miles per dollar." Charts can be obtained for free or they can be purchased for hundreds of dollars per region. The initial choice of a chart format and source has serious cost repercussions when it comes time to update your cartography.

Regardless of which charts you use, you should update them annually. NOAA charts are continually updated at no cost. NavSim BoatCruiser is unique in providing an auto-update feature (for a one-time fee) that compares and updates your free NOAA charts. Several certified NOAA chart distributors sell entire compilations for under $50. In contrast, some chart vendors provide new cartography through a repeat-customer update charge, an annual membership option, or in the most expensive scenario, a completely new full-priced purchase by region.

THE REVIEWS

In the chapters that follow, we review one viewer, one planner, and fourteen full-featured charting and navigation software packages. We begin by examining ease of installation and the user interface. We summarize technical support options, including the quality of user manuals, available forums, and the manufacturer's website and customer support. We evaluate the software's ability to handle chart data in terms of panning and scrolling, quilting and stitching, search functions, supported file formats, and exportability.

Once these basic features are reviewed, we turn to more advanced functionality. Can the software incorporate additional data such as tide and current predictions, street maps, or 3D bathymetrics? Does it include built-in transit calculators or route optimizers? Is the software able to import and/or export navigational data? Can it download and display weather information? How well does it integrate with external devices such as a GPS sensor, autopilot, radar, data recorder, or video camera?

First, take time to assemble your high-priority wish list. Then read the following reviews. With 81 feature comparisons and several thousand words each, it should be easy to separate the glitter from the gold. Your choice of an e-charting package is not only a cash investment, it's also a huge time investment. You don't want to waste either only to find that a feature you value, a chart format you need, or an instrument or data interoperability component is missing.

Chapter 12
CARIS Easy View

To understand CARIS Easy View you must first understand the company. CARIS was founded in 1979 by a survey engineering professor from the University of New Brunswick who worked on digital mapping programs in the basement of his home with his students. Its first commercial software product was called CARIS, which stood for Computer Aided Resource Information System.

Today, CARIS is a multinational company with offices in the U.S., Australia, and Canada. It develops software that stores and manages hydrographic data for use with paper or digital charts, often working under contract for federal agencies. For example, NOAA uses a CARIS application to update and create its Electronic Navigational Charts (ENCs). CARIS software is also used in agriculture, geology, forestry, and transportation.

CARIS created its original viewer, called EASY-ENC, so non-CARIS users could display their marine and hydrographic data products. Like its predecessor, Easy View is simply a viewer and not a full-featured navigation package (www.caris.com/products/easy-view). But it is an incredibly flexible viewer, able to display 18 data formats, including hydrographic data and 3D flight paths. It just so happens that two of these file formats, BSB/KAP and S-57 files, are of great use to recreational boaters. These are the file types associated with nautical charts in raster and vector formats.

CARIS may be the most surprised of all at how recreational boaters have discovered Easy View as a way to view both raster and vector digital charts. Sheri Flanagan, Marketing Coordinator at CARIS, estimates more than 600 people download Easy View each month and that as many as a third of those are recreational boaters.

Getting Started

Easy View can display the most common chart formats for U.S. boaters, but because it is only a viewer, you must obtain chart files from another source. You can download free raster or vector charts directly from NOAA or the U.S. Army Corps. You can also purchase charts on CD or DVD from commercial sources.

However, there are some files Easy View cannot display. For example, it won't open GEO/NOS chart files, a format used by SoftChart (recently acquired by Maptech), or Admiralty charts in ARCS format, which are popular with European and offshore cruisers.

CARIS is easy to load and use, although its manual is not particularly helpful. Easy View was not created for boaters, so the manual covers many non-boating features and focuses more heavily on Computer-Aided Design (CAD) users. Unless you are an engineer, it is not very intuitive.

The company does offer support via email and was very responsive to our requests for information. However, because CARIS provides this viewer as a freebie to large government agency customers, it would be unfair to pester them with your common PC problems. A better approach is to think of Easy View as a tinkerer's solution to free chart viewing and try to solve problems on your own.

Working with Charts

Bringing up a chart file is very straightforward in Easy View. After copying your chart files to your hard drive, simply choose Browse and select the file you want displayed. Once the chart is displayed, you can zoom and pan over the chart using your mouse. If you open a vector chart, you can hide or show its layers or click on an object to display its attributes.

But that is about the extent of Easy View features for the recreational boater. Once a chart file is opened, you cannot

CARIS Easy View

Pros: Free; displays a wide variety of data formats including raster and vector charts.

Cons: Only a viewer, not a full-featured charting and navigation application; cannot work with waypoints or routes; does not integrate with GPS sensors or other navigation equipment.

Coolest Feature: Extraordinary file format support.

Price: Free

Vista Capable: Yes

Version Tested: 1.0.4

System Requirements
PC with Intel Pentium 4 with 1.7 GHz processor
512 MB RAM
128 MB available hard disk space
128 MB video card (OpenGL support for 3D display)
Windows XP or Vista

CARIS
115 Waggoners Lane
Fredericton, New Brunswick E3B 2L4
Canada
www.caris.com

To open vector charts downloaded from the NOAA or USACE websites, expand the zip file, then drag the **ENC_ROOT** folder into a folder on your hard drive. You can then use Easy View to open the **.000** file in that designated chart folder.

move past the chart image on the screen to display the adjoining chart automatically. Nor can you click on an inset panel for a more detailed display in a larger scale. Easy View simply displays a graphical image. It does not know that chart 11512 adjoins chart 11511 to the south.

Unlike a true navigation program, it also does not automatically select an appropriately scaled chart, such as in the case where there is a small-scale chart and a large-scale chart available for the same area. You need to apply this intelligence, opening each chart file manually after figuring out the level of detail you want to see.

Because Easy View is not designed as a navigation program, it also does not connect to your vessel's electronics, such as a GPS sensor. You can view charts for planning but cannot show your vessel's position or create waypoints, markers, routes, or tracks.

Assessment

Because Easy View was created as a service to CARIS customers—not specifically for recreational boaters—it doesn't have many features boaters need. In many ways, it is a "look but don't touch" program: you can display nearly any kind of mapping or charting file, but you cannot add to the displays to incorporate any vessel information.

For that reason, Easy View is best for simply planning at home when you want to peruse a library of charts without the hassle or cost of a stack of paper. It may also have use in certain applications as a simple backup or augmentation to more sophisticated navigation software.

The boater who would be most comfortable setting up and using Easy View would come from a profession that uses CAD software, perhaps an architect or engineer. Other users will likely find the interface a bit cumbersome and counter-intuitive. If you are a CAD fan, then Easy View's can-opener ability to read a huge variety of file formats is a plus: you can use it to open those GeoTIFF or AutoCAD files your friends send you. But most recreational boaters want to do more than simply *look* at a chart, so would be better served by a simple, inexpensive marine-focused package.

Vector & Raster Viewer: *Of the huge assortment of file formats CARIS supports, S-57 vector charts (left) and BSB raster charts (right) will be boater favorites.*

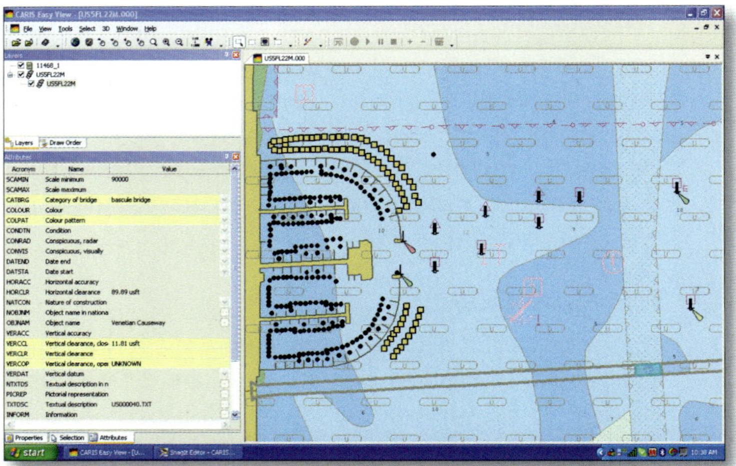

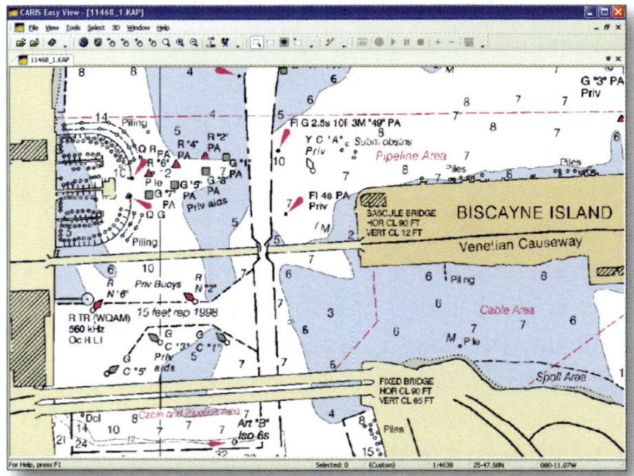

Chapter 13
EarthNC

More than 200 million people have logged on to Google Earth, the flashy Internet site that lets you virtually soar over satellite images of the world. Yet most people still use this tool as a novelty—downloading images to exclaim, "Look Honey, I can see our house!"

For Google Earth aficionados, however, this Internet resource has far more potential. Forums such as Google Earth Blog (www.gearthblog.com), moderated by cruising sailor Frank Taylor, bring together user-submitted ideas to make Google Earth a resource for travel, business, and recreation. With the familiar, popular, and free Google Earth resource already in place, why not integrate marine data?

That was the idea of former Air Force developmental engineer Virgil Zetterlind when he began coding EarthNC two years ago. His mission: to bring nautical content to the Google Earth platform. EarthNC, as in ENC (electronic navigational chart) and pronounced Earth 'N' Sea, is a software application that integrates nautical charts and other marine data sets with Google Earth.

The idea is simple and compelling: transform traditional nautical charts into Google Earth *layers* so that chart objects can be displayed on top of a Google Earth satellite image. In other words, your computer screen shows a satellite view supplemented by icons for navigational aids, depth contours, tide stations, boat ramps, and so on. The possibilities are endless as government offices, businesses, and users continually post new data sets, called *KML layers*.

Google Earth's imagery may be uneven—crystal clear in some places and flawed in others—but it can bring a collection of abstract chart symbols to life. In fact, the Merchant Marine Academy uses EarthNC to help midshipmen learn to read nautical charts. Entrance channels and sandbars stand out in full color, marked by red and green icons for navigation aids. Fishermen can view the actual bottom structure of bars, holes, or coral heads. You can even zoom in to see boat ramps, mooring fields, marinas, and docks. And of course you can zoom in for the obligatory, "Look Honey, I can see our boat!"

Setting aside the cool factor, EarthNC is a practical *planning* tool for navigation. We consider it a "planner-plus." The ability to connect a GPS sensor to view boat position makes it more than a planner, but it currently lacks the piloting tools of a fully-featured charting and navigation program.

What EarthNC does represent, however, is a new class of e-charting resource, just as Wikipedia redefined reference encyclopedias (www.wikipedia.org). Instead of a static reference, EarthNC is an online compilation of user-submitted data, brought together on the platform of Google Earth. And this type of tool has many uses, both at home and on the water.

GETTING STARTED

EarthNC has an enviable position as a new software application: many people already know how to use Google Earth. If that is your situation, there is very little additional learning involved in EarthNC. Most of your start-up will be familiarization with Google Earth's tools, such as layers, searches, push pins, and paths.

EarthNC is a separate software application that adds features and data to Google Earth. In order to use EarthNC, you must purchase a copy of the EarthNC software and have Google Earth (version 4 or later) loaded on your PC or Mac. A nice side benefit of the Google Earth platform is that it works on any operating system, PC, Mac, or even Linux. You can download Google Earth at no charge from the Google website (http://earth.google.com). If you already have Google Earth loaded, be sure to re-visit the website to upgrade to

EarthNC

Pros: Easy to learn; low-cost alternative to satellite cartography; varied and rapidly growing marine data overlays.

Cons: Not a full-featured e-charting application; lack of piloting tools, no instrument connectivity except GPS.

Coolest Feature: Nav aids on satellite imagery.

Price: $34.95 EarthNC (includes one region of U.S. raster charts); $49.95 EarthNC Plus (includes U.S. vector charts); $159.95 EarthNC Combo (includes U.S. vector charts, one region of U.S. raster charts, and a USB GPS).

Vista Capable: Yes

Version Tested: 2.4

System Configuration
PC with Intel Pentium (or equivalent), or Macintosh with OS X 10.4.5 or later
512 MB RAM
2 GB available hard disk space
3D-capable graphics card with 32MB of VRAM
1280 × 1024 color display
768 Kbps network speed
CD or DVD drive (DVD drive required for charts)
Windows XP or Vista or Mac OS X

EarthNC
1515 N. Swinton Avenue
Delray Beach, FL 33444
United States
www.EarthNC.com

EarthNC Plus vector charts are updated monthly so be sure to register to receive update announcements via email. EarthNC's Online Data links, containing additional point-of-interest, service, and weather information, are also updated each month.

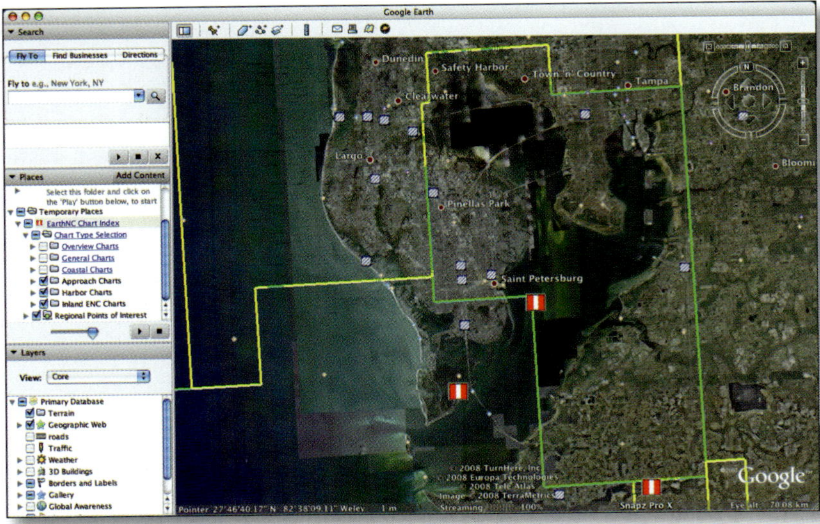

Chart Outlines: *EarthNC indicates available charts with outlines on the Google Earth projection. Clicking on the red-and-white icon links to the NOAA chart.*

For faster performance, keep your My Places folder in the Google Earth Places menu organized. If Google Earth is starting slowly, it is probably attempting to load too many layers. Uncheck layers you do not wish to load automatically. If you don't need a layer, delete it from the My Places menu.

the latest version. Each update includes important enhancements.

PC users can purchase a CD-ROM of EarthNC or download the application directly from the EarthNC website. The download option is not available for Macintosh users. EarthNC provides optional charts on CD-ROM, either in vector (ENC and IENC) or raster (RNC) formats. EarthNC Plus includes the software application plus all available U.S. vector charts on a single DVD. EarthNC Raster includes the software application and one region of U.S. raster charts on a supplemental single DVD. Raster charts are much larger—and expand in conversion to Google-readable format—so only a single region of the U.S. fits on a DVD. The five regions cover Southeast U.S., Northeast U.S., Great Lakes, West Coast, and Alaska.

Your choice of charts depends on your boating needs and geography. Since NOAA has converted only about 60 percent of its charts to vector format, purchasing EarthNC Plus by itself may leave you with coverage gaps. This can be solved by supplementing with a region of raster charts. Note that because EarthNC requires Google-formatted layers, you cannot download your own free raster charts from NOAA.

EarthNC offers a subscription service that includes monthly chart updates via its website. This online option also includes additional data layers as they become available. For example, recent releases include nautical points of interest in the Sea of Cortez, Gulf of Mexico hurricane debris, and Pensacola dive sites. When you purchase EarthNC, a one-year online subscription is included. Thereafter it is $50 per year.

For the whole enchilada, EarthNC offers a bundle for $159.95 that includes the software, vector charts for the U.S., one region of raster charts, a USGlobalSat BU-353 USB GPS

to show your boat position, and the GooPs Pro software needed to integrate the GPS with the display (PC only).

Installation of EarthNC was trouble-free. The vector charts on the supplemental DVD automatically load onto and thereafter run from the hard drive. The raster charts are designed to run off the DVD drive to save space on your hard drive. In other words, insert the vector DVD once to load your charts; insert the regional raster DVD each time you run EarthNC in raster chart mode.

If you do need help, EarthNC provides several resources through its website. Of course, since its platform is Google Earth, most questions can be answered through Google Earth's extensive Help menu and Help Center Website. For specific EarthNC questions, the EarthNC website has FAQs, email technical support, and video tutorials to get you started. The nine-page PDF Installation Guide is brief but covers the main features unique to EarthNC.

WORKING WITH CHARTS

Nautical charts are EarthNC's most important data layer. By converting NOAA charts into collections of objects in a KML layer, EarthNC places chart elements, such as navigation aids or depth contours, directly on a Google Earth satellite image. The nautical chart is not simply a transparency layer; it is fully integrated with the satellite view.

EarthNC is primarily intended for use with vector charts. As you zoom in on a Google Earth area, each available NOAA chart is represented with a yellow outline. Hovering over the chart's flag icon shows the chart number and its title (such as US4FL46M-East Cape to Mormon Key). Clicking on the chart flag opens a window with a View Chart option, which then loads the chart. Charts are also shown in a chart index drawer, organized by NOAA's groupings into Overview, General, Coastal, Approach, and Harbor Charts. Unfortunately, EarthNC does not auto-load charts as you pan and scroll.

Once a vector chart is loaded, you have all its information on your Google Earth display. Google Earth provides the true-to-life backdrop of sandbars, land forms, marinas, and channels. The chart overlay adds green and red navigational aids, lights, depth soundings, and colored contour lines. The effect is visually impressive—literally bringing to life the abstract flatness of a vector navigation chart.

Since each navigational icon is a separate vector object, you can query the icon for more information. Clicking on a navigation aid branches to icons showing aid and beacon light list information. Chart notes are accessible in nested folders in a sidebar window.

Nearly 200 object layers are available, including important navigational information such as bridge details, Coast Guard stations, and marine tide prediction stations. Blue-and-white logo icons show West Marine store locations. The possibilities are endless as any KML data layer—created and submitted by a government agency, a business, or even an individual—can be overlaid on the satellite backdrop.

While EarthNC is extremely data-rich, much of the information is deeply nested, making it a mouse-intensive interface. For example, in one case a single icon incorporated three layers of mouse clicks. Icons lead to sub-icons that display object information, but only one pop-up window can be viewed at a time. A navigation aid's appearance and flashing characteristics are two separate windows, reached through separate branching operations. Occasional users will not mind the extensive mousing, but power users or those who try to mouse while underway may find it cumbersome.

To manage the boundless possibilities for KML layers, you can customize your display through Google Earth, selecting only those layers of interest. Even slicker, you can selectively turn on or off layers at particular zoom levels. For example, you may want buoys and beacons to appear at a mid-level zoom, then shore structures, and finally depth soundings at a closer level of zoom.

EarthNC can also display raster chart overlays, an important option for areas without vector coverage. Although many boaters prefer the familiar look of a traditional raster chart, raster charts are a clumsier overlay. EarthNC has attempted to manage this by providing a slider bar to adjust the transparency of the raster chart, letting you overlay a raster and vector chart over a Google Earth satellite image. The ability to gain vector chart data is retained in the raster chart view: vector objects automatically float above the raster chart and can be queried to open a pop-up window with details on that object. In practice this overlay option is only feasible with thinly-annotated views.

For boaters who are Bahamas-bound, EarthNC is integrating newcomer Wavey Line charts beginning in 2008 (www.waveylinepublishing.com). Chart regions on CD-ROM, compatible with Google Earth, will be sold for the northwest, central, and southeast Bahamas for about $80 each. The Bahamas is a perfect geography for a Google Earth-supplemented view. Combining recent satellite images with nautical charts should reduce some of the stress in locating those narrow channels marked with tree branches and water jugs.

Naples Inlet, Florida: *Overlaying icons of navigational aids and tilting the view shows off the eye-candy nature of Google Earth.*

EarthNC Offline

Until now, you may have been following along, nodding and thinking, "Wow, online satellite imagery combined with electronic navigational charts." Sounds great at your desk with unlimited broadband Internet access. But what about when you're out on the water, either too far for a WiFi signal or unwilling to pay the cell minute tariff necessary to connect?

Fortunately, once you have downloaded Google Earth and EarthNC, EarthNC can be run offline. In other words, you don't need to be connected to the Internet to use the application. Part of Google Earth's design is the ability to save, or *cache*, a considerable number of images for offline viewing.

Zetterlind, EarthNC's founder and chief architect, points out that most users currently use EarthNC offline. However, he expects a trend toward *always-on* laptops in the future. More and more people now own portable air cards, typically for business travel, providing online access without per-minute charges. If you already have a wireless plan through a cell

In order to use a GPS with EarthNC, you must use a helper utility. If you're running a PC, use GooPs GPS (http://goopstechnologies.com). For Mac users, gps2geX is the way to go (www.grandhighwizard.net/gps2gex.html).

phone provider, then EarthNC is a low-cost and easy addition to your online arsenal.

Although you cannot access online extras such as weather, tides, or Wikipedia entries, you can cache the areas needed for your route. Google Earth stores up to 2 GB in its cache, which covers more geographic distance that most people would need for a day on the water. Longer-distance cruisers would need to plan a nightly wireless pit-stop. For example, one could cache some Bahamas coverage in Miami or Fort Lauderdale, then store additional islands at a full-service Bahamian marina, many of which now have WiFi.

Caching a route in advance is quite easy. Simply pan through your route in EarthNC and Google Earth automatically saves the files for later playback. You can adjust the level of detail as you pan the route: panning and scrolling at a zoomed level to store that level of detail in your cache. Conversely, if a portion of your route has very little navigational detail, panning at a lower level of zoom saves space. Google Earth seamlessly integrates your zoomed levels as you go through your route.

NAVIGATING WITH EARTHNC

Like Google Earth, EarthNC has a lot of flash and sizzle. But how would one use EarthNC on shore for planning or on the water for piloting?

EarthNC admits it is not intended as a primary tool for navigation. Although it integrates NOAA charts—the same cartography you navigate from with paper or an e-charting application—EarthNC is currently a planning application.

With EarthNC you can visually preview your navigation area to identify and understand your approach routes, marina locations, and shoal areas. There is no question that aerial or satellite images are welcome information to any navigator. Cruising guides with aerial images of harbor approaches have always been popular and most full-featured e-charting applications provide satellite data on CD-ROM or by incorporating C-Map or Navionics cards.

Although some boaters may be comfortable viewing these cached images while underway, we suspect most boaters need to keep their laptop dry below deck and their eyes peeled above deck. EarthNC does not replace a chartplotter, but provides a more visually intuitive view than a chartplotter's small and abstract representation.

Using Google Earth's tools for "pushpins" and "paths," you can create *mock* waypoints and *proxy* routes and transfer them to a GPS or a chartplotter. EarthNC has streamlined the transfer to later-model Garmin units with a transfer interface provided on their website (http://earthnc.com/triptool). Pasting a folder of Google Earth pins and paths into the transfer web page converts the data, with the option of sending to

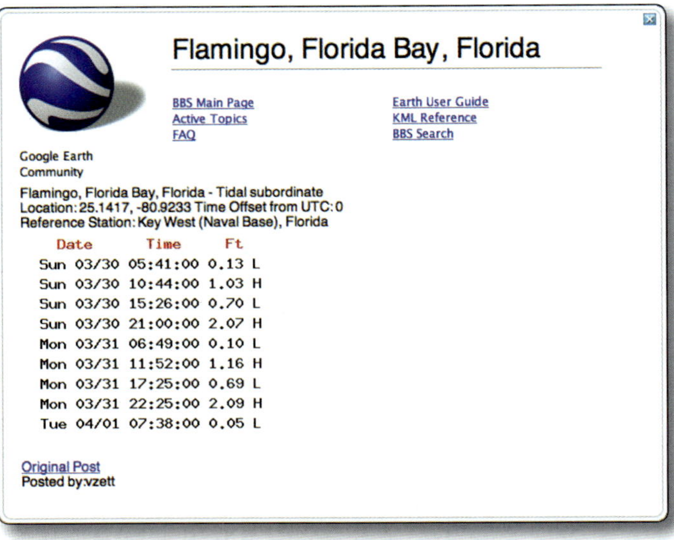

Flamingo, Florida Bay, Florida

BBS Main Page   Earth User Guide
Active Topics   KML Reference
FAQ   BBS Search

Google Earth
Community

Flamingo, Florida Bay, Florida - Tidal subordinate
Location: 25.1417, -80.9233 Time Offset from UTC: 0
Reference Station: Key West (Naval Base), Florida

Date	Time	Ft
Sun 03/30	05:41:00	0.13 L
Sun 03/30	10:44:00	1.03 H
Sun 03/30	15:26:00	0.70 L
Sun 03/30	21:00:00	2.07 H
Mon 03/31	06:49:00	0.10 L
Mon 03/31	11:52:00	1.16 H
Mon 03/31	17:25:00	0.69 L
Mon 03/31	22:25:00	2.09 H
Tue 04/01	07:38:00	0.05 L

Original Post
Posted by: vzett

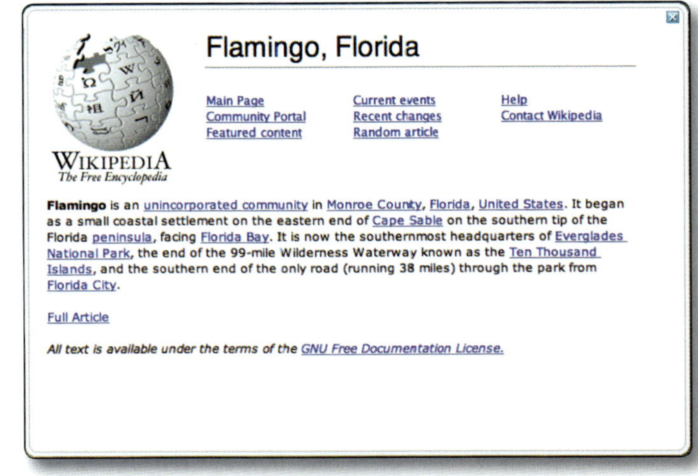

Flamingo, Florida

WIKIPEDIA
The Free Encyclopedia

Main Page   Current events   Help
Community Portal   Recent changes   Contact Wikipedia
Featured content   Random article

Flamingo is an unincorporated community in Monroe County, Florida, United States. It began as a small coastal settlement on the eastern end of Cape Sable on the southern tip of the Florida peninsula, facing Florida Bay. It is now the southernmost headquarters of Everglades National Park, the end of the 99-mile Wilderness Waterway known as the Ten Thousand Islands, and the southern end of the only road (running 38 miles) through the park from Florida City.

Full Article

All text is available under the terms of the GNU Free Documentation License.

a Garmin unit or saving as a GPX file. Waypoint and route data can then be transferred to other GPS or chartplotter units with a helper utility such as GPSBabel (www.gpsbabel.org).

Another option for planning with EarthNC is to print out annotated satellite snapshots of that day's trouble areas, such as channels, shoals, or the anchorage or marina approach. Google Earth includes built-in features to either print or email your annotated satellite shots. In effect, you can make your own annotated approach aerials to have at the helm, supplementing your chartplotter and cruising guide. Google Earth's email capabilities also let you email an annotated satellite view to communicate a rendezvous location to cruising club members or buddy boats.

Because EarthNC can integrate GPS data, it is more than a tool for planning. Adding a USB GPS sensor displays the position of your vessel on the EarthNC annotated display.

It is important to note that EarthNC does not include many of the piloting features that are now standard in e-charting applications. For example, although Google Earth's pins and paths can be converted to waypoints and routes, they are not used for navigation within EarthNC. For example, there is no autopilot connectivity, man overboard feature, GoTo or SteerTo functions, or cross-track error display or alarms. Until these and other piloting tools are added, EarthNC remains a visually sophisticated planner-plus software rather than a full-featured navigation and piloting application.

ADDITIONAL DATA LAYERS

Part of Google Earth's appeal is the infinite flexibility of data overlays, often called KML files for their *Keyhole Markup Language* format. Anything that is saved or converted to a KML file can be overlaid on a Google Earth image.

In fact, the creation of KML files has become a growth industry, with governments, businesses, and individuals creating publicly-available collections of whatever data interests them. In this sense, Google Earth is a dynamic, user-created resource like Wikipedia. The creators of these data sets want their information disseminated, and freely post it for others to use. EarthNC has emerged as the central collector of nautically-inclined KML files.

EarthNC includes several of these data layers as defaults. We already mentioned the blue-and-white striped logos to mark the West Marine store locations. Clicking on the icon brings up the store address, phone, fax, click-to directions, and click-to web page. Purple dots mark Wikipedia entries, such as a Wikipedia post on Flamingo, Florida. A tiny camera icon opens a geographically-referenced Panoramio photo, a user-created forum for photographs (www.panoramio.com). Another window displays tide station data.

Since anything can be added as a Google Earth KML layer, the ability to custom annotate your charts is endless. The amount of highly-specific data you can add to your chart is already staggering. For example, EarthNC includes data sets for boat ramps provided by state governments, wreck sites provided by NOAA, local tourism sites submitted by Hampton Roads, and canoe put-ins near Niagara Falls provided by a paddling buff.

EarthNC data can also be customized for member-only layers. For example, EarthNC is teaming up with America's Great Loop Cruisers' Association (AGLCA) (www.greatloop.com) to provide member-only layers of sponsor locations, membership directories, and cruising member locations. AGLCA members will be able to submit their boat position (by latitude/longitude, city/state, or waterway/mile) to automatically post their location on a shared cruising map.

It's impressive to have access to all this data. But going forward, quantity and quality will need to be monitored. User-driven information portals such as Wikipedia struggled to find the balance between open data submission and quality control. For example, icons to linked photos through Panoramio—a site where anyone can submit their photos as geo-referenced layers—sound very cool. But today the Florida-based photo assets are an almost worthless layer of amateur sunset pictures. Google recently bought Panoramio, which currently has over 3 million photos in its library and is grow-

Aerial Photos: *Google Earth includes links to many community-based Web assets. You can occasionally find a useful member-submitted photo in a Panoramio link.*

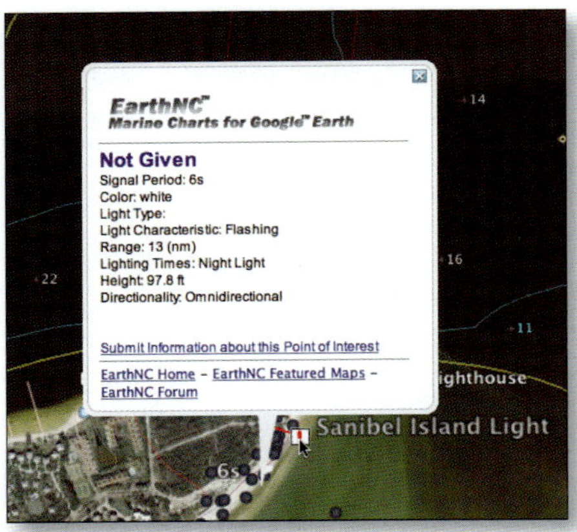

Light List Data: *Clicking on Sanibel Island Light displays a pop-up of abbreviated Light List data.*

ing rapidly. Personally, when navigating we'd rather have a single, non-artistic shot showing a bridge approach from the water than 100 tourist snapshots. In a move in the right direction, EarthNC recently added a layer of over 500 U.S. Army Corps pictures for inland waterways.

If you are the posting type, you can submit your own content through Google Earth's My Map feature. Simply create a My Map in Google Earth and EarthNC may consider adding it to its marine-based data sets. You can submit data for boating events or gatherings, dive or fishing spots, maps and charts, marinas and ramps, photos and videos, trip reports, web cams, or websites.

If you prefer to work with your own data, you can import your existing waypoints, chart marks, or routes into EarthNC by converting them to a KML file. The e-charting applications MacENC, Coastal Explorer, and Fugawi Marine ENC all exchange waypoints and routes as KML files automatically. We easily moved 2,400 anchorages, marinas, and mooring fields from MacENC to EarthNC. Be warned that Google Earth will load KML files multiple times without warning of duplicates, so it's easy to end up with redundant folders, resulting in replicated icons and slow load times.

ASSESSMENT

EarthNC is a new concept in e-charting that leverages Internet resources and user-created data. In this sense it's the Wikipedia of e-charting. Rather than purchasing a static encyclopedia, many people turn to online Wikipedia entries that are created by a community of users. Similarly, EarthNC uses Google Earth's platform to view user-submitted marine data in layers. If you are a blog, forum, or Wikipedia person, then EarthNC should have a lot of appeal to you.

Going forward, EarthNC will have to balance the quantity of data with quality and monitoring so it does not become bloated with specialized user-submitted information. But like any Internet resource, the information potential is staggering. Any marine data you need—whether dive sites in Pensacola or canoe put-ins in Upstate New York—can potentially be available as a layer on your satellite-enhanced nautical chart.

However, EarthNC should not at this time be viewed as a full-featured e-charting application. Although it can display your vessel's position using a USB GPS, it does not have the countless piloting tools that are fundamental to even the simplest e-charting application. The absence of these tools, such as range and bearing, planning and route calculators, autopilot control, GoTo with cross-track error, or instrument connectivity, make EarthNC a planner-plus application at best.

For boaters who already own an e-charting application, you may already have Google Earth capability. Several popular packages, such as Coastal Explorer and MacENC, include features that let you export your waypoints and routes as a KML layer to Google Earth. Fugawi Marine ENC goes even further with a new plug-in that integrates Google Earth as a split screen display. For boaters who want a full-featured e-charting application but like the dynamic satellite imagery of Google Earth, Fugawi Marine ENC is price-competitive and an extremely full-featured navigation package.

Also, recall that many marine cartography sets, including Maptech, C-Map, and Navionics, include satellite views, as well as marina, point-of-interest, port photographs, and other information not yet available through Google Earth. These assets, which you may already own as part of a chartplotter or e-charting application, are generally better than general-purpose Google Earth assets.

EarthNC may work best for the millions of boaters who already own a chartplotter but do not plan to purchase a full-featured PC-based e-charting package. Similarly, EarthNC is an easy-to-learn option for boaters who only use a free viewer with a handful of local NOAA electronic charts. If you choose EarthNC as your e-charting option, expect a visually intuitive planner with the potential for light navigational use onboard. EarthNC's current lack of planning and piloting tools will lead boaters who actively plan or navigate on computer to look elsewhere for a more full-featured navigation application. Yet EarthNC is a newcomer, giving them plenty of time to add navigation features to their catchy visual interface.

Chapter 14
SeaClear II

There is good news for boaters who want to test the waters in the ever-growing market of navigation software. SeaClear is a free charting and navigation software that is surprisingly full-featured. By combining a SeaClear download with free raster charts from NOAA, you can create a simple electronic charting system at no cost to explore the question, "Is e-charting right for me?"

SeaClear was created by Olle Soderholm, a Swedish recreation boater who was unhappy with the e-charting choices available back in 1995. He makes it sound easy, but he literally "made his own e-charting package."

More than a decade later, SeaClear has no commercial connection or support. Soderholm maintains the application in his spare time, releasing new updates at an impressive rate.

One major update, SeaClear II, was created to run on Windows 2000 or XP, replacing SeaClear for Windows 95. For boaters on older Windows operating systems, these versions are still available for download, which is a plus for boaters who are attached to their old (dare we say outdated?) equipment. So far, there is no version available that runs on Windows Vista.

GETTING STARTED

Since SeaClear is a free software application, it is only available for download from the Web (www.sping.com/seaclear). The download links are very clear, with choices available for your version of Windows. The full install (approximately 2 MB) includes an English language manual. You can also choose to download the manual in other languages, including Spanish, Swedish, German, French, Finnish, Dutch, and Hebrew.

The installation puts four items in your Program directory: MapCal II (a utility for chart installation), SeaClear II (the actual application), SeaClear II Manual (PDF), and SeaClear II on the Web (Web link). Part of the download includes small-scale base charts for the world. This is not a chart set you can use—it only shows continents with no data—but it lets the program display a backdrop when you first hook up your GPS sensor and begin loading charts for your region.

In order to get up and running with SeaClear II, you need raster charts in BSB/KAP or GEO/NOS (formerly SoftChart) format. SeaClear is designed to work with raster chart files only. Unlike many commercial applications, SeaClear does not include charts, nor does it include or display additional data files such as tides and currents, elevation maps, bathymetric data, street maps, or weather overlays. You can download BSB files from NOAA at no charge, or purchase regional CDs or DVDs inexpensively from several commercial sources.

Customer support for SeaClear is simple: you get what you pay for. Because you paid nothing, you get nothing. We tried to contact Sping multiple times by email with no response. That leaves customer support to you and the manual. We did eventually hear from Olle, who said he was out sailing for several weeks and replied briefly to some of our emailed questions. But you get the idea.

Unfortunately, the manual is poorly written and very thin. It's not written from a user's perspective, explaining "how to use" or "why one cares." Instead, it lists features from a software engineer's view. The manual has been translated into American English, with awkward sentences, grammatical mistakes, and unconventional terms (such as "position" rather than "waypoint"). Furthermore, although SeaClear was recently updated, the manual was not. Several of the screen menus do not match the documentation.

The best bet for troubleshooting is to post questions on one of the many recreational boater forums. Other recreational

To save cost, many vendors provide manuals as PDF files rather than printed documents. In order to view and print PDFs you must have a copy of Adobe's Acrobat Reader installed, available free at www.adobe.com.

GPS Setup: SeaClear does not include a Port Wizard, requiring you to migrate through Window XP's Device Manager to choose the appropriate COM Port and Baud Rate (in this case COM3 and 4800 baud).

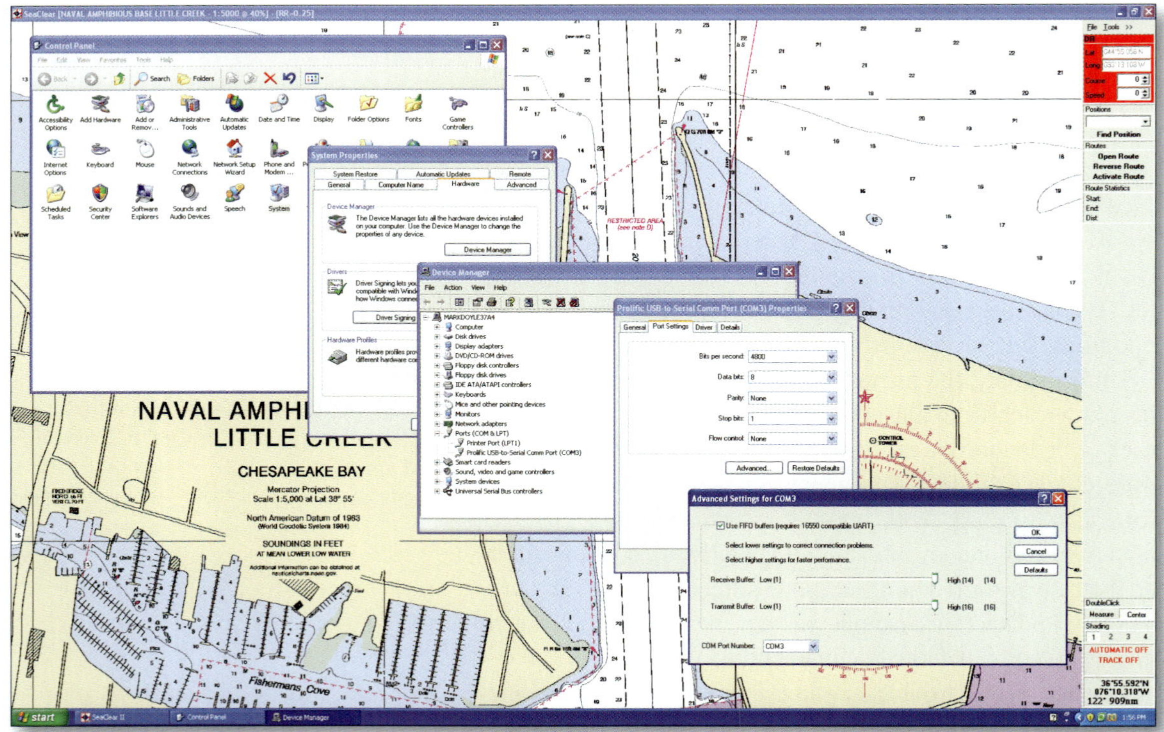

boaters using SeaClear may be able to help.

Unfortunately, the manual is particularly thin on the basic chart installation process for NOAA BSB files of U.S. waters. Even if you're familiar and comfortable with the Windows metaphor, the process is difficult to decipher. However, SeaClear comes with a separate map calibration and installation utility called *MapCal II* that helps with the install. The tricky part is that charts must be calibrated using a utility in MapCal before they can be read. Follow the instructions carefully.

There are two ways to install and organize chart files in SeaClear. You could simply drop any pair of BSB/KAP files into the C:\Program Files\SeaClear\charts folder. Although this method works, it leads to a cluttered and unmanageable chart directory. And that leads to sluggish laptop performance because all of your charts must be scanned and recognized by the application.

A real-world example is trying to unload 138 Chesapeake charts and load 200 ICW charts as you migrate south. A bet-

ter process is to keep the charts in separate regional folders, placing them in SeaClear's charts Folder as you need them. Fortunately, SeaClear can read nested folders, which allows you to organize your charts into sub-folders by regions. Lastly, in MapCal II, go to Tools>Autoload List>Scan for New Charts, which tells SeaClear to add those charts to its directory and makes them available for viewing and navigation.

SeaClear is designed to be used with a GPS sensor and you will get much more out of the software if you connect one. When you first start the application it prompts you for the GPS sensor—even before it asks for the chart files. If you don't want to connect a GPS, you can pass through these prompts to load the charts.

Port recognition between the GPS sensor and SeaClear was our first speed bump in getting the application running. SeaClear's default communication port for USB-to-serial is COM1, meaning it was looking on COM1 for the GPS. But the GPS driver's default was COM3. In the XP Device Manager, we had to change the communications port to COM3 and the transmission rate from 9600 baud to 4800 baud. Once your GPS is connected, the Dashboard shows green and displays your position data. A red circle shows your position on the chart. If you lose GPS connectivity, don't panic. The program automatically searches for the GPS device every few minutes and will return to that mode once it has re-located the sensor. (For a more thorough discussion of COM Ports and setting up your GPS sensor, refer to Chapter 5, *Instruments and Sensors*.)

LOOK AND FEEL

SeaClear gets high marks for its use of screen real estate, which is particularly important if you're on a laptop. Almost the entire screen is dedicated to the chart view. Most of Sea-Clear's functions are accessed with a right-click. There are also more than 20 preset keyboard shortcuts, such as Control-space to set a man overboard marker at your current position. This leaves only two non-chart items on display: a clean summary title bar along the top which lists the chart, scale, zoom level and ranging ring increments, as well as something called the *Dashboard*.

The Dashboard is a thin vertical menu that displays a wide variety of data, including a list of your waypoints and quick access to your route statistics. If you have a GPS sensor connected, the Dashboard displays your position, course, speed, and other data. Hook up your wind, depth, and compass and you can see the summary of this data too. You can also maximize the chart screen area by temporarily minimizing the Dashboard.

A negative of the Dashboard is its choice of content and display metrics. Throughout the program, there is a bit of a "translated" feel. For example, depths can only be shown in meters, not feet. The time is displayed in UTC rather than local time. Waypoints are called *positions*. Course over ground and bearings are in degrees True—you cannot change them to magnetic. Some of the displays don't have units, leaving you wondering what "Course 4" means. (The answer is 4 degrees True.)

Most navigation programs let you customize metrics and displays. Although some aspects can be customized, such as its *night mode* shading levels, SeaClear II is relatively rigid. We were a bit frustrated at not being able to change some of the interface or metrics in a way that made more sense to us. Unlike fully-featured and more expensive products, you don't have a lot of choice over what you're looking at and how the data is presented.

Working with Charts

There are four different metrics for chart display responsiveness: opening a chart, scrolling on a single chart, popping to an inset (a larger scale from the main chart panel showing more detail), and zooming.

SeaClear was highly responsive in all chart-display metrics. In fact, it loaded and re-displayed so quickly we were often behind with our mouse-clicks. Charts literally popped up on the screen. It was also very responsive when clicking-and-dragging to pan a chart. Using the *Outlines* option (File>Chart>Outlines) shows the extensions and insets, outlined with a rectangular box. Double-click on a chart outline and that chart panel comes right up. Double-click a chart margin to get back to the smaller-scale chart view. Use the mouse scroll wheel to zoom in and out to show more detail or a larger view.

SeaClear also has several nice features to help with navigating your chart files. If a GPS sensor is connected, right-clicking and choosing Charts>On Position brings up a window listing all charts spanning the current position. Single-click on a chart and it displays instantly. Right-clicking and choosing Charts>Find Boat centers you on the largest scale chart where the GPS says you are. You can also right-click and choose Charts>Best Chart to automatically view the largest scale chart for your position. If you don't have a GPS hooked up, choose File>Chart>List All and pick a chart.

There are some deficiencies too. SeaClear's screen images didn't appear as colorful and sharp as some other applications. SeaClear also cannot rotate charts, as you might do to show a course-up display. For all manufacturers, this is a trade-off between display speed and the limitations of the raster format. Unlike vector charts, which are comprised of data objects, raster charts are large picture images that are sluggish and more difficult to rotate. Although SeaClear cannot display charts vessel-up, it has a compromise option where you can set the boat icon toward the bottom of the screen, rather than the center, so that most of the chart display remains in front of your vessel.

Waypoints and Routes

SeaClear has many features for creating and saving routes (sequences of waypoints) and tracks (a record of your course over the chart). You can save an unlimited number of waypoints and routes, constrained only by the storage capacity of your system.

Waypoint entry was very easy using the chart display and the mouse-click menu: simply right-click, choose Position>Add and enter a waypoint name to mark that location. Waypoint

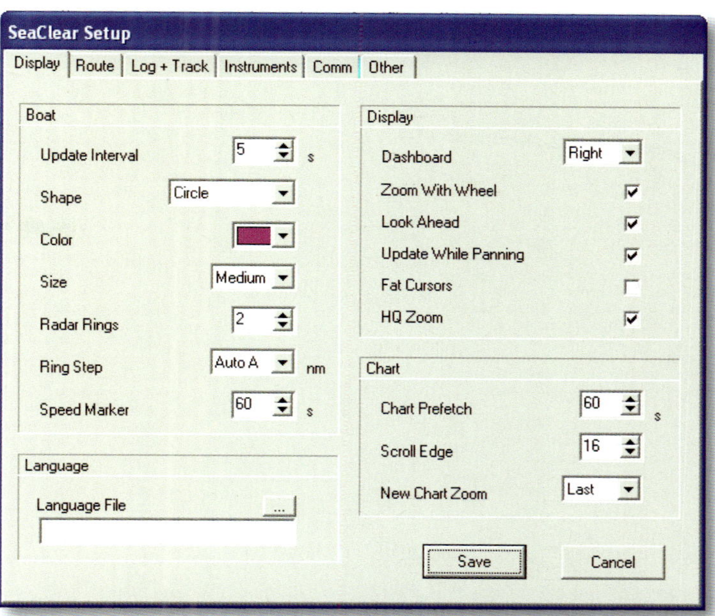

User Settings: *SeaClear II allows some control of user interface display and metrics through its Setup (Properties) tabs.*

SeaClear's manual does not include clear instructions for installing BSB chart files. Simplified loading directions are available at www.managingthewaterway.com/charts_help.html.

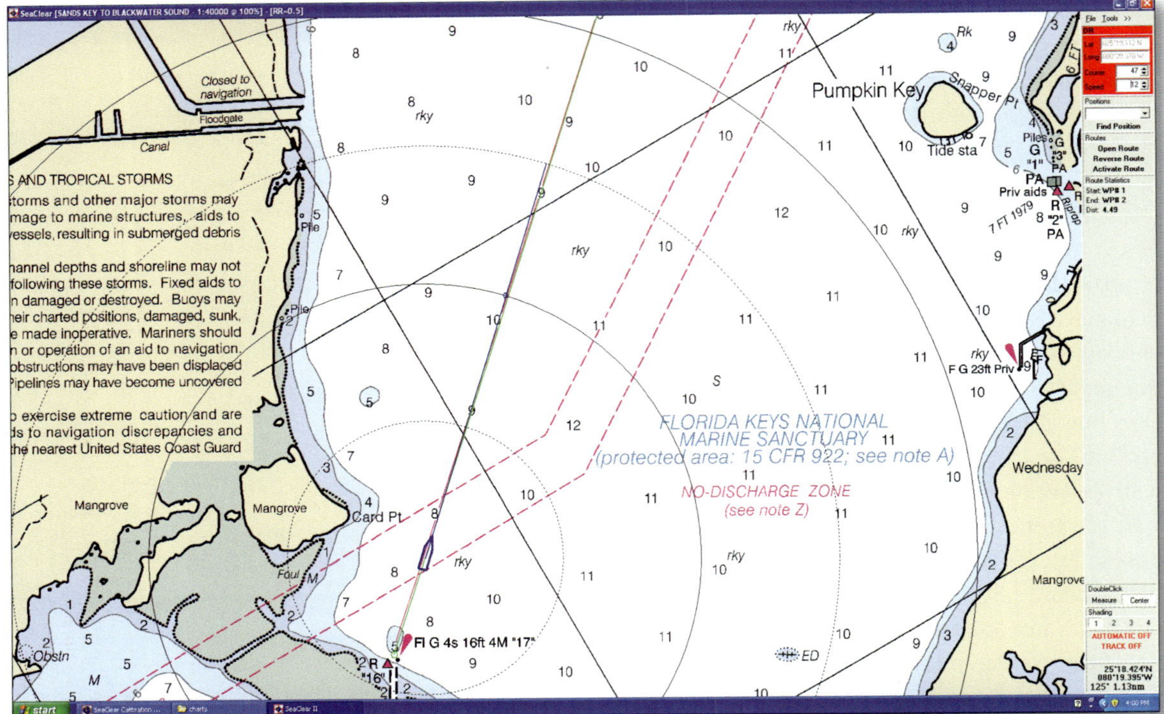

Chart Window and Dashboard: *SeaClear's uncluttered window features a Dashboard (right margin). The example shown is a DR route of 47 degrees True and 12 knots being followed northerly from Card Sound to Biscayne Bay (red box).*

names can include spaces and upper- and lower-case to make them easy to read and remember. For example, we entered the long waypoint label of *Rickenbacker Moorings*. Another nice feature is a comment field associated with each waypoint, letting you enter a lengthy personal note.

Tracks can be saved and plotted to create routes. You can join several shorter routes to create a longer route. It was easy to insert new waypoints into a route and to extend a route with additional waypoints. You can reverse a route with one click using Reverse Route in the Dashboard menu. In addition, the Dashboard displays your route statistics: start, end, and distance. If a GPS is connected, you can set a cross-track error alarm.

Waypoints, routes, and tracks can be imported and exported. You can transfer waypoint and route data from your PC to navigation units capable of receiving NMEA waypoint and route data. Your data can also be saved as a comma-delimited text file for export to other applications.

SeaClear also has some features reminiscent of the systems on commercial ships. For example, commercial vessels are required to keep an electronic log—essentially a *black box*—of their routes. Instead of only recording a track, Sea-Clear goes one step further and automatically keeps a log of date, time, position, heading, and speed. It also has a Logbook feature for manual entries. These records can be opened, edited, and saved as a text file.

Our primary complaint about SeaClear's waypoint, track, and route features is not in the options provided, but in the handling and organization of the data. We had quite a bit of trouble saving the waypoints in an organized way. In the end, it took several tries to understand the proper sequence of steps to save data—and this sequence was cumbersome and required a lot of screens and mouse-clicking for each data point. Even when we thought waypoints were correctly saved, we occasionally found some of our data either disappeared or was incorrectly stored in another regional waypoint folder when we restarted the application.

If you are a waypoint junkie (or cruising guide author), you ideally want multiple SeaClear *position files*, or collections of waypoints, within the SeaClear waypoint folder to organize your waypoint data. For example, you'd like a file for your Chesapeake Bay waypoints and a file for your Intracoastal Waterway waypoints. Unfortunately, the documentation is not much help here. It took about 45 minutes of experimenting to create regional waypoint files to organize our thousands of waypoints.

But after all that effort there was a catch: once a waypoint is saved to a file, your ability to edit it is seriously constrained. Although you can edit the attributes of an individual waypoint, you cannot do any advanced editing, such as moving waypoints from one regional file to another. Any future changes require deleting and re-creating the waypoint.

Ultimately, we were disappointed with the organization of waypoint and route data. If you are a light user, you probably should simply use SeaClear's default and store all your waypoints in one position file, such as a file named "My Waypoints." Unfortunately, the typical way to access a waypoint is through the Dashboard, which lists all waypoints alphabetically in a pull-down menu. Although it is straightforward to select the waypoint you want and then click Find Position, this search approach is cumbersome if you have many waypoints. You must scroll through an extremely long list of all your waypoints to choose the one you want.

Another problem we encountered was that, although all data files are supposed to be printable, we never could get printing of any kind to work. All options for printing, including charts, routes, and waypoint list data, resulted in a solid black printed sheet. Multiple computers, with multiple printers, on both wired and wireless networks all exhibited this bug. The workaround is to use a screen capture utility such as SnagIt (www.techsmith.com) or to export the data to a spreadsheet and print.

ADDITIONAL FEATURES

In addition to a GPS sensor, SeaClear integrates with your depth, wind, and compass data if you connect these NMEA instruments. The data displays on the Dashboard and a wind arrow displays on the chart showing the wind angle relative to your vessel. Similarly, autopilots and AIS instruments can be connected. Ranging rings (non-radar) may be overlaid on the chart display. Connecting an AIS device displays the target on the chart display with a target-name drop-down list on the Dashboard.

SeaClear also has a simulation mode, which is a feature available in many of the more expensive software packages. If you don't connect a GPS, SeaClear runs in what it calls *DR Mode*, maintaining a dead reckoning course based on your manual input of speed and heading. Your vessel is still shown with an icon, moving across the chart based on the DR calculations. A *Speed Marker* advances before your position to indicate where you will be in sixty seconds, five minutes, or one hour. You customize this speed marker in Tools>Properties>Display.

One of the unusual features of SeaClear is that you can scan paper charts, calibrate the files, save them in WCI, PNG, or BMP format, and import them for viewing and navigation in SeaClear. This is a great option for boaters heading to Croatia or Tonga. SeaClear's devotees include a community

of long-distance on-a-budget boaters who are scanning paper charts of non-U.S. waters to view using SeaClear. Since SeaClear doesn't read the international S-57 vector format—and these chart are also relatively expensive—scanning charts has become the discount workaround.

ASSESSMENT

Although SeaClear has many great features for boaters, including integration with your vessel devices, the documentation is so poor you'll pay for features in start-up time. The best target user for SeaClear II is someone who values "free" and is willing to slug it out with trial and error.

But there is such a boater. And you truly can put together a nicely-featured charting and navigation program, combined with free NOAA charts, for only your time and effort. If you have the time and know-how (or patience), this package is a great deal. It's also the perfect package for shoestring international boaters who want to scan paper charts to circumvent purchasing expensive international digital charts.

However, if you are easily frustrated by computers or are not very Windows-savvy, SeaClear may not be a bargain when you in factor in your time and stomach lining. One of the low-priced charting and navigation packages, which will include professional-level documentation and customer support, may be a better choice for you.

Customer Support: *A "help option" for all vendors is to Google for boating and e-charting forums. There are many boaters on the Internet using the same programs and facing the same issues as you.*

FEATURES AT A GLANCE

Each full-featured e-charting application has been evaluated according to the following 81 features. Some ratings are binary, "yes" or "no." Others are more subjective, evaluated on a scale from 1 to 5 (5 being "Excellent"). Any other comparison is defined in the Legend.

Out of Box Experience

Packaging	N/A
Documentation	2
Software install	3
Load charts	2
Load supplemental data	N/A
GPS hookup	2
Technical support	2

User Interface

GUI metaphor	NT
Screen space management	4
Program responsiveness	4
Chart display speed	4
Chart display quality	3
Customizable GUI	2
Customizable metrics	1
Split windows	No
Tabbed interface	No

Basic Features

One-button MOB	Yes
Waypoint creation	3
Route creation	2
Steer-to function	No
Track creation	3
Convert track to route	Yes
Waypoint and route management	2
Chart management	4
Search waypoint and route	3
Range/bearing tool	3
Chart annotation	No
Chart printing	1

Data Exchange and Networking

GPX support for waypoints	No
GPX support for routes	No
Waypoint exchange formats	None
Route exchange formats	None
Integrated GPS transfer	No
Multiple monitor support	No
Network application sharing	No

Cartography

Free U.S. rasters BSBs (RNCs)	Yes
Free U.S. vectors S-57s (ENCs, IENCs)	No
SoftCharts	Yes
CHS Canadian rasters	No
DNC	No
Seafarer	No
British Admiralty ARCS	No
International S-57s	No
International S-63s	No
Scan and geo-reference paper charts	Yes
Navionics cards	No
C-Map cards	No
C-Map CD/DVD	No
Satellite geo-referenced photos	No
Aerial informational nav photos	No
Topographic maps	No
Bathymetric data	No

Instruments and Sensors

GPS	Yes
Autopilot	Yes
AIS receiver	Yes
Wind	Yes
Depth	Yes
Water temperature	No
Radar	No
Heading sensor	Yes
Video camera	No

Advanced Features

Tides	No
Currents	No
Coast Pilot	No
POIs	No
Streets	No
Google Earth support	No
Advanced search	No
AIS	Yes
Buddy boat	No
Radar (ARPA/MARPA)	No
Radar (display overlay)	No
Fuel calculator	No
Transit calculator	No
Celestial calculator/Almanac	No
Auto route planning	No
Great Circle Route planning	No
Sail performance	No
GRIB weather integration	No
Weather options	No
Customizable bathymetric recorder	No

Legend

5	Excellent	CM	C-Map
4	Above average	NV	Navionics
3	Average	SC	SailCruiser
2	Below average	MT	Mr. Tides
1	Poor	PP	Plus Pack
CP	Chartplotter	C	CSV
Mac	Macintosh	E	Excel
NT	Non-traditional	G	GPX
Win	Windows	K	KML
N/A	Not applicable		

Chapter 15
DigiBOAT Software-On-Board

DigiBOAT proudly admits its charting and navigation software will never be Microsoft Certified, believing that software *on board* must be designed differently than software *on desktop*. Their mantra is to create "a navigator's tool on a computer, rather than a Windows program that does navigation."

The founders of this Australian company definitely have the sea-time to know what onboard software requires. Their collective experience ranges from motor to sail, dinghy racing to yacht regattas, and coastal cruising to bluewater passagemaking. In fact, our most recent discussions with Digi-BOAT's technical director Simon Blundell—one of three founding members and the sole programmer—occurred during a 6,000-mile "product test" from Thailand to Crete.

DigiBOAT's Software-On-Board, usually called SOB and pronounced as the word "sob" rather than the acronym "es-oh-bee," was first released as freeware in 2004. Blundell designed the software after an extensive research phase, during which he did some unconventional fieldwork, including following vessels around local Australian waters in a runabout and interviewing sailors with disabilities. By following vessels of all types—including racing yachts, fishing boats, commercial ferries, and recreational power boats—he hypothesized the information a helmsman would need in different situations or at different times.

This went beyond the obvious navigational queries such as position, speed or depth, and included information for more intelligent navigating—such as the time to round a point, the distance off a buoy or headland, or the relative position of a crossing vessel. From his interviews with a blind sailor, he developed *Talking Pilot*, a feature that narrates vessel and navigation data.

SOB now claims nearly a half-million downloads. It became commercial software in 2005, but remains a very inexpensive e-charting option. A generous three-month trial version includes free email technical support.

DigiBOAT offers four versions of SOB: Registered, LITE, Standard, and Pro. Registered is the full-featured program for a three-month trial. LITE is an unregistered free version. Some functions are disabled and only 20 waypoints can be displayed. Standard is full-featured and includes unlimited use on three computers; and Pro provides all the functionality while adding network support.

We reviewed the Standard version and found it to be a solid program that delivers many high-end features at a bargain price. While there is a learning curve to overcome, Digi-BOAT offers superlative documentation to help its users master the program.

However, while SOB is one of only four PC-based navigation applications making use of C-Map cartography, the company's plans to steer clear of raster chart support will cause pause for many U.S. boaters.

Getting Started

In the U.S., SOB is only available as a download from Digi-BOAT's website (with the option to obtain a CD-ROM via mail). SOB is sold through a few resellers in several European Union countries, Scandinavia, Singapore, India and Argentina. In keeping with DigiBOAT's international focus, the software is available in Norwegian, Dutch, French, and Italian. German, Russian, Portuguese, Chinese, Malay, Indonesian, and Taiwanese versions should appear in 2008 or 2009.

Registering with DigiBOAT initiates an email with the link to download SOB. You then reply by emailing your PC code, which returns up to three different unlock codes for use on three computers, such as your laptop, home, and office com-

DigiBOAT Software-On-Board

Pros: Excellent value for price; excellent documentation; clean interface.

Cons: Only supports C-Map cartography; does not support free NOAA or USACE charts; some stability issues with computer-intensive tasks.

Coolest Feature: Talking Pilot

Price: $57 (Standard download), $76 (Standard CD-ROM), $238 (Pro)

Vista Capable: Yes

Version Tested: 8.5

System Requirements
PC with Intel Pentium II 500 MHz processor
128 MB RAM
100 MB available hard disk space
Windows 2000, XP, or Vista

DigiBOAT
P.O. Box 400
Rose Bay, NSW, 2029
Australia
www.digiboat.com.au

DigiBOAT provides several Excel macros for converting waypoints and routes from lists or other e-charting programs. Email support@sob.com.au to request information.

Keyboard Shortcuts: DigiBOAT is optimized for touch-screen use. Menus, mousing, and keyboarding are minimized and screen real estate is maximized. [F1] displays the shortcuts you will need to memorize.

sOb - Keyboard and Mouse shortcuts

Keyboard

[F2] - Toggle "Chart Levels" toolbar
[F3] - Toggle "Chart Settings" toolbar
[F4] - Change Depth shading
[F5] - Toggle Depth Soundings
[F6] - Toggle "Level Mixing"
[F7] - Night Modes Form
[F8] - Toggle De-Clutter Mode
[F9] - Ship's Settings Form
[F10] - All Waypoints Form
[F11] - All Routes Form
[F12] - Perspective View
[I] [O] - Zoom In/Out
[0] - Reset Chart OverZoom
[1] - Show/Hide Main Toolbar
[2] - Show/Hide Statusbar

[Space] - Centre Ship. Refresh chart and tools
[Alt-Space] - Pan to exact Position, Level and Scale
[Enter] - Toggle "Info" mode
[Backspace] - Show PastTrack Form
[Arrow keys] - Pan chart. [Shift] to super-pan
[D] - Toggle Dead Reckoning mode
[L] [R] - Turn DR heading 10° left/right
[+] [-] - Increase/Decrease DR speed 2.5 knots
[4] - Show Depth Settings form
[Home, End, PgUp, PgDn] - Adjust Depth Shades
[N] - Show/Hide Raw NMEA Data form
[T] - Show/Hide AIS/ARPA Targets form
[C] - Show/Hide Cursor Box
[B] - Configure Display of Chart Borders
[W] - Show/Hide Wind tools

Mouse

[Left Click]   - Pan to clicked position, or
                 - select object under cursor.
[Ctrl Left-Click]- Selects object under ship.
[Right Click] - Show detailed Chart Data

[Middle Click] - Pan to clicked position
[Middle Drag] - Zoom to Window
[Wheel]   - Zoom In/Out large steps

Keyboard   QuickStart Card   User Manual   OK

puter. As an international "online" company, all communication with DigiBOAT is by Web or email. They do not maintain phone-based technical support.

The SOB download and installation was very easy. The program downloads with a C-Map worldwide background chart and 15 full-detail C-Map charts that allow you to sample the cartography before purchasing a C-Map Wide Chart Area or MegaWide Chart Area. These chart regions sell for $199 and $249 respectively. Since SOB can be used on several computers, DigiBOAT also recommends the C-Map USB Key, a $35 dongle that unlocks C-Map charts on any computer with the key inserted.

SOB can also read C-Map cartography on cards. This requires a C-Map USB Multimedia Reader. Unfortunately, DigiBOAT finds itself in the same boat as other e-charting companies, waiting for C-Map to distribute the support files for its latest USB 2.0 readers. In the interim, DigiBOAT's website requests users who already own the USB 2.0 version to email for a copy of their driver test file. You can use the USB 1.0 reader, but there is a quirk if you use your laptop's energy-saving options, which most boaters must do to conserve their battery while underway. If your PC goes to sleep, it loses the C-Map device and does not re-acquire it when the computer and application wake up. You must quit and restart SOB.

SOB's documentation is some of the best we've seen. The company philosophy is that any charting and navigation application has a learning curve. Realizing they cannot design every bit of novelty out of the interface, they work hard to get you started. SOB's installation places a folder on your C:Drive called SOBvMAX. This folder contains a gold mine of resources, including New User Information, a QuickStart Card, a User Manual, release notes, hardware installation guides, and more.

The 172-page User Manual, also available through their website as a PDF, is a fantastic resource. It includes full chapters on networking and connecting NMEA devices. This document was clearly written in-house by experienced boaters, not sub-contracted to a technical writer. It doesn't simply list the buttons and knobs; it often explains when you would use them and why.

LOOK AND FEEL

With no menus and few actions that require dragging a mouse, SOB has a slightly different look-and-feel. In fact, SOB is designed for future use with a touch screen, either directly or through a wireless device. Mouse support is considered an additional functionality until touch screens become more mainstream. But Blundell points out that mousing, particularly clicking on small icons or dragging-and-dropping objects, is difficult in a moving environment.

Instead, SOB focuses on a toolbar and an extensive set of shortcut keys. The toolbar contains icons for intuitive actions such as zoom in and zoom out, creating a waypoint, getting range and bearing, centering the boat, auto-panning the chart, dropping a man overboard marker, and locating ports and services.

DigiBOAT's goal is an interface that answers a variety of detailed questions with one or two finger *taps* (or today, mouse clicks). For example, the distance between a buoy and a headland is obtained by clicking on the dividers icon, clicking on the chart buoy icon, and clicking on the headland. A single click on the paper-clip icon brings up the ViewPanels window, which includes pertinent data such as your vessel's heading and speed, the depth, and any saved messages or notes. The Quick Navigation Box displays a condensed overview of navigation to a designated point. Both Great Circle and rhumb line distances are included in this display if they differ by more than five nautical miles, a handy detail for long-distance voyagers.

SOB's minimal-mousing goal is more successful with some actions than others. Most notably, waypoint or route creation requires more levels of clicking than some other navigation applications. In addition, because windows are not resizable, sometimes you have to move the vertical and horizontal elevator bars to show important data.

Relying on shortcut keys is another common way to avoid mousing. Because of its "no-menu-bar" interface, SOB uses shortcuts heavily. Of course, the flip side of shortcuts is having to memorize which key triggers which functions—and there is a fair bit of memorization with SOB. But to be fair, they provide a handy cheat sheet to get you started and many of their shortcuts are intuitive, such as I for Zoom In, 0 for Zoom Out, and C for Show Cursor.

SOB also has a very useful status bar at the bottom of the screen that contains lots of relevant data. You can interactively read the latitude and longitude of the cursor position on the chart; the distance of that point from your boat; and its true, magnetic, and relative bearing (such as 4.4 nm 132°T [138°M] [21° Stbd]). The status bar also shows information about your current chart, such as its number, scale, and depth metrics. Most importantly, the status bar notes whether the chart is displayed at "real chart" scale or at "over zoom," in which case you have over-zoomed the chart image and compromised accuracy.

As Americans, we found the learning curve to be a few degrees steeper than it might be for boaters from Canada, Europe, or Australia. Some of SOB's feel is a bit different from what American boaters are used to finding when they fire up a piece of software. Most of the differences are metrics you can change, such as setting the program to use feet instead of meters. Some are simple word choice differences, such as *Centre* instead of the commonly-used *GoTo* command. But there are also some tougher challenges. For example, to optimize (or shall we say, optimise) the chart display for your screen, you must enter your screen size in centimeters. Americans think in terms of inches. If DigiBOAT were to make a few of these easy adjustments in advance, it would help SOB break into the U.S. market.

Working with Charts

Several e-charting applications leverage proprietary vector chart resources from C-Map and Navionics. SOB is one of four PC-based applications—along with BoatCruiser, Nobeltec, and MaxSea—that use C-Map cartography. (Of course, many chartplotters use C-Map cards.) C-Map charts have international coverage and embed additional information such as tide and current predictions, animated navigation aids, aerial photographs, and port and marina information.

In fact, SOB is currently only compatible with C-Map cartography, able to load the charts either from a C-Map CD-ROM or from C-Map cards. SOB is compatible with C-Map FP, NT, and MAX charts. It does not currently support C-Map MAX Pro charts. If you already own C-Map cards, such as for your chartplotter, SOB can read them using a C-Map USB Multimedia Reader, which is required to un-encrypt the files.

Users of SOB should have no trouble loading and managing their C-Map charts. With a 28-page chapter in the User Manual, DigiBOAT provides more content and detail on C-Map charts than C-Map themselves. Within SOB, you can easily view a list of your C-Map assets to confirm all your files are properly installed and recognized.

SOB does not support—and does not intend to support—raster charts, focusing on the intelligence that comes with the embedded information native to vector charts. In other words, common raster formats such as NOAA or Maptech BSBs, ARCS, and Seafarer charts are not supported. SOB also does not support user-scanned charts because the company considers the process of scanning paper charts to be an unsafe navigation practice. Other proprietary formats such as MapSource (Garmin), BlueChart (Garmin), SoftChart, and Navionics are also not supported. Although SOB does not currently support standard S-57 charts, such as those available free

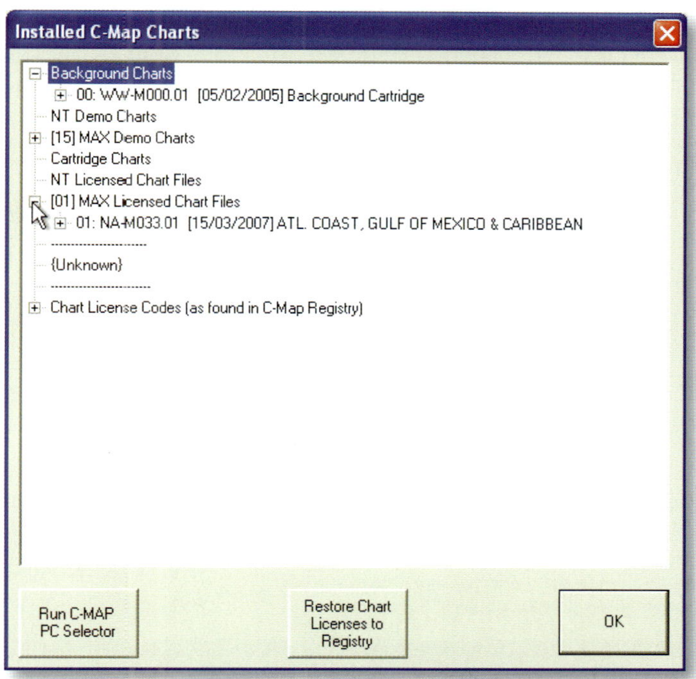

C-Map Chart Library: *Checking to make sure all of your C-Map licensed chart files and cartridges are recognized is easy with* About SOB>C-Map Chart Info.

SOB's isomagnetic variation graph is an excellent example of a nautical reference that can now be brought aboard *digitally* rather than in print. SOB can overlay a graph of isogones (lines of equal magnetic variation); clicking on any location on the underlying world map shows local magnetic variation.

Perspective View: *Looking westward over Government Cut (Miami), C-Map cartography can display a perspective view with toggleable tides and currents, bottom conditions, soundings, chart boundaries, and other important chart features.*

from NOAA, this project is reportedly in the works.

SOB's charts and features are largely a function of the cartography. Chart rendering is good but not exceptional. Charts easily pan and scroll, with much faster responsiveness when loaded from CD-ROM than from an external USB 1.0 multimedia reader. There were occasional display artifacts. For example, moving objects results in a blurred after-image across the screen, like a fanned deck of playing cards.

Charts can be printed directly from a toolbar icon. However, we consistently experienced print artifacts in the form of solid black rectangles striping across the images.

The use of C-Map cartography does allow SOB to take advantage of the intelligence of vector charts and C-Map's additional assets. For example, depth areas can be custom shaded. C-Map also provides the capability to animate the color and rhythm of navigation lights. Additionally, to help you calibrate what you see in the real world with your PC display, only lights within the nominal range of your vessel show as illuminated. Unfortunately, we were unable to verify this C-Map functionality in SOB. The company says SOB supports animated lights but we were at a loss to get this feature functioning properly.

An Anti-Grounding function can check a defined area ahead of your boat for potential navigation hazards—at least those that appear on the chart—every two seconds. A green look-ahead triangle displays if no obstructions are found on the chart, turning red if a potential hazard is located. You can customize the feature to match your draft and set the look-ahead distance and degree width. This feature was temporarily disabled by the company, but is scheduled to return in a future release.

To remove non-essential chart elements, such as text, tools, and symbols, you can choose *Declutter Mode*. For example, your boat's label becomes transparent, route and range/bearing lines are simplified, and target names (but not their tracks) are hidden. For night viewing, charts can be set to gray, red, or black. These settings not only alter the luminance of the chart, they adjust all other components such as borders, buttons, toolbars, and scroll bars.

C-Map charts can also be displayed in *perspective* view, providing an aerial view of an inlet, channel, anchorage or port. Perspective charts can only be displayed at the scale they were digitized—in other words, you can't zoom continuously between scales. But they are digitized at several scales. We did occasionally get an error message when working with perspective views or other computer-intensive displays, warning us to "Close an unresponsive program." At this point the application must be restarted, losing any unsaved waypoints and routes.

Waypoints and Routes

Waypoints are easy to create in SOB. The Waypoint tool icon creates the mark, which easily can be moved by mouse-clicking at the old and new locations.

Unlike many other e-charting applications, SOB has waypoint management. You can move or copy waypoints to a new file, creating separate files for grouped waypoints. Although SOB certainly gets credit for including waypoint management, the implementation is a bit cumbersome. A more intuitive "transfer" or "drag-and-drop" interface would be a welcome improvement.

When SOB creates a waypoint, it becomes a "temporary waypoint," which is unsaved until you assign it to a waypoint folder. When you quit the program, SOB doesn't warn you of these unsaved items, deleting them unless you pay attention and manage your waypoints carefully.

Although SOB does not have a designated search field, you can select a waypoint from the All Waypoints form [F10], double-click, and then choose the Centre icon on the Waypoint Detail form. Although a combined search and center form would be more efficient, the current implementation provides a useful way to move around charts.

Creating routes is not quite as easy as creating waypoints. In order to create routes beyond the current chart display, Auto-panning must be turned on. Unfortunately, Auto-panning slows the program so severely that conventional *rubber-band* route creation is not feasible. The better approach when making a large route is to rough-in a route on a small-scale chart, then zoom in and make any necessary adjustments on a more detailed chart.

One particularly nice feature is SOB's one-touch range and bearing lines. With SOB you can set *multiple* bearing lines, which many other programs don't allow. Using this tool makes it easy to sight bearings or even create parallel line routes to maintain a safe distance from a hazard.

SOB also creates automatic search routes, like you might use during a man overboard search and rescue. Waypoints and a corresponding route (in a circular or grid pattern) are instantly created over a pre-set search area.

Many SOB files, including default waypoints, tracks, AIS data, buddy boats, and user-created messages, are stored using an innovative file naming trick. These files begin with an exclamation mark (!), which sorts them to the top of the alphabetical listing, making them easy to locate.

Many of these files are created automatically by SOB, such as the data files !LastTrack and !PastTrack. The LastTrack file is a backup of the track currently displayed. PastTrack is an accumulation of all PastTrack data. About five times a minute,

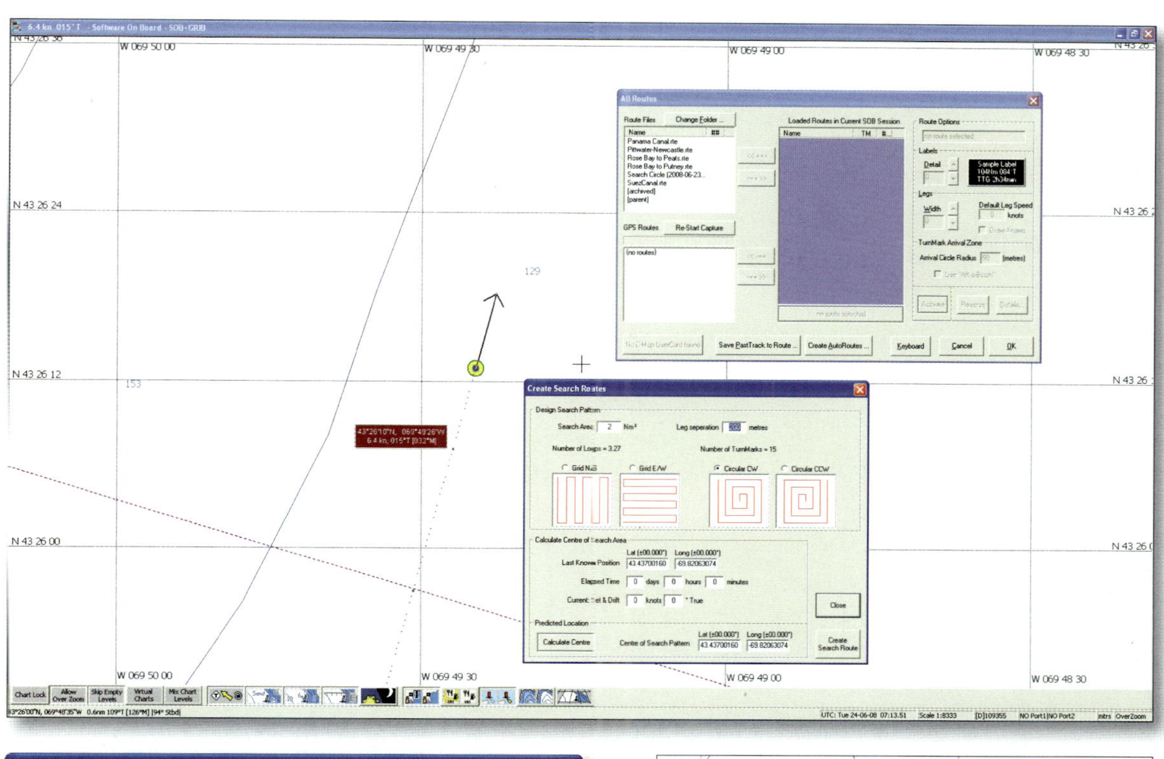

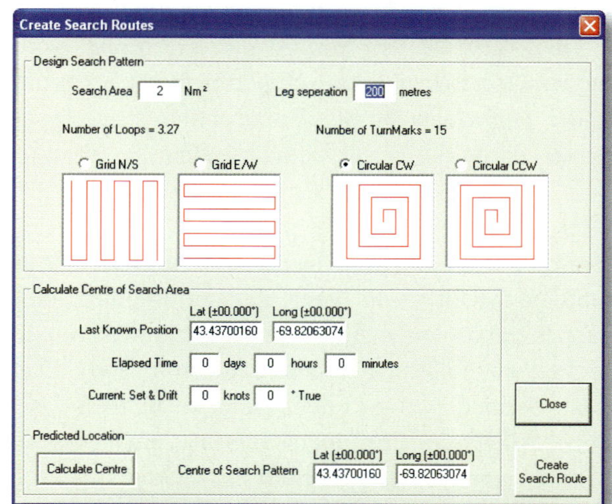

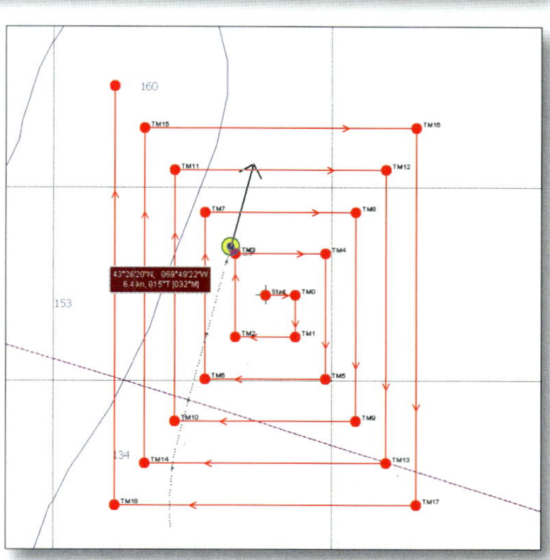

Search and Rescue: *A conscious crewmember goes into the water during a night passage from Provincetown to Boothbay Harbor. It is flat calm and there is a full moon but the water is very cold. While the deck watch initiates a standard MOB procedure, the navigator prepares for a worst case scenario, a lost sighting. DigiBOAT's Auto-Search Route capability is an easy three-step process. Step 1: From the All Routes menu [F11], choose Create AutoRoutes... (top). Step 2: Define search parameters such as search area size, grid spacing and layout, and additional data such as last known position and set and drift (lower left). Step 3: Follow the search route, waypoint-to-waypoint, closely adhering to cross-track error (XTE) (lower right).*

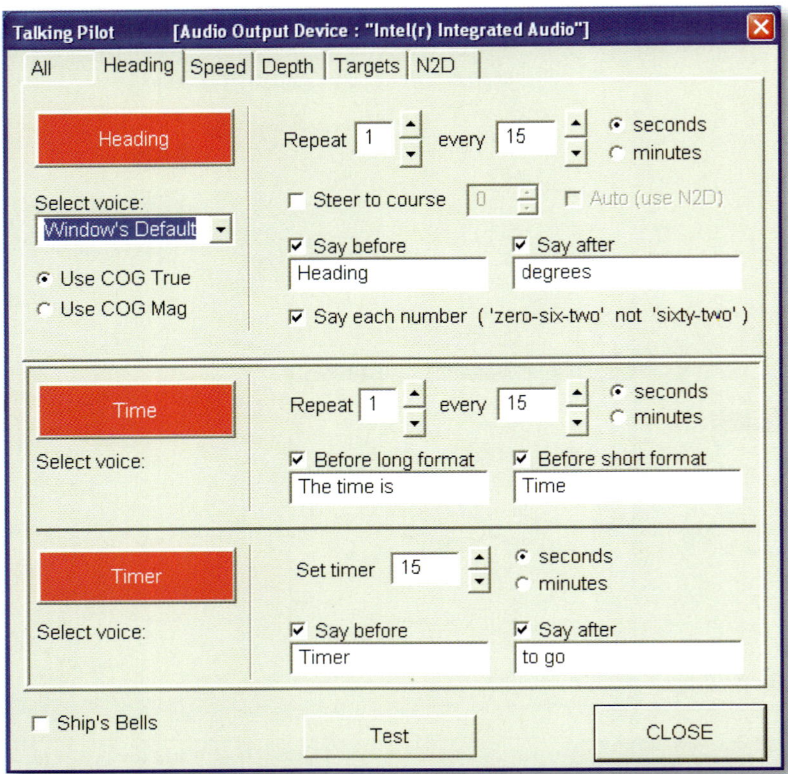

Talking Pilot: *Using built-in Windows speech utilities, DigiBOAT's Talking Pilot is a great aid to disabled boaters as well as navigators and tacticians who benefit from spoken boat data.*

To improve the display of your vessel's vectors (such as COG, SOG, BoatSpeed, or BoatHeading), use the Dampen & Calibrate form.

SOB drops a dot on the chart display to create a bread crumb trail. Each PastTrack point records all ship's data provided by any connected instruments. *Voyage Replay Mode* lets you replay the NMEA data from your GPS, depth sounder, or other instruments. This record is great for reviewing depth profiles or water temperatures for fishing. (Or if you ever happen to find yourself in Admiralty Court!)

Waypoints can be imported and exported, but with a bit of work. SOB doesn't support GPX, instead working from TXT (text) or CSV (comma-separated value) files. Importing waypoints from a Garmin GPS requires a utility such as EasyGPS or GPSBabel (www.easygps.com) (www.gpsbabel.org) or transfers using a C-Card multimedia reader. Unfortunately, we were unable to successfully import the standard CSV file we use for testing.

ADDITIONAL FEATURES

SOB supports a wide range of instruments, including AIS, GPS, autopilot, ARPA radar, depth sounder, knot/log, electronic compass, and wind instruments. These can all be connected and either monitored or controlled from SOB. DigiBOAT has provided two excellent chapters in their User Manual on NMEA and instrument interface, including details on wiring connections and data formats. Our experience installing a GPS was painless. The Raw NMEA Data window has an Identify PORTS selection that automatically scans all COM ports, letting you select the likely port for your GPS.

In addition to the standard AIS target information, SOB also has a buddy boat feature. If your buddies are also trans-mitting—which may be rare, since most recreational boats only receive AIS data—you can simply add your buddies' MMSI numbers to the Target Friends list. Any vessel on this list will show up on the chart with a user-specified color.

SOB integrates tide and current predictions through the C-Map cartography. Click on a T icon on the chart, designating a tide station, and a tidal prediction window opens. Currents can be displayed as a text table, a line graph, or a 24-hour polar graph.

SOB now includes GRIB weather overlays. Data is requested by email through the *Saildocs* server. Choosing Request a GRIB File constructs an email with the appropriate request command string. You can choose your area, forecast horizon, and type of data, including wind, pressure, temperature, geopotential height, and precipitation.

SOB also includes features specifically designed for sailors, such as Show Shadow and Show Laylines. After entering your sailboat's maximum upwind sailing angle, a sector of the wind circle displays where your boat *cannot* sail. DigiBOAT warns that using this feature extensively may bog down the program, and suggests using its Show Laylines features on a more regular basis to show closest point of sail. Alternatively, you can show the Waypoint Wind Shadow, the region in which your sailboat should remain to optimize the upwind leg to the waypoint.

One of SOB's most innovative features is its *Talking Pilot* for sight-impaired users. SOB works with existing built-in Windows utilities to make the display more readable or to even speak navigational data. The Talking Pilot feature uses the Windows speech system (also known as text-to-speech or TTS) to narrate selected ship data. It can repeat data of your choice at a given interval, such as announcing a countdown timer every minute for a race start or announcing your vessel's heading or speed every five minutes. You can even configure the Windows operating system to use Spoken Commands to activate Macros that are linked to SOB's features. For example, the voice command "Pan-East" can be set to click to the right of the display center. The voice command "Show-Ships-Form" can figuratively press the [F9] key, displaying the Ship's Setting Form.

DigiBOAT's most extensive software package, SOB Pro, adds the ability to network vessels, computers, and devices over a distance. For example, if you have NMEA instruments that are far away from your PC, you can use the network features to maintain a wireless connection. Networking also allows the use of a wireless tablet. And, you can connect your instruments to multiple computers. With networking, you can even track a particular boat or monitor a fleet of boats in real-time from anywhere in the world. DigiBOAT's User Manual includes an extensive chapter on networking for Pro.

ASSESSMENT

DigiBOAT's Software-On-Board Standard version is an interesting e-charting package at a bargain price. Although it has some small quirks in both design and function, it also boasts many features usually found only in much more expensive packages. Their ease-of-use claims are not warranted but, more importantly, their documentation is excellent, helping users cope with the initial learning curve.

SOB is one of a handful of PC-based charting and navigation applications that support C-Map cartography. If you like (or already own) C-Map charts, then SOB is a good low-cost choice. However, in order to keep SOB as sprightly as possible, we recommend avoiding reading charts from the external USB 1.0 multimedia reader.

The flip side of DigiBOAT's monogamous relationship with C-Map is that it doesn't read free NOAA or USACE raster and vector charts. But, if all goes as planned, you should be able to display NOAA and USACE vector charts in a future SOB release. However, recognize these would be redundant, and have fewer integrated features than comparable C-Map offerings.

DigiBOAT states that SOB will never support raster charts such as NOAA BSBs or British Admiralty ARCS. This is a serious impediment to boaters in U.S. waters considering SOB. The availability of free—and more important, frequently updated—raster and vector charts of U.S. waters is too great a resource to ignore. In fact, it may be a deal-breaker for many American boaters.

DigiBOAT Software-On-Board
FEATURES AT A GLANCE

Each full-featured e-charting application has been evaluated according to the following 81 features. Some ratings are binary, "yes" or "no." Others are more subjective, evaluated on a scale from 1 to 5 (5 being "Excellent"). Any other comparison is defined in the Legend.

Out of Box Experience

Packaging	N/A
Documentation	4
Software install	3
Load charts	3
Load supplemental data	N/A
GPS hookup	4
Technical support	2

User Interface

GUI metaphor	NT
Screen space management	4
Program responsiveness	3
Chart display speed	3
Chart display quality	3
Customizable GUI	3
Customizable metrics	4
Split windows	No
Tabbed interface	No

Basic Features

One-button MOB	Yes
Waypoint creation	3
Route creation	2
Steer-to function	No
Track creation	3
Convert track to route	Yes
Waypoint and route management	3
Chart management	N/A
Search waypoint and route	3
Range/bearing tool	4
Chart annotation	No
Chart printing	1

Data Exchange and Networking

GPX support for waypoints	No
GPX support for routes	No
Waypoint exchange formats	C
Route exchange formats	C
Integrated GPS transfer	No
Multiple monitor support	No
Network application sharing	No

Cartography

Free U.S. rasters BSBs (RNCs)	No
Free U.S. vectors S-57s (ENCs, IENCs)	No
SoftCharts	No
CHS Canadian rasters	No
DNC	No
Seafarer	No
British Admiralty ARCS	No
International S-57s	No
International S-63s	No
Scan and geo-reference paper charts	No
Navionics cards	No
C-Map cards	Yes
C-Map CD/DVD	Yes
Satellite geo-referenced photos	No
Aerial informational nav photos	No
Topographic maps	No
Bathymetric data	No

Instruments and Sensors

GPS	Yes
Autopilot	Yes
AIS receiver	Yes
Wind	Yes
Depth	Yes
Water temperature	Yes
Radar	Yes
Heading sensor	Yes
Video camera	No

Advanced Features

Tides	CM
Currents	CM
Coast Pilot	No
POIs	CM
Streets	CM
Google Earth support	No
Advanced search	No
AIS	Yes
Buddy boat	Yes
Radar (ARPA/MARPA)	Yes
Radar (display overlay)	No
Fuel calculator	No
Transit calculator	No
Celestial calculator/Almanac	No
Auto route planning	No
Great Circle Route planning	Yes
Sail performance	2
GRIB weather integration	3
Weather options	No
Customizable bathymetric recorder	No

Legend

5	*Excellent*	**CM**	*C-Map*
4	*Above average*	**NV**	*Navionics*
3	*Average*	**SC**	*SailCruiser*
2	*Below average*	**MT**	*Mr. Tides*
1	*Poor*	**PP**	*Plus Pack*
CP	*Chartplotter*	**C**	*CSV*
Mac	*Macintosh*	**E**	*Excel*
NT	*Non-traditional*	**G**	*GPX*
Win	*Windows*	**K**	*KML*
N/A	*Not applicable*		

Chapter 16
TIKI Navigator Pro

The TIKI Navigator software line is the brainchild of Norwegian sailor and programmer Fred Jenssen, who named it after the most famous of Norwegian boats, Kon-Tiki, a raft that crossed the South Pacific. The name is more apt than simple Norwegian patriotism. The Kon-Tiki expedition remains famous decades later because its commander, Thor Heyerdahl, went against conventional boat-building wisdom, re-thinking trans-ocean passages based on his fundamental notions of the sea and boat materials.

Jenssen also admits he is going against the flow with his TIKI Navigator software. He used his 30 years of sea experience to re-think how a computer can and should be used on board. He tossed out established Windows metaphors, moved away from traditional menus, limited the use of a keyboard, and altered the mouse interface. As Jenssen puts it, his goal was a simple scheme that could be used at sea by "boaters who don't think like computers."

Challenging a standard PC user interface is fraught with risk. In fact, we often criticize software developers for deviating from established PC standards. Jenssen hasn't just tinkered with a Windows layout, he fundamentally redesigned it—and we think he pulled it off! As evidence of his success, only two years after its first release TIKI became Norway's best-selling navigation software for raster charts. He is now attempting to break into the U.S. market.

TIKI Navigator is available in three versions: Planner, Basic, and Pro. TIKI Navigator Planner ($37) is a planning tool, able to display charts, create waypoints, and run in simulation mode only. It does not show boat position and is not for real-time navigation.

TIKI Navigator Basic ($74) adds GPS navigation, showing boat position, course, and speed on the chart. You can create routes of ten or fewer waypoints. However, to obtain features like split windows, tracks, logbooks, collision avoidance, autopilot control, and anchor watch, you need TIKI Navigator Pro ($148).

Here we review TIKI Navigator Pro ($148), the full-featured program. For users who don't want to initially pay for Pro, one can run TIKI in demo mode for 30 sessions or purchase a license for Planner or Basic, and then upgrade to Pro easily through TIKI's website.

GETTING STARTED

All TIKI versions are available only by downloading the installation software through TIKI's website. In order to run the software you must purchase a license through the site for the version you want (Planner, Basic, or Pro). The license is emailed to you, allowing you to open the application beyond the trial period.

The TIKI download includes a collection of demo charts, including some charts of the Seattle area and an NMEA 0183 data file showing real AIS traffic. You must purchase or download your own raster charts. TIKI reads all raster BSB charts up to Maptech's 4.0 format, as well as SoftCharts (GEO/NOS format) and the new European raster format (BSB/EAP). When you download TIKI, you can opt to download free U.S. raster charts or order a DVD of charts for $39.95. Raster charts for other parts of the world require a purchase from a retailer such as Maptech or NV-Verlag.

TIKI does not currently display vector charts in S-57, S-63, or proprietary formats such as Navionics or C-Map. However it does display raster topographic maps and aerial and satellite image files if you have them.

All our BSB raster charts, GEO/NOS raster charts, aerial and satellite image files, and raster topographic maps loaded automatically when we first opened the application. In fact,

TIKI Navigator Pro

Pros: Intuitive, creative, and efficient user interface; excellent use of chart screen area; very responsive; excellent image quality; minimal system requirements.

Cons: Only reads raster (not vector) charts; no tides and currents; no weather download; no integrated print function.

Coolest Feature: Digital Magnifier

Price: $148

Vista Capable: Yes

Version Tested: 6.0

System Requirements
PC with 500 MHz processor
256 MB RAM
800 x 600 monitor
DVD drive (required for charts)
Windows 2000, XP, or Vista

TIKI Navigator As
Ovre Hon Terrasse 10A
1384 Asker
Norway
www.tiki-navigator.com

You'll love TIKI's Digital Magnifier tool. To view a magnified section of the chart, simply click-and-hold the right mouse button.

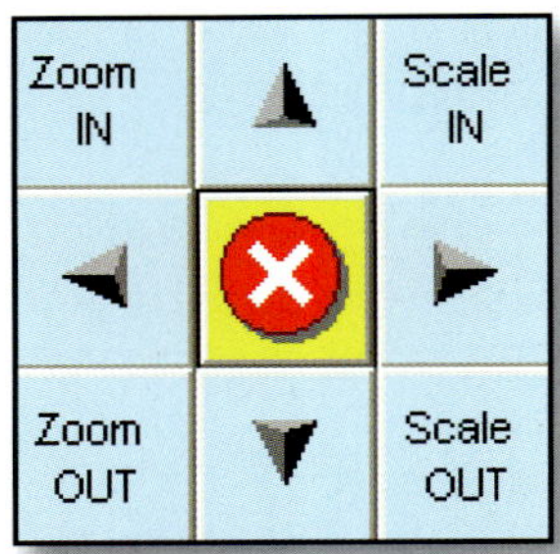

Navigating Charts: *Right-click anywhere on a chart and the Chart Panel pop-up window appears under your cursor.*

Split Windows: *TIKI can simultaneously display one, two, or three chart windows. In this example, a chart overview (left) and zoom (top right) are complemented by a satellite photo (bottom right).*

since our extensive Maptech chart library was already on our hard drive, TIKI immediately located all these assets by looking in the default chart folder location.

The program start-up has a slick user interface: a striking graphic of a hand holding a glass globe with your chart titles scrolling by like movie credits. The chart loading display is a great idea, and Jenssen has taken efforts to buffer the chart list to make this a quick and painless review of your potentially lengthy list of loaded assets.

We also had no trouble connecting our GPS sensor. Although TIKI does not include a Port Wizard to locate the appropriate communications port, a series of feedback windows easily guide you through the process.

The first three times TIKI is opened, it prompts you to begin a tutorial. Take the time to do it! The tutorial is excellent and consists of only 20 short windows. Without the tutorial, you won't find all the menus, which are intentionally hidden to keep the screen clear for the chart.

To use all the available features of TIKI Navigator Pro, you should consider a few additional software plug-ins and accessories. If you want to upload waypoints and routes to your GPS, you need the EasyGPS utility (www.easygps.com). To use TIKI's BuddyTracking feature, you must have a general-purpose GPS utility program called GpsGate ($39.95) (http://franson.com/gpsgate). If you want to print charts you'll want a screen capture application such as SnagIt for $39.95 (www.snagit.com). Finally, TIKI Navigator intentionally has a cursor-based, rather than a keyboard-based, user interface. In fact, most operations are accessed through contextual, mouse-based menus that require a mouse with both a right- and left-click button.

We had no trouble either installing or using TIKI. If you do need technical support, and the FAQ page doesn't answer your question, TIKI asks that you email through its website, leaving your telephone number for a return call. At press time TIKI did not have a printed or online English manual, relying instead on its integrated pop-up help windows. An older-version English manual had been removed from TIKI's website.

LOOK AND FEEL

When Jenssen began the TIKI project, he sought to create an intuitive user interface that broke away from a traditional PC Windows format. But why re-invent an established interface that much of the world already uses? Jenssen points out there are many reasons, including rough sea conditions that make keyboarding and traditional mousing difficult; chart displays on laptop screens that are too small; type so small it is illegible on a moving vessel; and graphical user interfaces with insufficient contrast for a marine environment.

TIKI's user interface is definitely different—but it works. Gone are the tiny pull-down menus that consume screen real estate and are too small to read underway. Gone are the small icons and graphics that clutter the screen. In fact, TIKI's screen at first appears to be all chart display. Where are the menus and choices?

There are a lot of menus—including a Main Menu, a Chart Select pop-up window, a QuickButtons pop-up window, and several contextual mouse menus—but you only see them when you need them. Menu objects are large enough to read and to target with your mouse cursor in a seaway. TIKI's menus can be displayed in large format since they automatically disappear when not in use.

The Main Menu opens as a strip window along the right edge of the screen when the cursor is moved to this edge. When the cursor is brought back onto the chart, the Main Menu panel disappears. The Main Menu shows a series of "rolling" displays, accessed by left-clicking on the menu's title bar. For example, you can scroll between customizable strips of data for waypoints, routes, autopilot, sailing, or racing. Each of the eight *Framesets* contains a collection of *Databoxes*, which you customize with your choice of instruments or data.

A different pop-up appears when the cursor hovers over the top left corner of the screen. The Chart Select pop-up window lets you click to change chart scale, to split the display screen, or to flip between daytime, dusk, or nighttime screen illumination. Again, the pop-up disappears as you move the cursor away, returning the majority of the screen to the chart.

Hover the cursor over the bottom left of the screen and the QuickButtons pop-up window appears. This strip menu displays twelve QuickButtons with their text descriptions, such as dropping an MOB marker. Buttons are numbered from one to twelve and match the function keys on your keyboard for even faster access.

TIKI includes useful shortcuts covering many of the functions. Pressing [F2] brings up the Keyboard Help pop-up window, which shows a list of actions and their corresponding shortcut. Alternatively, typing [F3] brings up the Shortkey Strip pop-up window, listing all Shortkeys alphabetically.

Although TIKI uses menus and pop-up windows, it is fundamentally a mouse and cursor interface. Most features can be accessed easily using the mouse, with keyboarding only for function keys, shortcuts, or naming waypoints or routes.

TIKI relies on contextual menus, where the menu choices vary depending on the cursor's location. For example, a mouse-click on a route waypoint brings up the Route Menu pop-up window, with contextually relevant choices to edit, rename, delete, or cancel the route. Right-clicking anywhere in the chart window brings up the Chart Panel pop-up window,

with buttons to zoom for more detail on the current chart, scale in or out to change the chart scale, or center the boat on the best chart.

We were initially concerned about remembering all these right-click and left-click commands, but we found they quickly became second nature. Remember, the price of removing menu clutter from the screen is that you need to memorize common cursor commands. To help you accomplish this, TIKI includes integrated help, called *Mousehelp*, which can be toggled on or off with the [F1] key. We recommend turning Mousehelp on when you start with TIKI, and at the beginning of each boating season.

With Mousehelp on, hovering anywhere in the program— on the chart, any graphic item, menu or button—displays what left and right mouse functions are available.

Although TIKI is a heavy mousing program, there are several "smart mousing" design elements, including large hover zones, *lock mouse*, and intelligent cursor placement. The large display areas are designed to not only read easier underway, but to be an easier cursor target in rough conditions. If you need even more mouse control, the lock mouse option locks the mouse inside the main and chart panels, preventing the cursor from slipping and sliding all over the screen. We also really appreciated TIKI's intelligent cursor placement, a subtle mouse-friendly bit of programming that places the cursor automatically on the default button of a window prompt. This is a fantastic detail in lieu of moving the mouse for each window. Once you become familiar with the window choices, rather than playing the familiar game, "click…chase the window…click," you can simply double click without physically moving the cursor.

Working with Charts

TIKI's biggest limitation is that it currently only reads raster format charts. This means many of the intelligent features of vector charts—such as integrated alarms and navigation queries—are not available. However, TIKI intends to add Navionics vector chart support, and eventually S-57 vector support, to accommodate the needs of its international customers.

TIKI gets very high points for chart display. We've already

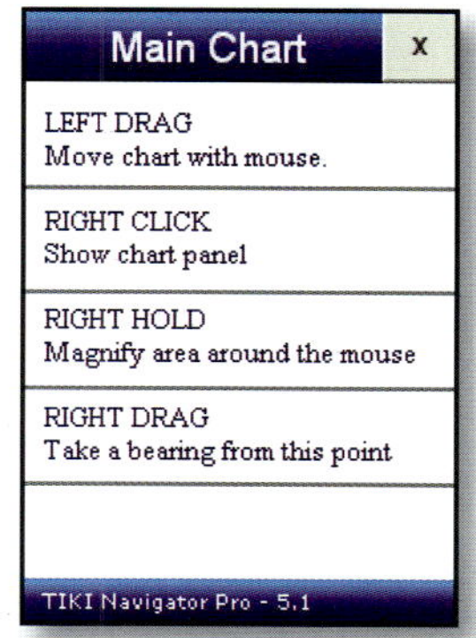

MouseHelp: *Mousehelp is activated by toggling the [F1] key. Hover over any menu or chart display to quickly see what left and right button functions are available.*

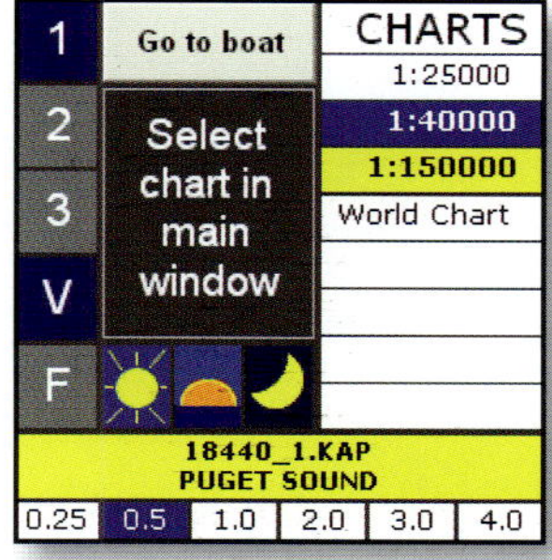

Chart Select Pop-up Menus: *The Chart Select pop-up window lists all charts in the displayed area. Additionally, you can choose a single, double, or triple split-window view; horizontal or vertical chart display; day, dusk, or night lighting; and six zoom levels.*

Clicking the Esc button at the top right of the screen, or pressing the Esc key on the keyboard, aborts a process and immediately returns you to normal mode with the chart centered on your boat position.

AIS Targets: *Once an AIS target is selected, TIKI opens a panel allowing you to continually assess any threat of collision.*

Range and Bearing: *How far is it from the marina to a restricted bridge? How far from an anchorage to a restaurant dinghy dock? Rather than build a route, TIKI's innovative range/bearing tool allows you to quickly extend ranges to answer these common serpentine questions. Q: How far from the Palm Beach anchorage to Lake Worth Inlet (below)? A: 1.66 nm.*

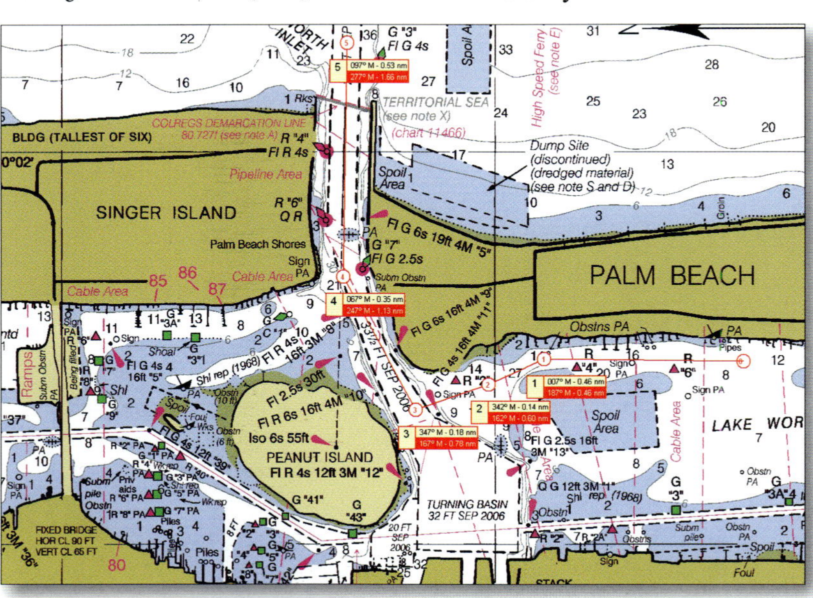

mentioned its maximized chart area. In our assessment, this feature is very important for all charting and navigation programs. If you can't see the chart on a laptop screen there is not much point in laptop charting. TIKI's chart images were also exceptionally sharp—in fact, some of the best looking we've seen. Remember these are the same BSB charts used by all applications, so Jenssen has achieved this by putting quite a bit of programming work into chart rendering.

More impressively, the screen detail was accomplished without sacrificing display speed. The charts were very sprightly to pan and scroll. We never noticed any lag or blurring. As you pan across a chart edge, TIKI automatically loads the next chart. However, charts are not seamlessly quilted: you see the edge of the chart and then flip to the next chart.

TIKI is designed to navigate chart displays primarily through the Chart Panel pop-up window. You can right-click at any position on the chart and zoom or scale in or out, centering the chart at this position. After the first click on zoom or scale, the chart center position does not change. With the window's arrow buttons you can pan the chart east, west, north or south. You can click once or continue to depress the button to continue panning. The pop-up window's red center button instantly centers the chart on the boat position. (You can also display the boat with maximum view ahead, which shifts the boat position to show more chart area ahead of the predicted course.) If you left-click the red center button for about one second, TIKI selects the best chart for the boat position.

Charts can be opened in up to three chart windows, using a split-screen display, showing a chart overview, a zoom, and a satellite or aerial view. TIKI also lets you link your own image file to a chart, such as a photo of an anchorage.

TIKI Navigator Pro includes a *Digital Magnifier* feature, similar to the Spyglass Tool included on NavimaQ navigation software for Apple Macs. Clicking and holding the right mouse button invokes the Magnifier tool, which zooms in to the chart scale showing the most detail.

WAYPOINTS AND ROUTES

We had no trouble creating waypoints and routes with TIKI Navigator Pro. The Pro version accommodates an unlimited number of routes, each with 500 waypoints. Each route shows the cumulative distance at each waypoint, which is a nice detail. The Pro version of TIKI also includes features such as logbook, tracking, and creating routes from tracks.

If you already have a large collection of waypoints and routes, you'll want to import these using GPX. You can also export waypoints and routes to your GPS using the freeware EasyGPS we mentioned above.

TIKI recently added annotation capability and excellent waypoint, route, and chart management, admitted weaknesses in earlier versions. And in keeping with TIKI's superior looks and usability, all mark labels can be adjusted for font, color, and placement.

TIKI has a sophisticated waypoint search panel. Waypoints can be located by type of mark, keyword search, or distance from your boat position or chart center. Once a search filter is applied, the results can be displayed alphabetically or by distance. The search field is also a software navigation tool, letting you quickly center on, GoTo, edit, send, or delete a particular waypoint.

ADDITIONAL FEATURES

TIKI Navigator Pro integrates with several vessel instruments, including wind speed and direction instruments, knot/log, compass, depth sounder, autopilot, and AIS receiver. However, it does not have the capability to display tide and current predictions or weather downloads. Like many applications, tide and current predictions will likely come from Navionics cards. GRIB weather incorporation is underway and scheduled for a future release.

When connected to a wind instrument, true and apparent wind direction can be displayed as graphic arrows on the chart. TIKI has several alarm features, including depth alarms when connected to a depth sounder. The Anchor Watch pop-up window shows your anchoring data, options to set an alarm radius, and a zoom of the chart showing your boat location. TIKI Navigator Pro can also send steering data to your autopilot. Again, implementation is streamlined: create a route, press the autopilot menu button, and TIKI steers the boat via your autopilot, giving a warning as the vessel reaches a waypoint on a route and when the autopilot changes course.

TIKI Navigator Pro recently added AIS support and a feature called *BuddyTracking*. Although most charting and navigation applications now include AIS—the latest trend in e-charting—Jenssen has clearly put quite a bit of thought into TIKI's AIS interface. AIS targets change color if they are predicted to become a hazard according to time and distance parameters set by the user. Each target shows a line representing the vessel's course and speed. An additional line to the left or right indicates if the vessel is turning. You can roll the display forward between two and ten minutes to simulate courses and proximity, assuming no changes to course or speed. Selecting a target displays detailed information on that vessel through an AIS Watch pop-up window.

BuddyTracking lets you exchange position data with your friends and see them on the chart with course line and speed indicators. This feature requires an additional piece of software called GpsGate and an Internet-enabled device such as a cellular phone. If your buddy boat's location is off your main chart display, then TIKI's split screen capability automatically displays a small separate chart window with their location. This feature is great for yacht club cruises or long-distance cruisers who travel in company.

ASSESSMENT

TIKI Navigator Pro is an innovative charting and navigation program that is extremely intuitive, easy to learn, and includes many great features for the price.

TIKI gets very high points for a creative and efficient graphical user interface. It is easy to use—but not because it

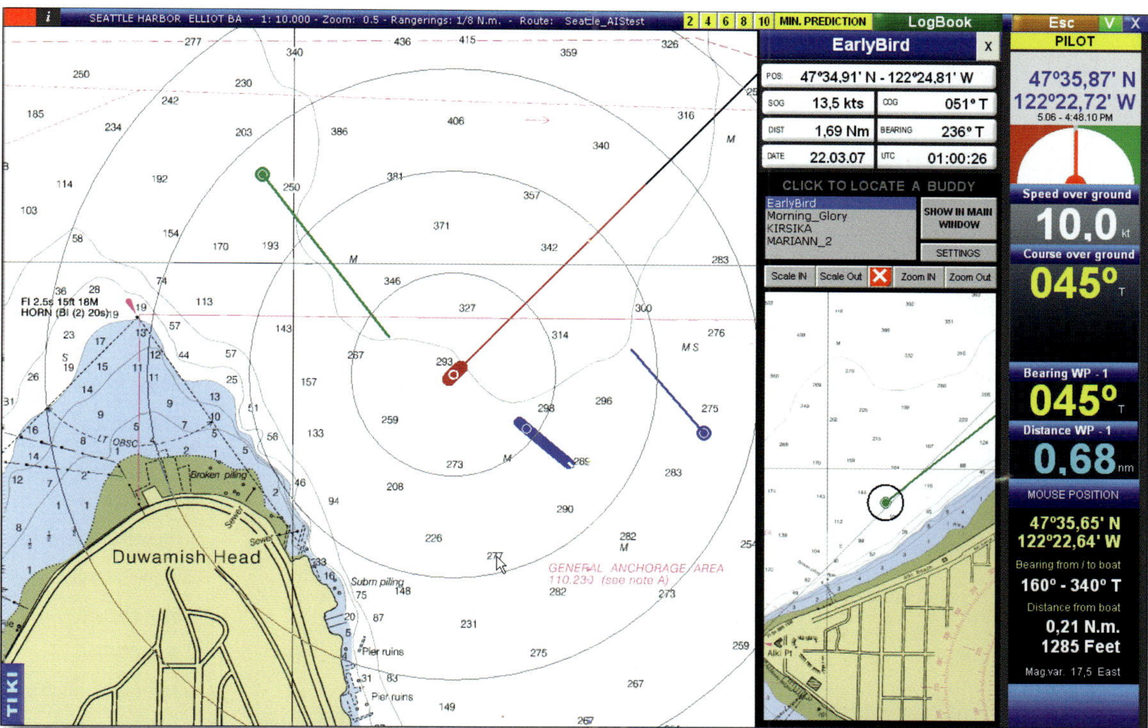

BuddyTracking: *AIS targets (blue), buddy boats (green), and your vessel (red) are visible in the main chart display. BuddyTracking even allows you to keep track of buddy boats that are off your main chart view in a separate window (bottom center).*

is de-featured or stripped-down. Although it does not include some very important features, such as vector chart support, weather downloads, or tide and current predictions, it does more than the basics and does them all well.

Its clean and simple interface focuses on the important features that a real boater needs. It also contains extras such as AIS and BuddyTracking. Recent additions such as rich waypoint, route and chart management and support of GPX for waypoint and route portability are major improvements.

TIKI's creator is refreshingly in touch with his product and market demands. He has a realistic assessment of upcoming features to add to TIKI, with plans to integrate Navionics chart support, GRIB weather downloads, and S-57 vector capability in future releases. When Jenssen completes these additional features, TIKI Navigator Pro will be a world-class product. But like its namesake vessel, TIKI is already a program that can weather the storms and go to sea.

When you click on *any* route waypoint, a popup menu gives you the option of calculating the distance from the boat, time to go, and estimated time of arrival for that waypoint.

TIKI Navigator Pro
FEATURES AT A GLANCE

Each full-featured e-charting application has been evaluated according to the following 81 features. Some ratings are binary, "yes" or "no." Others are more subjective, evaluated on a scale from 1 to 5 (5 being "Excellent"). Any other comparison is defined in the Legend.

Out of Box Experience

Packaging	N/A
Documentation	2
Software install	3
Load charts	3
Load supplemental data	3
GPS hookup	3
Technical support	4

User Interface

GUI metaphor	NT
Screen space management	5
Program responsiveness	4
Chart display speed	4
Chart display quality	4
Customizable GUI	2
Customizable metrics	3
Split windows	Yes
Tabbed interface	No

Basic Features

One-button MOB	Yes
Waypoint creation	4
Route creation	4
Steer-to function	Yes
Track creation	3
Convert track to route	Yes
Waypoint and route management	4
Chart management	3
Search waypoint and route	5
Range/bearing tool	5
Chart annotation	4
Chart printing	No

Data Exchange and Networking

GPX support for waypoints	Yes
GPX support for routes	Yes
Waypoint exchange formats	G
Route exchange formats	G
Integrated GPS transfer	No
Multiple monitor support	No
Network application sharing	No

Cartography

Free U.S. rasters BSBs (RNCs)	Yes
Free U.S. vectors S-57s (ENCs, IENCs)	No
SoftCharts	Yes
CHS Canadian rasters	Yes
DNC	No
Seafarer	No
British Admiralty ARCS	No
International S-57s	No
International S-63s	No
Scan and geo-reference paper charts	No
Navionics cards	No
C-Map cards	No
C-Map CD/DVD	No
Satellite geo-referenced photos	Yes
Aerial informational nav photos	Yes
Topographic maps	Yes
Bathymetric data	No

Instruments and Sensors

GPS	Yes
Autopilot	Yes
AIS receiver	Yes
Wind	Yes
Depth	Yes
Water temperature	No
Radar	No
Heading sensor	Yes
Video camera	No

Advanced Features

Tides	No
Currents	No
Coast Pilot	No
POIs	No
Streets	No
Google Earth support	No
Advanced search	No
AIS	Yes
Buddy boat	Yes
Radar (ARPA/MARPA)	No
Radar (display overlay)	No
Fuel calculator	No
Transit calculator	No
Celestial calculator/Almanac	No
Auto route planning	No
Great Circle Route planning	No
Sail performance	No
GRIB weather integration	No
Weather options	No
Customizable bathymetric recorder	No

Legend

5	Excellent	CM	C-Map
4	Above average	NV	Navionics
3	Average	SC	SailCruiser
2	Below average	MT	Mr. Tides
1	Poor	PP	Plus Pack
CP	Chartplotter	C	CSV
Mac	Macintosh	E	Excel
NT	Non-traditional	G	GPX
Win	Windows	K	KML
N/A	Not applicable		

Chapter 17
Fugawi Marine ENC

When the U.S. government announced the release of its entire digital chart library for free, ironically it was a Canadian software company that first stepped forward to take advantage of the new resource.

Northport Systems, which began making GPS mapping software, had already created products such as Fugawi Marine ENC and Fugawi Global Navigator for marine applications, so the appeal of charts was obvious. Today, Fugawi Marine ENC remains one of the major offerings in the market and continues to add new functions and data sets far faster than many of its competitors. You might think of Marine ENC as the "Swiss army knife" of navigation software. It is designed to view many different types of data, from topographic maps and satellite images to nautical charts and street data. It is also highly compatible across platforms, meaning you can use it on a computer, a smartphone, or a PDA.

Fugawi is the perfect package for world adventurers who want different maps and charts to facilitate a wide range of hobbies: boating and canoeing, offshore and lake fishing, hiking and climbing. And for those interested in viewing many types of cartography on many different devices, it is an excellent option.

Getting Started

Marine ENC is the Fugawi product directed more toward the marine market, although it still contains many features for hikers, small boaters, and lake fishermen. Another Fugawi product, Global Navigator, is a more general offering covering land, sea, and air for use on a PC, GPS, or PDA.

Fugawi Marine ENC is available as a boxed set, directly from Northport Systems or through retailers. The $279.95 set includes an installation CD, two Data Pack DVDs, a 40-page Getting Started Manual, and a Navionics Multi Card Reader with documentation. Our first impression was extremely positive. The software is professionally packaged and had us excited to try all its features and the included extras.

When you purchase Marine ENC, you choose one of four Fugawi Bonus Data Packs—U.S., Canada, Europe, or International. The U.S. pack includes all raster and vector charts from NOAA, as well as vector charts from the U.S. Army Corps of Engineers covering U.S. inland waterways.

Coastal water bathymetric data, elevation data, street overlay data, and landmark data for the U.S., as well as a digital version of the MPC Boaters Directory and a world background map, are also included. The Canadian Data Pack is comparable. The Europe and International Data Packs include primarily street, landmark, and elevation data.

However, there is another bonus: the Navionics Multi Card Reader lets you read Navionics chart cards with Marine ENC. If you already own Navionics digital charts on a Compact-Flash, Secure Digital, or MultiMedia card, such as Platinum or Gold charts, you're in luck.

We had no trouble installing the software or reading in our Gold+ charts using the Navionics card reader. The software took over with window prompts and the installation was very straightforward. A set of Navionics charts, stored on a cartridge and accessed through the card reader, are a worthwhile extra. For example, a single $199 16XG Gold+ card gives coverage of the entire U.S. East Coast, Florida Keys, Bahamas and Bermuda, and much of the West Coast of Florida. In addition, they include street name overlay data, tides and currents, and marina locations—all with a trouble-free data installation.

Don't get frustrated if you have some difficulty with the initial loading of your own charts or the data on the Bonus Data Pack. Fortunately, Fugawi has extensive first-rate cus-

Fugawi Marine ENC

Pros: Competitive price for a multitude of features and data extras; good data exchange with handheld GPS units; excellent waypoint management; customizable printing.

Cons: Difficult set-up process; some stability issues; outdated graphical user interface.

Coolest Feature: Sophisticated waypoint and route management.

Price: $279.95 (includes Navionics Multi Card Reader)

Vista Capable: Yes

Version Tested: 4.5.18

System Requirements
PC with Intel Pentium (or equivalent)
512 MB RAM
250 MB available hard disk space
Graphics accelerator (for 3D viewing)
CD or DVD drive (DVD drive required for Data Pack)
Windows XP or Vista

Northport Systems, Inc.
95 St. Clair Avenue West
Suite 1406
Toronto, Ontario M4V 1N6
Canada
www.fugawi.com

Fugawi Marine ENC supports an extraordinary assortment of chart formats, which may be useful in certain applications or common only in specific regions. For a complete list, see http://www.fugawi.com/web/products/fugawi_marine_enc-ca.htm#charts.

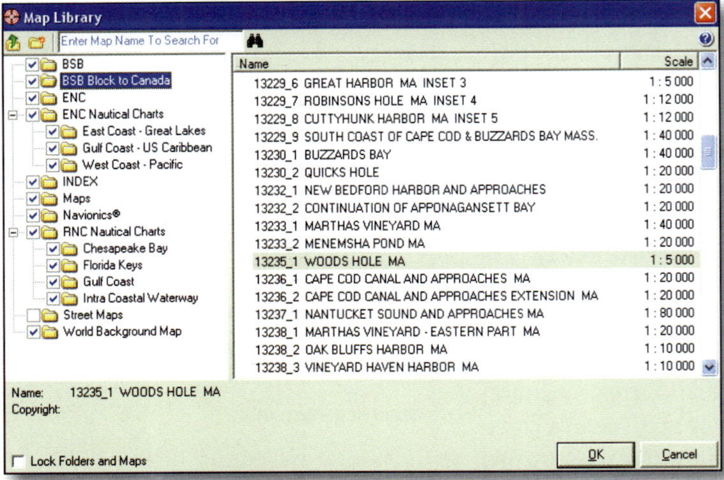

Good Housekeeping: *The Map Library provides flexible organization of charts and maps. NOAA RNCs and ENCs can be cataloged by region. Note also the separate folders for specific digital assets such as Maptech BSB4s, Navionics charts, and street maps.*

If the Bonus Data Pack will not install, go to FAQ #4 (www.fugawi.com/web/support/faq/faq_0004.htm). If it still won't install, call technical support for a quick fix.

If the Navionics Multi Card Reader fails to read your chart card, go to FAQ #5 (www.fugawi.com/web/support/faq/faq_0005.htm), which directs you to a firmware update.

tomer support. The software package includes a helpful 36-page Getting Started Manual, but don't neglect the detailed 142-page User's Guide which is included as a PDF document on the setup CD. We recommend immediately printing this PDF and creating a binder. Its format—brief step-by-step directions, coupled with lots of screen shots—makes it easy to follow along chapter-by-chapter. Fugawi Marine ENC also includes On-Line Help Files, which you access by pressing the [F1] key from any point in the program.

Technical support is available to all registered users, but you must register within 30 days of purchasing your software license or you lose this support option.

Fugawi's website has a page for Troubleshooting & Frequently Asked Questions (FAQs). The latter addresses five issues, including configuring a GPS, port problems with a serial-to-USB converter, and the Data Pack import problems noted above. For users who hate to read manuals, this page includes online training videos to learn how to use Marine ENC. Many important and relevant technical specifications are also detailed here, such as a list of supported GPS and PDA models. Another nice contribution is their NMEA cable diagram to help you hook up your onboard GPS.

Northport Systems has gone even further by providing additional troubleshooting software utilities and documentation as free downloads (www.fugawi.com/web/support/utilities_and_docs/utilities_index.htm). For example, these resources help you convert file formats, obtain and set up drivers for third party devices, and provide sample data files to test AIS or GRIB weather features. The company also provides a monthly newsletter of tips and tricks.

Before loading additional charts, you should develop an organizational strategy to manage your chart files. Loading newer versions of charts you already have will likely create duplicate charts in the Chart Library. You must isolate or re-

move potential duplicates manually, checking for the most current edition. The program does not automatically identify and display the most recent chart edition. For example, if your chart folder contains three charts of Miami Harbor, the software simply takes the first one in the folder, regardless of whether it is the most recent chart. If you are a folder slob, mend your ways or be very careful using this program.

LOOK AND FEEL

Marine ENC uses a standard interface based on menus and icon toolbars. You can easily pan over charts with the mouse and zoom with the scroll wheel. Functions can be accessed through a main menu or through toolbars that place the cursor into a *mode* to perform certain functions, such as creating a waypoint or measuring distance and bearing.

In addition to the main menu across the top, a Status Bar shows information such as GPS location, cursor location, and chart datum. Two toolbars include icons for functions such as opening charts, panning and zooming, and shortcuts for routes, tracks, and waypoints.

The interface is heavily based on icons, many of which are not very intuitive. It took some effort to memorize them all. The fact that some of the icons had to be labeled with letters, such as WP for waypoint and RTE for route, suggests that words may be a better choice than the current icon system. The user interface also has a dated Windows feel. Icons are pixelized and limited to primary colors, and the floating windows utilize designs and features reminiscent of older software programs.

In addition, the frame of menus and toolbars cramps the chart display area, which is often obscured by working windows, such as the Waypoint Library and Route Library. There were some adjustments that helped. You can make your own window *drawers* with the Waypoint Library and Route Library windows by pulling them off the bottom edge of your screen, letting you read the window title but clearing the screen for the chart display.

Fugawi Marine ENC is a mouse-intensive program. Because its user interface is based on cursor modes, a user must commute back and forth to the toolbar to switch modes for different operations. Keyboard shortcuts, the typical way to avoid extensive mousing, were limited. The program can be extra effort on the hands and wrists if you use it extensively.

On the positive side, Marine ENC wins high marks for a customizable display and experience. You can set imperial, metric, or nautical units. Soundings can be set in meters or feet. One particularly nice customized setting is depth area shading. You can set the chart display colors by depth, customizing safe water boundaries for your vessel's draft. It also has a dusk and night vision display. Charts can be displayed north-up, map-up, course-up, route-up, or (uniquely) by degrees of rotation.

Working with Charts

One of Marine ENC's strengths is its compatibility with a wide variety of chart formats, including NOAA raster and vector charts, U.S. Army Corps of Engineer IENC vector charts, BSB4 or earlier charts from Maptech, files in GEO/NOS format (formerly SoftChart), and nv.digital files from NV-Verlag for Europe, Scandinavia, and the Caribbean. It can also read user-scanned paper charts and, with a card reader, Navionics Silver, Gold, Gold+, Platinum, Platinum+, Fish'N'Chip, and HotMaps charts. Many charting and navigation programs are weak when it comes to international coverage. With the Navionics and NV-Verlag options, Marine ENC gives you access to digital charts for much of the world.

Overall, Marine ENC was very responsive. It was most responsive when displaying raster charts. Some operations were noticeably slower, particularly 3D viewing, but this is to be expected given the demands placed on your machine by 3D data. Another speed bump was that panning across vector charts sometimes resulted in a lag with a motion-blurred display, where depths and contours smear across the screen until the display catches up.

It seems straightforward, but many charting and navigation programs fall short on the seemingly-simple feature of printing chart excerpts. With the exception of one bug, Marine ENC has excellent printing features. You can select an area of a chart to print by clicking and dragging a rectangle, or set a print area based on a list of waypoints or a route. A particularly useful print feature lets you scale the waypoint data labels on your printout. You can customize the printing of waypoint labels to make the text bigger if you have trouble reading them, or smaller if densely-placed waypoints are obscuring chart features.

Unfortunately, we could not print *vector* charts at high resolution with our HP 1200 DPI printer and 380 MB of memory. We also tried our Epson C88+ color printer but this didn't solve the problem. Vector charts did print in "draft mode," but were coarse and unreadable.

Waypoints and Routes

Marine ENC receives high marks for operations such as creating routes, transferring data, and managing navigational assets. It is less smooth in creating waypoints and managing waypoint icons.

Creating a waypoint falls victim to the over-mousing interface we mentioned earlier. You can't simply double-click on a chart to create a waypoint or double-click on a waypoint icon to edit its name or description. However, waypoints can include a comment, be hidden from the display, be locked to prevent unintentional changes, or be linked to an image or sound file.

Routes are exceptionally easy to create by selecting and dragging waypoints. By displaying the Route Library, Display Route, and Waypoint Library windows at the same time, you can drag-and-drop collections of waypoints directly from your Waypoint Library to the Route Details window and a route is instantly created from those waypoints. Use shift+click to highlight a selection of waypoints and control+click to select a collection of non-contiguous waypoints.

Alternatively, routes can be created in the traditional manner, by linking a sequence of existing waypoints. You can also freehand draw a route on the chart with your cursor in Route Mode. The course you freehand trace on the chart is automatically converted to a series of straight line segments, the "granularity" or number of which is determined by your cross-track error setting.

Each waypoint, route, or point of interest can be marked with one of literally hundreds of icons—perhaps too many

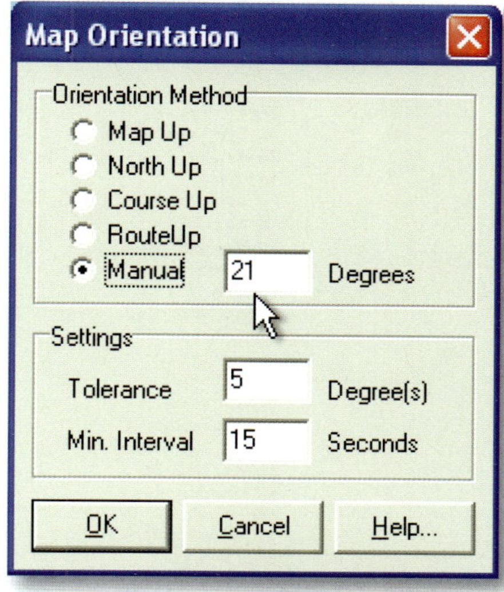

Which Way is Up? *The Map Orientation window allows a user to choose from the standard array of chart display orientations, as well as a unique option for manually defined chart rotations.*

To increase chart real estate, pull windows off the bottom edge of the screen, leaving the window title visible but clearing the screen for your chart display.

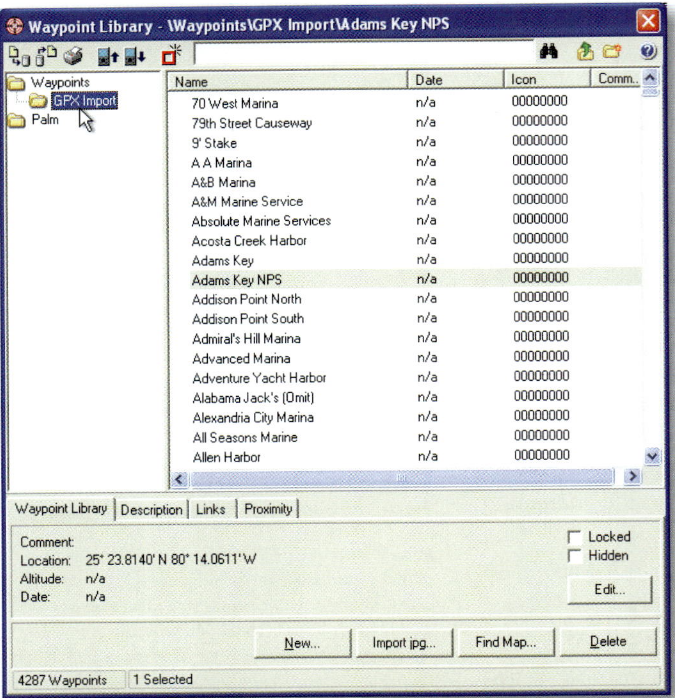

GPX Import: *Easy as 1-2-3. The Waypoint Library showing a seamless GPX import of 4287 waypoints originally created in another charting and navigation application.*

The Import feature puts BSB charts into a generic BSB folder and ENCs into a generic ENC folder. Instead of this "bucket option," keep your chart files organized by adding charts to the Map Library using the **Install Map and Data** option, which maintains the original folder name, such as Bahamas, Maine, etc.

icons. One feature we would have liked is the ability to customize a subset of about ten common icons, such as marina, anchorage, fuel, or restaurant. Unfortunately, each time you set an icon you must scroll through the full collection, sorting through icons for buses, hikers, airplanes, and helicopters.

Having said that, waypoint and route management is excellent using the Waypoint Library and Route Library windows. You can create subfolders to organize your waypoints and move waypoints between folders simply by clicking-and-dragging. To help you visualize your waypoint folders, the nested organization is shown with a tree structure view. This excellent and very organized treatment of waypoint and route data is an important feature for boaters with many waypoints for different regions.

Unfortunately, unlike many charting and navigation applications, Marine ENC does not have a way to search for a waypoint and then quickly go there on a chart. There is no "Go To..." menu choice and no way to type in the first few letters of a waypoint name and ask to go there. Instead it is a cumbersome panning and zooming search process, typically beginning with the world map.

But Marine ENC was a champ when it came to importing and exporting waypoint data. We easily imported 4287 East Coast waypoints using GPX. (For an explanation of GPX, see Chapter 8, *Networking and Data Exchange*.) Marine ENC does not limit descriptions and all our information for each waypoint imported successfully. Although the software also imports and exports in ASCII text format, we recommend always using GPX transfer if it is available.

Indeed, Northport has made GPS data transfer a priority—not surprising for a company that began making GPS software. You can import and export routes between a handheld GPS and your PC simply by clicking command buttons, and

data can also be shared with Navionics-compatible plotters via flash memory card.

ADDITIONAL FEATURES

When it came to hardware, Marine ENC worked well with our external USB GPS sensor and the setup was straightforward. Consistent with Fugawi's emphasis on portability and cross-platform use, you can also run Marine ENC on your Pocket PC or Palm devices.

The software also integrates with AIS information if you connect an AIS receiver to your PC, and Marine ENC added extensive AIS features as part of the version 4.5 update. Vessels that are transmitting an AIS signal are displayed color-coded to indicate whether they are underway or stationary. The outline of the vessel is drawn to scale and shows its name, type, track, and current heading.

Marine ENC also includes additional data on the Bonus Data Pack DVDs, which are included with the software. The discs include elevation data, street name overlays, landmark data based on the MPC Boaters Directory, and bathymetric depth data. Tide and current data is available to users with Navionics chart cartridges.

The elevation data is used for 3D displays. Once you have the data loaded, simply choose File>Open>3D>Show 3D Window. You can seamlessly fly over this landscape using your mouse or a collection of 13 keyboard shortcuts. Marine ENC also has bathymetric data, but depths do not display in 3D.

The Places Menu lets you search for cities and streets. As you zoom in on a chart, smaller street names continue to fill in. This street map overlay is handy for boaters, either to identify an anchorage (which are often at street ends) or to locate services using an address.

The MPC Boaters Directory of the U.S. is a searchable directory of 1.8 million places and 25,000 marine services. These can be plotted as an optional overlay on a chart. Unfortunately, because this directory is an advertising-based listing of marine services, and therefore not comprehensive, you generally can't rely on it as the definitive source on services.

Neither the street data nor the directory data are as high in quality as one would find on a dedicated street atlas program or in the POI (points-of-interest) data contained in other marine-only software packages. But if you have third-party BSB format charts, such as those from Maptech or NDI DigitalOcean, you can view their marine facilities and photo location data using Marine ENC.

Fugawi Marine ENC includes the ability to download and display GRIB weather data. Simply select ENC>Weather (GRIB)>Retrieve Data Via Email from the main menu and a window opens prompting you for the desired area and time. A GRIB file with data for surface wind, 500 millibar height, wave, surface temperature, or air temperature arrives immediately in your email inbox as an attachment.

Marine ENC recently provided a free Google Earth plug-in that lets you automatically download Google Earth images of your current location. A particularly nice feature is the split screen display, letting you view an aerial image—showing actual sandbars, channel openings, and coral reefs—next to a marine chart of the same area.

You'll need a fast Internet connection to download these images—Google Earth is a notorious bandwidth hog—but Marine ENC allows you to save them to your hard drive so they are ready to use later when underway. Waypoints and routes can be uploaded to Google Earth with a single mouse click, displaying your points on a satellite view. You can also send this data to others to communicate location or for use as a float plan.

If you like cool—but maybe less practical—features, you'll like Marine ENC's option of linking digital sounds and photos to a waypoint. You can link a WAV file of the red wolf howl from the Alligator River in North Carolina or link a JPEG file showing the scenery at your favorite Maine anchorage. The feature is very easy: highlight the waypoint and click on the Links tab, selecting the file to attach.

Marine ENC also includes some quick and handy tools. For example, there is a tool to calculate distance, including both rhumb line and Great Circle distances; a tool to calculate magnetic variation; an odometer tool; and a tool for sunrise and sunset predictions.

Assessment

Overall, Marine ENC is ideally suited for people who use multiple platforms in multiple outdoor settings. It is fundamentally designed to be a cross-platform, cross-activity tool. You can move from a PDA to a handheld GPS for hiking or to your boat's chartplotter for cruising—all using the same software package. This orientation is reflected in its capabilities to import and export data and manage waypoints and routes across devices.

The flip side of this "Swiss army knife" approach is that it may do too many things and contain too much data for certain boaters. It is alluring to have topographic maps, hiking tracks, 3D views, and street maps, but some users will only use a fraction of the features, and may become frustrated by the need to push aside unwanted menu choices and data.

However, for people who want the portability and extra data and options, Fugawi Marine ENC is about half the cost of many other charting packages. The best bet for marine applications is to purchase the convenient optional Navionics cards, which fuels Marine ENC with top-notch nautical charts and marine data.

In order to take advantage of Marine ENC's Google Earth feature, you need two downloads. Google Earth can be downloaded for free at http://earth.google.com. You must have version 4 or later. Then download the free plug-in from Fugawi at www.fugawi.com/web/support/google_earth_plugin.htm.

Bird's Eye View: *We applaud vendors who leverage free Internet resources. Here a Google Earth side-by-side helps define a poorly-aided Bahamian swash channel.*

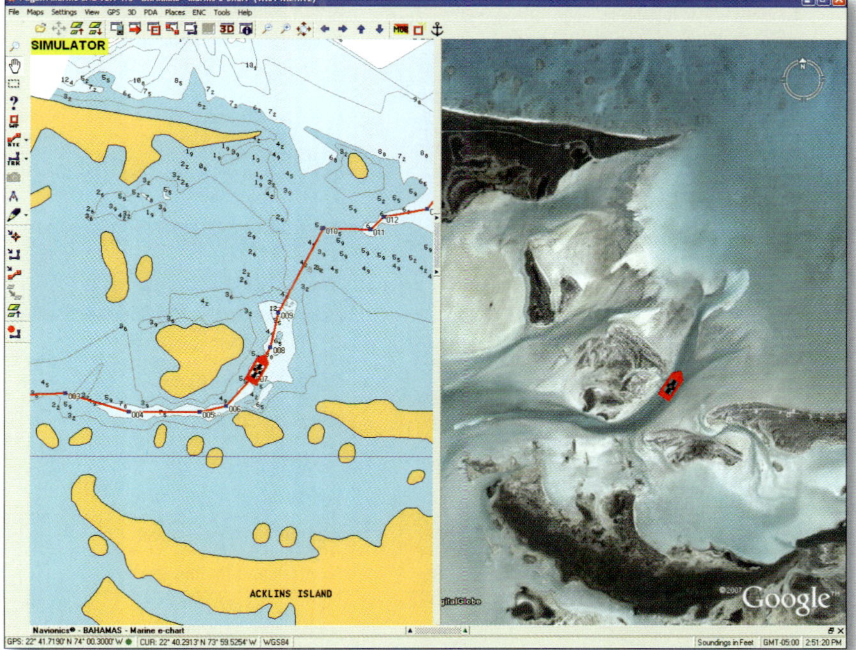

FEATURES AT A GLANCE

Each full-featured e-charting application has been evaluated according to the following 81 features. Some ratings are binary, "yes" or "no." Others are more subjective, evaluated on a scale from 1 to 5 (5 being "Excellent"). Any other comparison is defined in the Legend.

Out of Box Experience

Packaging	3
Documentation	3
Software install	3
Load charts	2
Load supplemental data	2
GPS hookup	3
Technical support	3

User Interface

GUI metaphor	Win
Screen space management	2
Program responsiveness	3
Chart display speed	3
Chart display quality	3
Customizable GUI	3
Customizable metrics	3
Split windows	No
Tabbed interface	No

Basic Features

One-button MOB	Yes
Waypoint creation	3
Route creation	4
Steer-to function	No
Track creation	3
Convert track to route	Yes
Waypoint and route management	4
Chart management	4
Search waypoint and route	2
Range/bearing tool	3
Chart annotation	2
Chart printing	4

Data Exchange and Networking

GPX support for waypoints	Yes
GPX support for routes	Yes
Waypoint exchange formats	GCK
Route exchange formats	GCK
Integrated GPS transfer	Yes
Multiple monitor support	No
Network application sharing	No

Cartography

Free U.S. rasters BSBs (RNCs)	Yes
Free U.S. vectors S-57s (ENCs, IENCs)	Yes
SoftCharts	Yes
CHS Canadian rasters	Yes
DNC	No
Seafarer	No
British Admiralty ARCS	No
International S-57s	No
International S-63s	No
Scan and geo-reference paper charts	Yes
Navionics cards	Yes
C-Map cards	No
C-Map CD/DVD	No
Satellite geo-referenced photos	Yes
Aerial informational nav photos	Yes
Topographic maps	Yes
Bathymetric data	Yes

Instruments and Sensors

GPS	Yes
Autopilot	Yes
AIS receiver	Yes
Wind	No
Depth	Yes
Water temperature	No
Radar	No
Heading sensor	No
Video camera	No

Advanced Features

Tides	NV
Currents	NV
Coast Pilot	No
POIs	Yes
Streets	Yes
Google Earth support	Yes
Advanced search	Yes
AIS	Yes
Buddy boat	No
Radar (ARPA/MARPA)	No
Radar (display overlay)	No
Fuel calculator	No
Transit calculator	No
Celestial calculator/Almanac	No
Auto route planning	No
Great Circle Route planning	Yes
Sail performance	No
GRIB weather integration	3
Weather options	No
Customizable bathymetric recorder	No

Legend

5	*Excellent*	**CM**	*C-Map*
4	*Above average*	**NV**	*Navionics*
3	*Average*	**SC**	*SailCruiser*
2	*Below average*	**MT**	*Mr. Tides*
1	*Poor*	**PP**	*Plus Pack*
CP	*Chartplotter*	**C**	*CSV*
Mac	*Macintosh*	**E**	*Excel*
NT	*Non-traditional*	**G**	*GPX*
Win	*Windows*	**K**	*KML*
N/A	*Not applicable*		

Chapter 18

NavSim BoatCruiser

BoatCruiser and its sibling SailCruiser are two relatively-new charting and navigation applications from NavSim Technology, a Newfoundland company specializing in computer applications for the commercial piloting and shipping industry.

NavSim is best known for its NavCruiser PRO software, a professional package currently used by more than 75 percent of the pilot boats on the Great Lakes and all pilot boats serving the St. Lawrence Seaway. Originally a spin-off from the National Research Council's Institute for Marine Dynamics in Canada, they continue to work on the development of an advanced marine autopilot system for the shipping industry.

Several years ago, NavSim began to adapt its technologies and leverage its software code base for recreational products. BoatCruiser was introduced in 2004, followed in 2006 by a sailboat edition called SailCruiser.

Today NavSim offers four software packages: NavSim Viewer, BoatCruiser or SailCruiser ($399), and NavCruiser PRO ($1199). The NavSim Viewer is available free with the purchase of two charts from the NavSim MapServer (www.navsim.com/products/mapserver), which we'll discuss later, but it only views and prints charts. At the other end of the continuum, NavCruiser PRO is their product for professional or commercial vessels. BoatCruiser is targeted for the recreational power boater, incorporating features such as the Fuel Cost and Consumption Calculator. SailCruiser is built on BoatCruiser's platform, with sailing-specific features such as *SailTimer*.

NavSim's current challenge with BoatCruiser, and its sailboat edition SailCruiser, is the transition from professional and government sales to the retail recreational market. Boaters are just beginning to see these products in major trade publications and retail outlets.

NavSim has a strong marketing team and the product looks professional and reads even better. But we found it fell a bit short on some ease-of-use and product feature claims. However, at only three years old, NavSim does an excellent job of listening to users and responding with updates, so we feel it is only a matter of time until they complete their documentation and round out the feature set.

Getting Started

NavSim has chosen C-Map to be the principal provider of charts for its software. In fact, BoatCruiser may be purchased as a bundle with C-Map charts, including a MAX Wide Chart Area for an extra $199 or a MAX MegaWide Chart Area for an additional $249. Bundled packages can be ordered through NavSim's website. They are also available at select marine retailers, such as Bluewater Books & Charts (www.bluewaterweb.com) and The Nautical Mind (www.nauticalmind.com).

If you already own cartography, or don't want C-Map charts, a software-only edition of BoatCruiser ($399) may be purchased online through NavSim. If you already own C-Map MAX *cartridges*, these can be read only with a C-Map USB 1.0 multimedia reader, available directly from C-Map. Note that the newer USB 2.0 multimedia reader offered by C-Map is currently not supported by NavSim; you must buy a dated PC-Planner kit containing the older reader and software for $179.

If you choose a C-Map option, the North America BoatCruiser package includes three C-Map discs and an unlock code for one region. (All three regions are provided in advance for convenience, in anticipation that you will later purchase additional unlock codes.) Although NavSim promotes this as a "free unlock code for new users," you do not receive an additional unlock code. Additional unlock codes, through

NavSim BoatCruiser

Pros: Very customizable layout; free automatic software updates; TimeMachine; Fuel Cost and Consumption Calculator; SailCruiser edition with SailTimer.

Cons: Difficult installation and set-up; not all features work as described; lack of documentation.

Coolest Feature: Automated NOAA and CHS chart updating

Price: $399 ($598 with one C-Map MAX Wide Chart Area; $648 with one C-Map MAX MegaWide Chart Area).

Vista Capable: Yes

Version Tested: 2.0.19.8

System Requirements
PC with 500 MHz processor
128 MB RAM
100 MB available hard disk space
800 x 600 monitor with 16-bit color
Internet Explorer 5.01 or higher (required for upgrades)
Windows 98 SE, ME, 2000, XP, or Vista

NavSim Technology, Inc.
P. O. Box 12093
Arctic Avenue
St. John's, NL A1B 3T5
Canada
www.navsim.com

Whenever you are connected to the Internet, take advantage of NavSim's free automatic chart updates. With NavSim open and an active Internet connection, a dialog box will appear if an update is available.

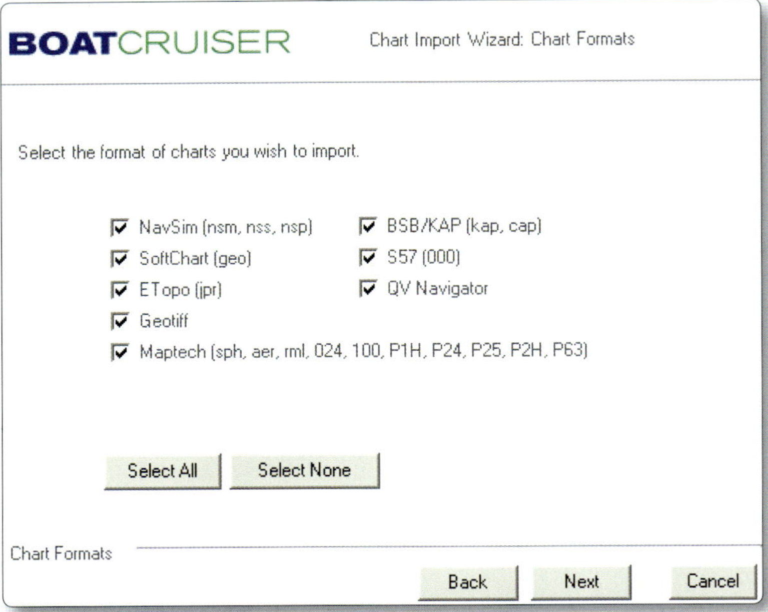

Chart Import Wizard: *In addition to supporting C-Map cartography, NavSim imports multiple raster, vector, topographic, image, and photo formats.*

NavSim provides a downloadable Help Handbook for its software at **www.navsim. com/products/help/Help.pdf**. The document is a collection of screenshots.

C-Map, cost $199 for a Wide Chart Area, such as the U.S. East Coast and the Bahamas or The Great Lakes and the Maritimes. A MegaWide Chart Area, for an additional $249 per unlock code, covers an area such as The North American Atlantic Coast, Gulf of Mexico, and the Caribbean. NavSim is now preparing to offer a *MAX Pro Add-On* package for $149, adding the ability to use C-Map's latest MAX Pro cartography. C-Map MAX Pro charts must be purchased separately.

BoatCruiser is not easy to get started, particularly installing C-Map and Maptech charts. Connecting the GPS is easier. The Device Finding Wizard scans all ports and all baud rates to make instrument hook-up painless. On the positive side, NavSim is a company that makes frequent updates to its software, automatically checking for and downloading the latest changes. Future versions should continue to streamline the installation process.

BoatCruiser would benefit from more user documentation, currently only including a thin User Manual with eight pages of content. Printing BoatCruiser's Help Handbook is not an acceptable substitute for a User Manual, since it consists of an amalgam of discrete help menus with no index, table of contents, or sequence. Although the PDF is 266 pages, most pages consist of only a large screen shot with very little text.

LOOK AND FEEL

BoatCruiser uses a standard Windows interface with traditional pull-down menus. A modifiable toolbar across the top displays 30 icons.

BoatCruiser's interface is highly customizable, combining elements similar to TIKI's Frameset customization and RayTech's Excel-like tabbed interface. With the Layout Manager, a user can design the entire screen layout, including the toolbar and screen contents. Although BoatCruiser does not have traditional split-window chart viewing—standard in packages at this price level—it does allow for multiple *navigation consoles*. Each of these display boxes, chosen from a list of 40 choices, can be customized for size and location.

The flip side of this feature is the tendency to cram too many data boxes onto your laptop screen. To see the information you'll need in a given scenario—be it cruising, racing, or fishing—it's easy to create a busy patchwork display. Although you could insert a two-by-two-inch camera view, your instrument data, and a chart, you need to be a judicious designer. Every additional display box subtracts from the chart display. However, with NavSim's customization options, the user has the option to maximize the chart display and even hide the toolbar.

WORKING WITH CHARTS

BoatCruiser and SailCruiser support several different chart formats, but are optimized for C-Map cartography, including C-Map NT+, C-Map MAX, and soon, C-Map MAX Pro. These proprietary vector charts also include tide and current predictions, flashing navigation aids, perspective views, and harbor inlet and marina photos.

BoatCruiser supports several other chart formats through a chart import scheme. (C-Map cartography runs native so does not need to be imported.) Of these import choices, BSB/KAP and S-57 charts are most relevant to mariners. Note that although the many Maptech choices look intriguing, non-Maptech applications typically don't integrate the files into their chart display. In other words, BoatCruiser can read and display those Maptech aerial photos you purchased separately. But since they are not linked to a chart location, the chart does not visually indicate when and if you have an aerial navigation photo available. All files are accessed through a single "bucket" of Maptech files. The exception is Maptech satellite images, which are geo-referenced within BoatCruiser.

BoatCruiser's chart rendering is very good. Chart images are sharp and easy to read. Although loading time for charts is slow (because of the conversion to NavSim's internal NSM format), the benefit becomes apparent when BoatCruiser promptly displays the images. Both BSB and S-57 format files were very sprightly. Unfortunately, the C-Map data using C-Map's USB 1.0 reader is noticeably sluggish. C-Map now ships a new, faster USB 2.0 reader, which BoatCruiser says it will support beginning in late 2008.

Panning and zooming across charts was intuitive and responsive. Panning uses a traditional *grabber hand* with an arrow option as the cursor nears the edge of the screen. Zooming uses the mouse's scrolling wheel. Charts are seamlessly displayed and may be rotated to course-up, chart-up, true north-up, or magnetic north-up. Charts also can be displayed in dusk or night modes. The Goto Boat option in the toolbar quickly pulls up the chart for your vessel's location.

BoatCruiser allows extraordinary customization of the chart display. It has choices over a wide range of units, including statute miles—needed for two-thirds of the U.S. Eastern Seaboard—which we often use as a litmus test for ultimate flexibility. Users have near infinite control over unit metrics for speed, distance, or depth. A convenient feature lets you globally set all units to either metric, nautical, or imperial for the entire application.

Although BoatCruiser touts high resolution chart printing, we had difficulty with the printing feature. We could print charts, but were unable to print routes as claimed. Sometimes a chart would print directly from the application; sometimes it would only print from Preview. Regardless, the print quality was poor with ragged fonts and fuzzy chart images. The NavSim and BoatCruiser page footers consumed nearly one-quarter of each printed page.

WAYPOINTS AND ROUTES

It is easy to create waypoints, markers, and routes in Boat-Cruiser. In fact, you can even drag-and-drop photos or documents directly onto the chart as a way to geographically organize related files. You can turn off the linked file icon if you need to reduce screen clutter.

BoatCruiser's routes are very easy to build and very visually appealing. You simply string a sequence of clicks. As you move to the next waypoint, a line *rubber-bands* with you,

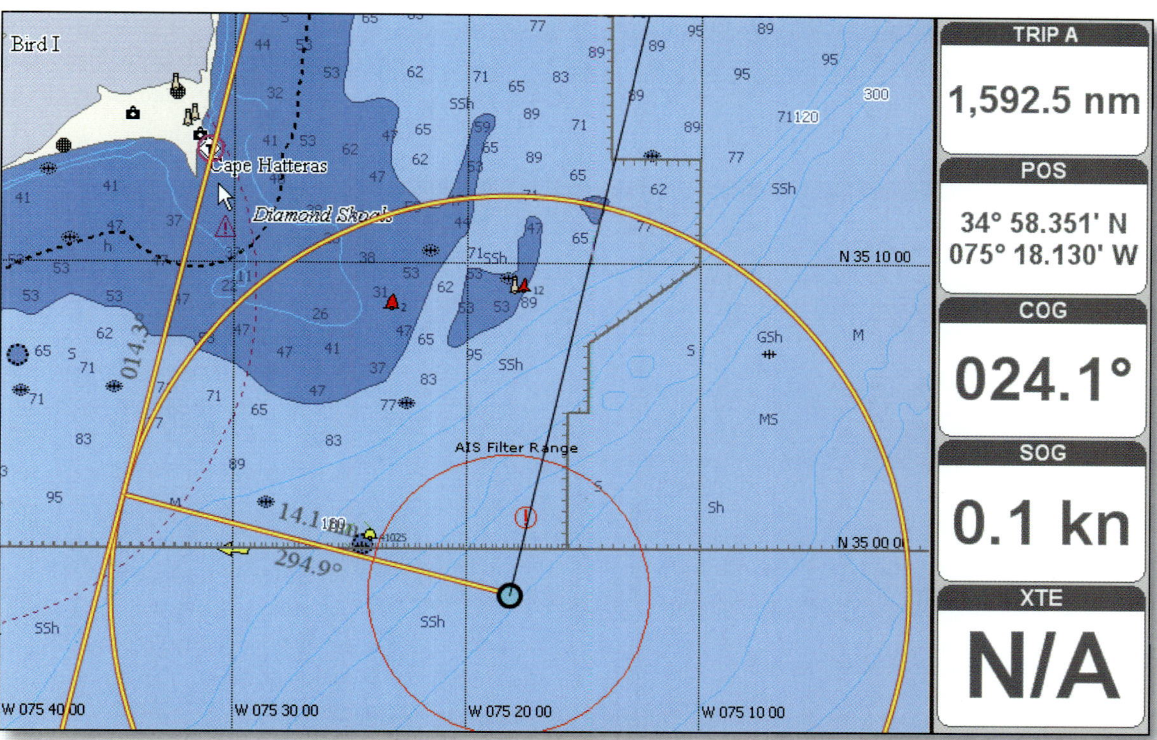

showing the distance and bearing. When you click, the next numbered waypoint is created. Waypoints are marked with vivid easy-to-see yellow circles. The route shows direction, distances, and bearings, and can be easily reversed.

A particularly nice feature is the *Parallel Index Line*. This feature gives you the ability to visually affix a route leg in parallel to the leg you are traveling. This is useful in several scenarios, such as when you are trying to maintain a safe distance offshore when rounding a headland.

All your navigational objects, including charts, routes, markers, zones, and annotations, are managed in one central location, called the *NavManager*. This high-level window is the portal to your navigational assets. NavManager not only lets you locate your objects, it allows you to organize them. For example, you can copy or move charts into different folders, such as a folder for New England and a folder for the Great Lakes. To reduce clutter, you can display only those charts for your current region with the hide and show buttons.

Parallel Index Line: *Rounding Cape Hatteras northbound, a 14-mile Parallel Index Line can be set to assure safe passage.*

You can link photos and documents to a chart by browsing to the file location, then dragging-and-dropping the file directly onto the chart.

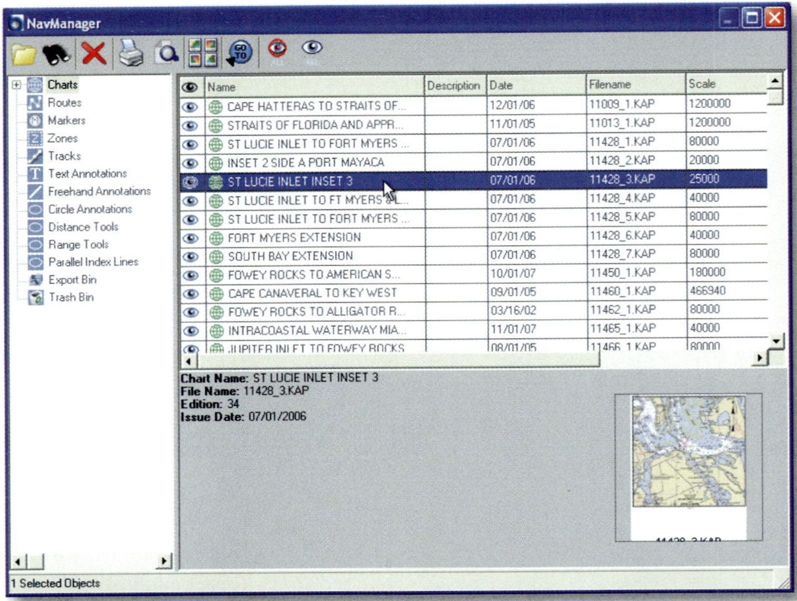

NavManager: NavManager is the main view for charts and user-created objects such as routes, marks, zones, tracks, and annotations.

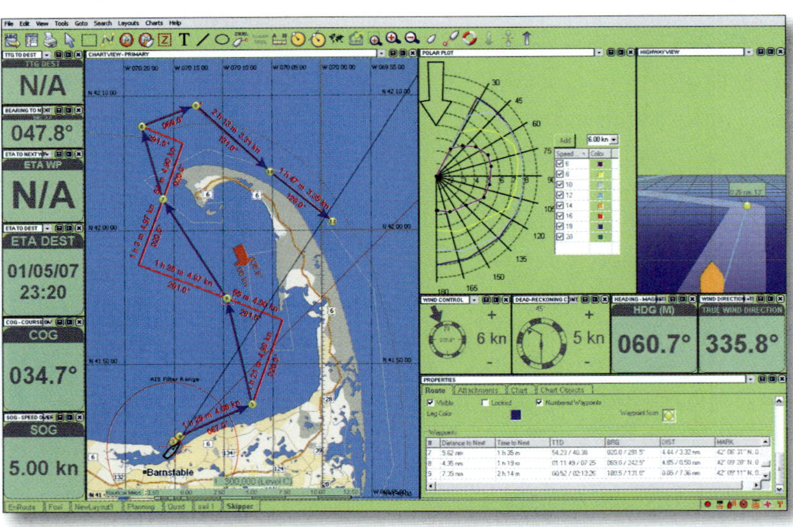

SailTimer: NavSim's sailing version displays the best tacking choices for given waypoints or race marks. Polar plots and boat instrument data complement chart and highway views.

NavManager is an important feature, omitted in many other charting and navigation packages.

If you already have a large collection of waypoints and routes on your GPS or another e-charting package, a fair bit of work is required to import them into BoatCruiser. BoatCruiser supports only two import/export exchange formats: NSO (NavSim Object file) or CSV (comma separated value file). GPX (GPS Exchange Format) is not currently supported. Any import or export process will likely require an additional software utility such as EasyGPS or GPSBabel (www.easygps.com) (www.gpsbabel.org), or transfers using a C-card multimedia reader.

ADDITIONAL FEATURES

BoatCruiser can connect to GPS, AIS, autopilot, radar with ARPA or MARPA, depth sounder, or video camera. In the case of SailCruiser you can also connect wind instruments. You can set alarms for either entry into or exit from an alarm zone, and can drop a Man-Over-Board (MOB) marker.

Tide and current predictions, marina locations, and points of interest data are not provided by NavSim but are integrated through C-Map cartography or by importing Maptech data files. For example, C-Map MAX charts have the option of overlaying colored arrows showing direction and intensity of currents. Unfortunately the C-Map granularity is not very helpful at higher current velocities. Although the color scheme begins with one-knot increments (yellow, darker yellow, orange), all currents between three and ten knots are lumped together with a red arrow. In most vessels, there is a huge difference between a three-knot current and a ten-knot current. Tide stations are shown with a T icon; double-clicking brings up a tidal chart. BoatCruiser does not support GRIB weather downloads.

In addition to the basics, BoatCruiser has several innovative additional features, including their TimeMachine, Fuel Cost and Consumption Calculator, SailTimer (SailCruiser only), and Auto Chart Updater.

TimeMachine is basically a simulation feature that lets you run a course backward or forward in time. With BoatCruiser in Playback mode (as opposed to Real Time mode), trips that were recorded can be reviewed. You can also simulate a passage in a future time frame. The TimeMachine displays daylight information, tidal data, trajectories, and any alarms you have set. A time frame is attached to every event or action that is relevant to navigation. You can quickly alter the time scale of your simulation from a span of months to seconds by scrolling with the mouse or zooming in or out with the "+" or "-" keys.

The Fuel Cost and Consumption Calculator is a great feature for power boaters. You begin by entering your vessel's fuel consumption information and local fuel prices. If you are not sure of your fuel use over all speeds, you can enter your gallons per hour as points on an X-Y graph. BoatCruiser will interpolate between the points to create an estimate of fuel consumption. Estimates of your fuel use and costs are then available by reviewing a route's Properties.

SailCruiser replaces the Fuel Cost and Consumption Calculator with SailTimer, a module developed by Craig Summers of Indepth Navigation (www.indepthnavigation.com). This feature automatically calculates the optimal tacking routes and angles given wind direction and boat speed. The tacking routes are clearly displayed on the chart, with tacks shown as red laylines. SailTimer also calculates Tacking Time to Go to Destination (TTG) and Estimated Time to Arrival (ETA). Polars are also displayed, beginning with default calculations averaged across all sailboats. If you know the performance characteristics of your sailboat in different wind conditions, you can input that information to increase the accuracy of the polar plots.

If you are a sailboat that also motors on long cruises, a recent upgrade includes an option for switching routes between sailing and motoring. When sailing, TTG and ETA are based on tacking and wind conditions. When motoring they are based on straight line calculations.

The Auto Chart Updater is a new feature based on NavSim's MapServer (www.navsim.com/products/mapserver). MapServer started in 2004 as a way for NavSim users to obtain individual BSB charts rather than purchasing an entire Maptech CD. When NOAA and the USACE began distributing those same charts for free through their websites, MapServer lost much of its value.

But, capitalizing on MapServer's utility, NavSim has recently integrated free automated BSB raster and S-57 vector chart updating with their software products. This $49 optional feature directs BoatCruiser to compare loaded chart editions with updated chart offerings and automatically download new editions.

ASSESSMENT

Recreational boaters demand e-charting software that is trouble-free. They are out on their boats for recreation—and do not take kindly to any computer hassles during their limited free time.

BoatCruiser should eventually reach this goal. It is strongly positioned as one of the few navigation packages that supports C-Map cartography and their Auto Chart Updater feature is a blessing. Marketing may be a bit ahead of engineering on many of BoatCruiser's features, but we think in time they'll tune their features to match the chutzpah.

Our optimism stems from NavSim's exceptional responsiveness to user feedback. Frequently, NavSim incorporates user feedback from online forums and emails into a new software release (www.navsim.com/products/history.php). These updates are available free in *perpetuity* for the product! Each release builds on BoatCruiser and SailCruiser, fixing bugs, improving existing features, and adding new functionality. With these constant improvements—and hopefully the commitment to a bona fide User Manual—BoatCruiser should evolve into a world class e-charting product.

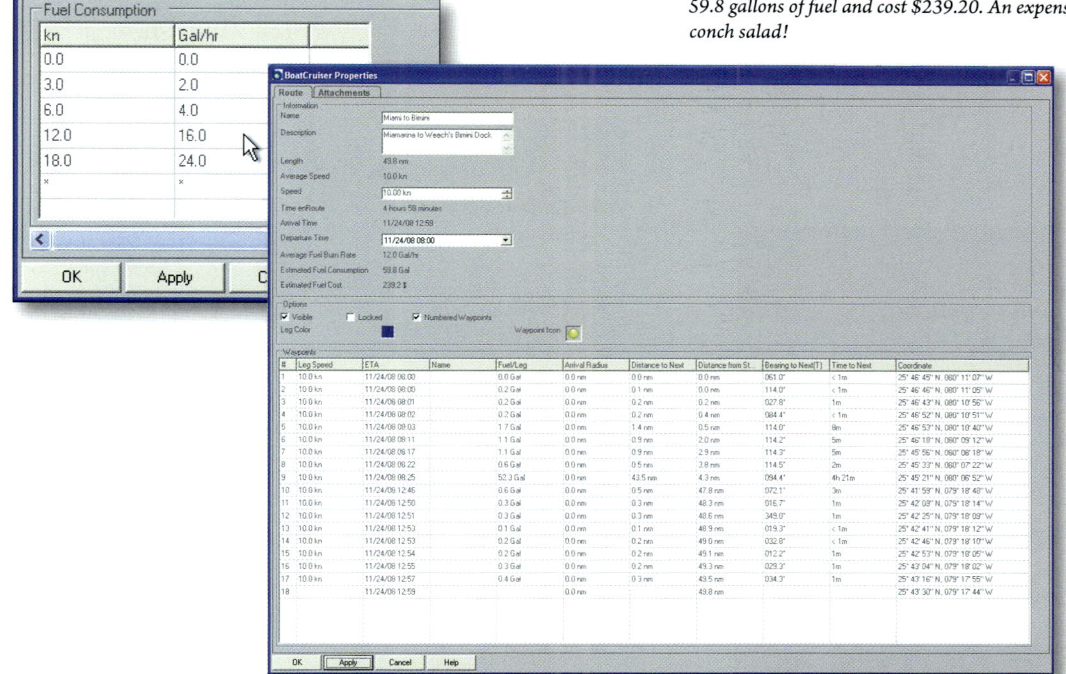

Fuel Data Input: *Fuel consumption may be entered on an X-Y graph or in an Excel-like spreadsheet (left). Both fuel consumption and cost ($6.00 in our example) is entered in* Tool>Options.

Route Detail and Fuel Consumption: *Motoring at an average of 10 knots, the 49.8 nm trip from Miamarina to Weech's Bimini Dock would consume 59.8 gallons of fuel and cost $239.20. An expensive conch salad!*

NavSim BoatCruiser
FEATURES AT A GLANCE

Each full-featured e-charting application has been evaluated according to the following 81 features. Some ratings are binary, "yes" or "no." Others are more subjective, evaluated on a scale from 1 to 5 (5 being "Excellent"). Any other comparison is defined in the Legend.

Out of Box Experience

Packaging	4
Documentation	2
Software install	3
Load charts	2
Load supplemental data	2
GPS hookup	3
Technical support	3

User Interface

GUI metaphor	Win
Screen space management	4
Program responsiveness	3
Chart display speed	4
Chart display quality	3
Customizable GUI	4
Customizable metrics	4
Split windows	No
Tabbed interface	Yes

Basic Features

One-button MOB	Yes
Waypoint creation	3
Route creation	4
Steer-to function	No
Track creation	3
Convert track to route	Yes
Waypoint and route management	3
Chart management	3
Search waypoint and route	4
Range/bearing tool	4
Chart annotation	4
Chart printing	2

Data Exchange and Networking

GPX support for waypoints	No
GPX support for routes	No
Waypoint exchange formats	C
Route exchange formats	C
Integrated GPS transfer	No
Multiple monitor support	No
Network application sharing	No

Cartography

Free U.S. rasters BSBs (RNCs)	Yes
Free U.S. vectors S-57s (ENCs, IENCs)	Yes
SoftCharts	Yes
CHS Canadian rasters	Yes
DNC	No
Seafarer	No
British Admiralty ARCS	No
International S-57s	Yes
International S-63s	No
Scan and geo-reference paper charts	No
Navionics cards	No
C-Map cards	Yes
C-Map CD/DVD	Yes
Satellite geo-referenced photos	Yes
Aerial informational nav photos	Yes
Topographic maps	Yes
Bathymetric data	Yes

Instruments and Sensors

GPS	Yes
Autopilot	Yes
AIS receiver	Yes
Wind	Yes
Depth	Yes
Water temperature	Yes
Radar	Yes
Heading sensor	Yes
Video camera	Yes

Advanced Features

Tides	CM
Currents	CM
Coast Pilot	No
POIs	CM
Streets	CM
Google Earth support	No
Advanced search	4
AIS	Yes
Buddy boat	No
Radar (ARPA/MARPA)	Yes
Radar (display overlay)	No
Fuel calculator	Yes
Transit calculator	Yes
Celestial calculator/Almanac	No
Auto route planning	No
Great Circle Route planning	No
Sail performance	SC
GRIB weather integration	No
Weather options	No
Customizable bathymetric recorder	No

Legend

5	Excellent	CM	C-Map
4	Above average	NV	Navionics
3	Average	SC	SailCruiser
2	Below average	MT	Mr. Tides
1	Poor	PP	Plus Pack
CP	Chartplotter	C	CSV
Mac	Macintosh	E	Excel
NT	Non-traditional	G	GPX
Win	Windows	K	KML
N/A	Not applicable		

Chapter 19
Coastal Explorer

It's a beautiful thing for e-charting when software engineers are also experienced boaters—and that's exactly how Rose Point Navigation Systems, the maker of Coastal Explorer software, was born.

Rose Point was founded in 2003 by Brad Christian, an avid boater and 17-year Microsoft alum, who built the first prototype of Coastal Explorer while cruising Puget Sound. His efforts set a company standard for clean, stable, and efficient code (the entire application takes up just 3.5 MB).

Rose Point continues to be a company of high-tech boaters. Three of the company's principals are long-time boaters—with an average boat length of 60 feet—and they regularly use their own product on the waters of Puget Sound. In fact, Rose Point's development cycle is geared toward testing and modifying the software during the Pacific Northwest's peak boating season.

Unlike many competing companies, Coastal Explorer is capable of serving a wide range of boaters. Weekend or short-season skippers will appreciate the easy-to-use interface. Long-distance cruisers can take advantage of advanced features, such as the ability to read international charts and connect an AIS receiver or MARPA-equipped radar.

In addition to selling Coastal Explorer, Rose Point also licenses its code to other companies, notably Maptech, which sells it under the brand name *Chart Navigator Pro*. Maptech, a large and established navigation company, adds value by in-corporating its extensive chart and data libraries and charges more ($499) for its version.

But the underlying program is the same. In fact, whether you are running Rose Point's Coastal Explorer or Maptech's Chart Navigator Pro, requests for software updates and weather data are routed through Rose Point's servers in Redmond, Washington.

GETTING STARTED

Coastal Explorer can be purchased directly from Rose Point Navigation (http://rosepointnav.com). Alternatively, it is also sold through dealers and select retailers, such as Landfall Navigation, Seabreeze Books & Charts, and Armchair Sailor. A complete list of retailers is available on the Rose Point website (http://rosepointnav.com/CoastalExplorer/Dealers). Versions are available in European Spanish, South American Spanish, French, Dutch, and U.K. English.

The software comes in a boxed set with a CD, DVD, and a 56-page *Exploring Coastal Explorer* booklet, which serves as a Getting Started guide. The CD contains the installation software. The DVD has raster and vector chart data, including all available NOAA raster charts. The vector chart library includes all available NOAA vector charts of coastal U.S. waters; all available vector charts from the U.S. Army Corps of Engineers for U.S. inland waterways; and 250 Digital Nautical Charts (DNCs). According to Rose Point, Coastal Explorer is the only charting and navigation package that includes DNCs, an alternative vector format produced by the National Geospatial-Intelligence Agency that supplements NOAA's incomplete vector coverage.

Although Coastal Explorer includes all available vector charts for the U.S., it does not include NOAA's raster library. Raster charts must be purchased or downloaded separately (see Sources of Electronic Charts in Chapter 6). However, version 2009 (to be released in late 2008) will include a chart update and installation service, which will update or download NOAA raster or vector charts directly through Coastal Explorer's Chart Portfolio.

If you want data extras such as 3D contour maps, shoreline topographic maps, or aerial and navigation photos, then con-

---

Coastal Explorer

Pros: Extremely stable; strict Windows compliance; excellent creation, management, and transfer of waypoints and routes; excellent search capabilities; integrated guidebook information.

Cons: Does not include 3D contour maps, shoreline topographic maps, or aerial and navigation photos; less-than-perfect image quality and chart rendering for raster charts.

Coolest Feature: Gazetteer

Price: $399

Vista Capable: Yes

Version Tested: 2009 Edition

System Requirements
PC with Intel Pentium III
128 MB RAM
500 MB available hard disk space
800 x 600 monitor
CD drive
Windows 2000, XP, or Vista

Rose Point Navigation Systems
16150 NE 85th Street
Suite 210
Redmond, WA 98052
United States
www.rosepointnav.com

Coastal Explorer's default is to automatically load all chart and supplemental data. If you don't need some charting regions or data, review the installation prompts and only load desired regions.

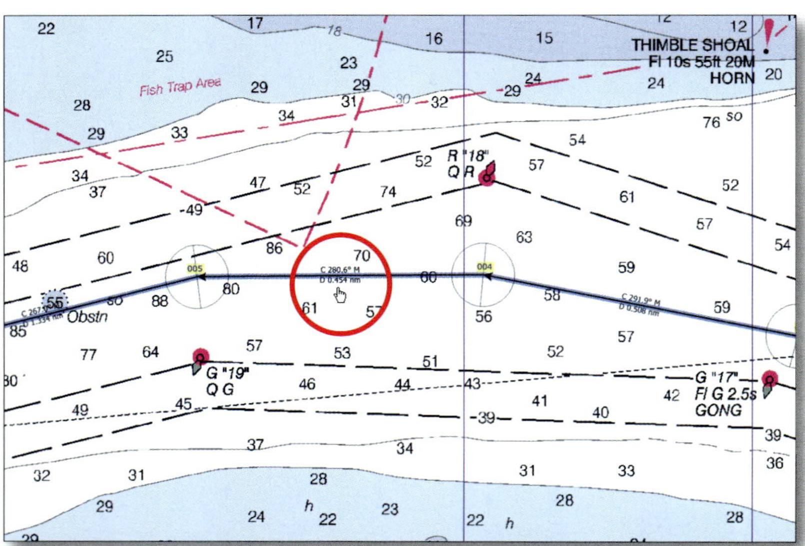

Bearing and Distance: *Following a route entering the Chesapeake Bay from seaward, Coastal Explorer labels the legs between waypoints with course and distance information (red circle and grabber hand).*

sider the branded Chart Navigator Pro version from Maptech. This package includes the same software as Coastal Explorer, with Maptech's additional chart and data libraries. Alternatively, announced updates to version 2009 are scheduled to include support for Navionics cartography, which bundles all these data extras.

If you value a hassle-free installation process, you'll enjoy working with Coastal Explorer. We had no trouble installing the software, databases, or our external devices. In fact, Coastal Explorer's installation is so easy and automatic it attempts to auto-load everything "nautical" on any disc you insert. If you are short of hard drive space—or just want to keep your system lean and mean—intervene and don't load all the chart and data files.

Coastal Explorer does not include an extensive printed or PDF manual. However, there are hot-linked topics through the Help menu. Although it is a bit tedious and time-consuming, we recommend opening and printing each of the eight topics. Other than the manual, Rose Point's website is a bit thin and outdated, particularly compared to Maptech's website on the branded Chart Navigator Pro product.

The foundation of Coastal Explorer's customer support strategy is an application that is extremely easy to use and seemingly impossible to crash. We intentionally pushed its features and never experienced even a lock-up. We suspect their technical support team may feel a bit like the lonely Maytag repairmen. However, if you do need to call, they can be reached by phone or email. Software tips, descriptions of Coastal Explorer installations on user boats, and links to technical articles are also posted on the Coastal Explorer User Network (http://coastalexplorer.net).

Look and Feel

Coastal Explorer is an elegant and carefully-crafted application that pays attention to details. Its careful design is evident in uncluttered screen layouts, quick access to information, and an easy-to-use interface.

There are no floating toolbars or overlapping chart windows, a common problem of many charting and navigation applications. This means your chart remains unobscured as you access features and functions. You don't have to constantly "clean up" your screen display as part of your workflow.

To make the user experience even more efficient, Coastal Explorer operates in two display modes: *Planning Mode* or *Cruise Mode*. Even slicker, the display changes automatically to the most appropriate, based on your vessel's movement (or lack thereof). You can also toggle between the two displays by pressing the [F12] key.

Planning Mode is the normal mode, providing tools for planning such as creating routes or getting information from the guide books. It uses small Windows-style menus suitable for working at home or at the slip.

In Cruise Mode, the toolbars and the windows change to a chartplotter-like display, with windows showing instrument displays and position data (such as range, bearing, speed over ground, course over ground, and cross-track error). These displays are customizable in the Data Console. Cruise Mode displays large easy-to-read buttons and windows of virtual analog-style instruments.

Unlike many charting and navigation applications, where every task seems to require multiple clicks and window displays, in Coastal Explorer most information is literally a mouse-click away. Those of us who use computers extensively have come to expect a streamlined user interface. Coastal Explorer has brought this standard of software engineering into the marine computing environment.

For example, when viewing routes, many e-charting applications require you to physically measure range and bearing from one waypoint to another or have a large route summary window open. But why have a dedicated window of data cluttering an already-too-small laptop screen? Coastal Explorer displays your magnetic bearing and distance directly along your course line, just as you would record it in pencil on a paper chart.

Similarly, all included NOAA paper chart notes are integrated into the chart display. One click on a small black rectangle containing the word NOTE opens that object and alerts you of a cable area or special anchorage.

Like any sophisticated and mature program, Coastal Explorer uses a lot of shortcuts. You can view a list of keyboard shortcuts at the end of the *Exploring Coastal Explorer* booklet. More welcome, Coastal Explorer uses many standard Windows shortcuts. Standardization is particularly important in an application that is used seasonally or only on weekends. Coastal Explorer spares you re-learning a slew of new shortcuts or icons at the start of each boating season or after five days at work using other computer applications. And during those first two weeks of the boating season—when you're still warming up—the multi-level undo capability is great for those of us who click first and ask questions later.

WORKING WITH CHARTS

Coastal Explorer is compatible with the major chart formats, including NOAA RNCs (raster charts), NOAA and international S-57 ENCs (vector charts), and U.S. Army Corps of Engineers IENCs (vector charts of U.S. inland waterways). Coastal Explorer also reads all Maptech digital charts, SoftChart digital charts, NDI Digital Ocean charts, and PhotoNavigator products. Later updates to version 2009 should support Navionics cartography.

We had no trouble panning and zooming over a chart. Coastal Explorer's "chart quilting" seamlessly moves across chart files of different scales. Charts can be displayed north-up, course-up, or heading-up.

Laptop navigating tends to become porthole driving, where you only see the small chart area around your vessel. In addition to the detailed, full-screen chart display, a small window in Coastal Explorer's upper right screen shows the larger overview area. And we particularly like the split chart display option. You can choose to split the main chart display into two or three panels.

Our favorite chart display feature is Coastal Explorer's

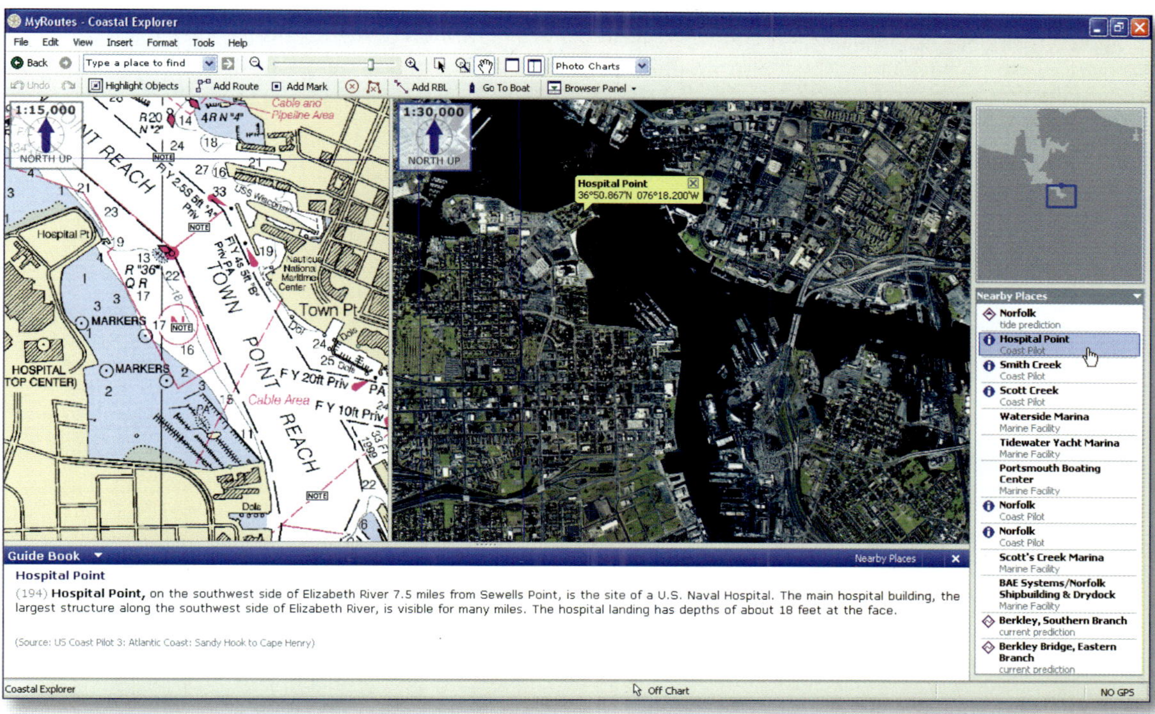

Gazetteer: *Split window functionality allows Coastal Explorer users simultaneous access to raster charts (left), satellite images (right), and Coast Pilot information (bottom).*

search capability. Search functions tend to be the Achilles' heel of many e-charting applications we've reviewed so far. Coastal Explorer allows a complete search of all objects in its database—a staggering five million places and points-of-interest (POIs). For example, type any word or string that relates to data on a chart. We typed "Elliott Island" and the chart for Elliott Island, Maryland popped up. In case we didn't mean this Elliott Island, a small window displays similar options, letting you click on an alternative location.

WAYPOINTS AND ROUTES

The creation of waypoints and routes is very sophisticated in Coastal Explorer. Obviously you can do all the regular tricks: create unlimited waypoints and routes, set boundary circles and areas with alarms, make chart annotations, draw range and bearing lines, and so on. But let's focus on some "wow" features that require getting down to the details.

To make a route, you can simply click along a string of waypoints. This is certainly convenient—and a lot better than

Sometimes NOAA charts do not quilt together very well. To see the chart you want (rather than the one Coastal Explorer has chosen for you), right-click on the chart and choose **Charts Here...** Then click on the chart that you do *not* want included in the quilt.

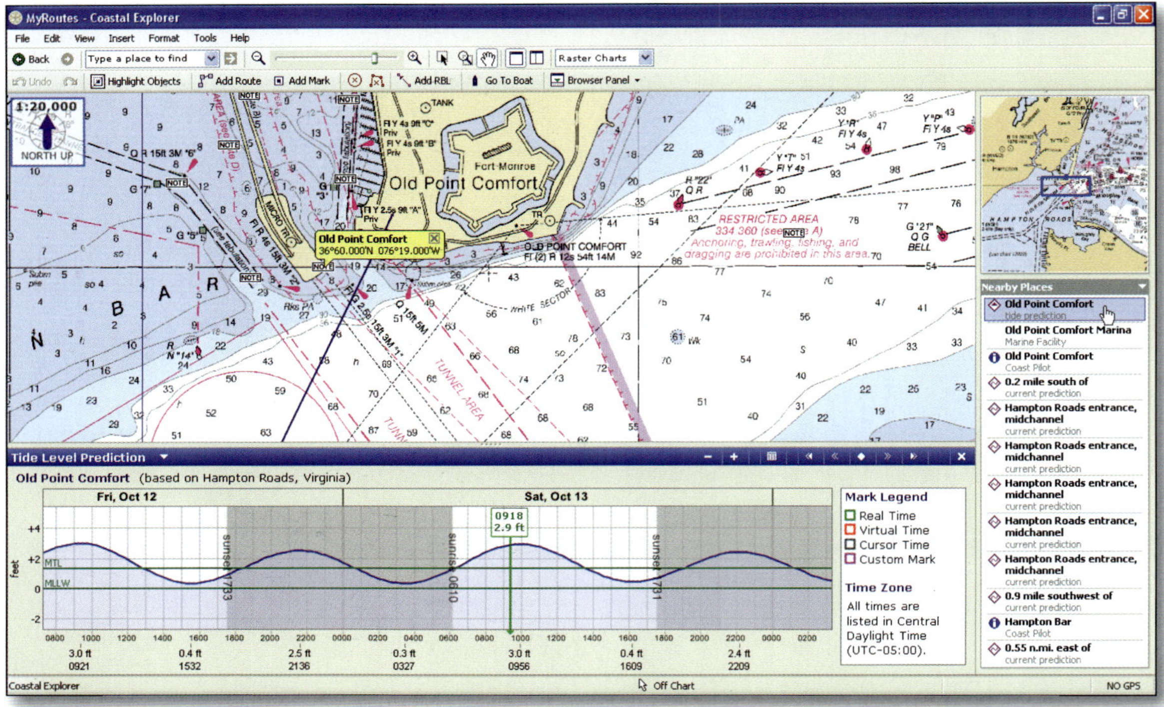

Tidal Predictions: *Current, historic, and predictive tide level information is available and richly displayed in Coastal Explorer.*

Use Coastal Explorer's Route Forking feature when you want to create a new route that uses most of the waypoints in an existing route.

klutzy alternatives such as dragging and dropping waypoint names into a "new routes" folder. But even better, as you string together waypoints and approach the edge of the chart, the cursor flips to a white arrow. Nudge the edge and the chart automatically scrolls so you can continue creating your route seamlessly across displays and charts.

A nice safety feature is the Check Route for Obstacles function. When you create a route, Coastal Explorer can read the data from its navigation object vector database and alert you if your proposed route encounters obstacles such as navigation aids or wrecks. Even more impressive, this feature works even if you are viewing a raster chart. Remember, a raster chart is simply a picture of a paper chart and does not contain any navigation aid, depth, or obstruction linked data. Coastal Explorer allows you to view a raster chart but have access to the navigational object vector database at the same time.

The new Route Timeline Planning Tool, part of the 2009 release, automatically calculates travel time estimates. The timeline tool plots a route on a time scale, incorporating dis-

tance and tidal current predictions. Assigning a waypoint to a particular time automatically adjusts the times of all other waypoints in the route, letting you easily determine the best departure time.

After creating a route, an option called Fork Route lets you reuse this information to build additional routes. This a perfect example of an option designed by boaters who have clearly put in time on the water. For example, you probably have two or three common routes that split or "fork" after you exit your home marina or mooring field. Or, when you enter an anchorage, you may enter from the south and depart to the north. Fork Route lets you create a route, then use that route to build a new route that splits off from any waypoint in a new direction.

ADDITIONAL FEATURES

Coastal Explorer is designed to work with nearly any NMEA 0183 device with an appropriate PC interface. For example, you can connect a GPS, autopilot, depth sounder, water temperature sensor, knot/log, wind speed and direction sensor, rudder angle sensor, AIS receiver, and radar (including ARPA and MARPA equipped models). Version 2009 adds video camera support and expands radar sensor capabilities.

Coastal Explorer does not include the extensive chart and data libraries of the Maptech-branded Chart Navigator Pro version. However, it does include the most commonly used non-chart databases, including tides and currents, weather, and guide book information.

Rather than including a plug-in or separate companion application, tides and currents are fully integrated into Coastal Explorer. A horizontal window shows the tidal current prediction displayed as a graph and marking slack water times and maximum ebbs and floods. We like this display because it is visual and includes specifics for flood or ebb in knots, date, time, degrees true of current flow, time of slack, and maximum ebb and maximum flood.

Obtaining weather data is also exceptionally easy with Coastal Explorer. Rose Point maintains a server with GRIB weather files, allowing you to download weather data directly rather through an email attachment. Simply select your choices of location and data from the Weather Browser window and click Download. You can choose to download all the weather data for your location (GRIB files are not very large), then toggle to display specific information such as sea temperature, wind, or air temperature. The weather data is stored on your hard drive so you can view it underway without an Internet connection.

Coastal Explorer currently includes a free "one year" subscription to its weather server, although the free access has currently been in place for two years. In other words, weather downloads are currently free but in the future there may be a cost associated with renewing the weather subscription.

Coastal Explorer is the only program that integrates NOAA Coast Pilots, geo-referencing this important navigational information so it can be accessed directly from the charts. Some applications include this data, but in the less useful form of PDF documents viewable with Adobe's Acrobat Reader. Coastal Explorer also integrates guide book information on ports-of-call. Click on a harbor with the *Guide Book* browser open and a window opens with a short cruising-guide paragraph on that location. Integrating this data with the charts is what fuels Coastal Explorer's powerful and extremely valuable search engine.

Rose Point's connection with Microsoft comes through in Coastal Explorer's concept of a *documents file*. All waypoints, routes, tracks, and event marks—called "Navigation Objects"—are stored in a special user documents file. This file is not simply an ASCII waypoint text file—a common feature of many e-charting applications—but a file that packages and saves all your navigation data for export.

With the documents file, you can send your navigation information to other users of Coastal Explorer or Chart Navigator Pro simply by selecting Send To. This navigation document is a great way to transfer information to other boaters. For example, a flotilla captain could develop a rich float plan with waypoints, routes, marinas, and attractions, then distribute all this information as a package by email or CD to other flotilla boats using Coastal Explorer or Chart Navigator Pro.

Version 2009 launches *Coastal Explorer Network*, an online service harnessing some of the new Internet trends of user-generated sites. For example, Coastal Explorer Network will include a blog-version Ship's Log, a Community Guidebook, and eventually a Vessel Registry. The Ship's Log lets you drop text or images onto your current track, with the option of making them available to family, friends, or other Coastal Explorer Network users. The Community Guidebook collects photos, text, or Web links for sharing among Coastal Explorer users. And the Vessel Registry, scheduled for a future update to version 2009, let's you associate text or photos with an AIS target.

Assessment

With Coastal Explorer, we've entered the world of full-featured—and more expensive—charting and navigation packages. For example, Coastal Explorer costs twice as much as Fugawi Marine ENC—but in fairness, only a third as much as Nobeltec Admiral MAX Pro or MaxSea Time Zero. But the price difference is not simply a brand name mark-up: Coastal Explorer is a robust, well-designed, and incredibly stable program.

Coastal Explorer is appropriate for a wide range of boaters. Weekend or short-season boaters will love Coastal Explorer's easy-to-use interface. Its standardized and intuitive design lets you minimize time re-familiarizing yourself with the software every time you use your boat. At the same time, Coastal Explorer is full-featured enough for long-distance cruisers, with its ability to read international charts and connect to AIS receivers and MARPA-equipped radars.

If you must have the fancy data extras, such as the 3D bathymetrics, aerial and satellite photos, or shoreline topographic maps, then you'll want the slightly more expensive Maptech-branded version or wait for Coastal Explorer's upcoming support of Navionics cartography. But don't assume that Chart Navigator Pro is exactly like Coastal Explorer. In addition to different chart and data libraries, the differences in technical support, product range, and corporate infrastructure make these two very different product offerings.

NMEA's low baud rate can produce an error message if you try to *repeat* the NMEA information and send it out from your computer. Avoid this problem by limiting what data is sent using NMEA sentence filtering, available in **Tools>Options>Instruments>Filter> Sentences**.

Although Coastal Explorer runs under Microsoft Vista, some information is not saved due to Vista's security settings. You can download a fix for this problem at http://coastalexplorer.net/articles/coastal_explorer_with_vista?qx=1mp43pw.3.

Coastal Explorer
FEATURES AT A GLANCE

Each full-featured e-charting application has been evaluated according to the following 81 features. Some ratings are binary, "yes" or "no." Others are more subjective, evaluated on a scale from 1 to 5 (5 being "Excellent"). Any other comparison is defined in the Legend.

Out of Box Experience

Packaging	4
Documentation	3
Software install	4
Load charts	4
Load supplemental data	4
GPS hookup	5
Technical support	4

User Interface

GUI metaphor	Win
Screen space management	3
Program responsiveness	4
Chart display speed	3
Chart display quality	2
Customizable GUI	3
Customizable metrics	3
Split windows	Yes
Tabbed interface	No

Basic Features

One-button MOB	Yes
Waypoint creation	4
Route creation	5
Steer-to function	Yes
Track creation	3
Convert track to route	Yes
Waypoint and route management	4
Chart management	3
Search waypoint and route	5
Range/bearing tool	4
Chart annotation	4
Chart printing	3

Data Exchange and Networking

GPX support for waypoints	Yes
GPX support for routes	Yes
Waypoint exchange formats	GK
Route exchange formats	GK
Integrated GPS transfer	Yes
Multiple monitor support	No
Network application sharing	No

Cartography

Free U.S. rasters BSBs (RNCs)	Yes
Free U.S. vectors S-57s (ENCs, IENCs)	Yes
SoftCharts	Yes
CHS Canadian rasters	Yes
DNC	Yes
Seafarer	No
British Admiralty ARCS	No
International S-57s	Yes
International S-63s	No
Scan and geo-reference paper charts	No
Navionics cards	No
C-Map cards	No
C-Map CD/DVD	No
Satellite geo-referenced photos	Yes
Aerial informational nav photos	Yes
Topographic maps	Yes
Bathymetric data	No

Instruments and Sensors

GPS	Yes
Autopilot	Yes
AIS receiver	Yes
Wind	Yes
Depth	Yes
Water temperature	Yes
Radar	Yes
Heading sensor	Yes
Video camera	Yes

Advanced Features

Tides	5
Currents	5
Coast Pilot	5
POIs	5
Streets	No
Google Earth support	No
Advanced search	5
AIS	Yes
Buddy boat	No
Radar (ARPA/MARPA)	Yes
Radar (display overlay)	Yes
Fuel calculator	No
Transit calculator	No
Celestial calculator/Almanac	No
Auto route planning	Yes
Great Circle Route planning	No
Sail performance	No
GRIB weather integration	5
Weather options	No
Customizable bathymetric recorder	No

Legend

5	Excellent	CM	C-Map
4	Above average	NV	Navionics
3	Average	SC	SailCruiser
2	Below average	MT	Mr. Tides
1	Poor	PP	Plus Pack
CP	Chartplotter	C	CSV
Mac	Macintosh	E	Excel
NT	Non-traditional	G	GPX
Win	Windows	K	KML
N/A	Not applicable		

Chapter 20
The Capn

You can't understand The Capn software—the first computer navigation program and long an industry standard—without understanding the story behind it. As with all "Captains," there is a story, and like many salty tales this one sets sail on the coast of Maine.

It begins with Dennis Miller, a boater and former journalist who wanted to create a computer program to help mariners calculate and display lines of position from celestial sights. This was nearly 20 years ago, when most people didn't own a personal computer. The project literally began as a computerized version of the *American Practical Navigator*, so he called it CAPN, short for Computerized American Practical Navigator.

At the same time, the National Oceanic and Atmospheric Administration was beginning to experiment with electronic versions of their paper charts. As described in Chapter 2, *The History of Global Position Finding*, NOAA was looking for a private company partner to release the charts for sale to boaters. After considering 37 bids, the winner was a company in Bangor, Maine, named BSB Electronic Charts, who produced spiral-bound collections of paper charts called ChartKits. (If you think this sounds a lot like Maptech, you're correct; BSB Electronic Charts was eventually acquired by Maptech.)

With charts now available in electronic format, Mills added a revolutionary feature to his CAPN program. He included the ability to plot a vessel's position directly on an electronic chart display. His company, Nautical Technologies, was one of the first companies to take advantage of NOAA's electronic charts. And The Capn became the first commercially available Windows-based charting and navigation program.

Today, more than fifteen years later, The Capn is an aging player in a field full of modern products. But there are still about 8,000 registered Capn users, and the program is used by the U.S. Coast Guard, Navy, and Marines, as well as pilots, tug operators, merchant vessels, and commercial workboats.

It also was very popular among pirates. The Capn was created early in the PC revolution—long before today's technology of registration, licenses, keys, and dongles. Retired Coast Guard boaters introduced recreational boaters to The Capn, which was then the only show in town. Lacking anti-piracy features, recreational boaters freely copied and passed around the software and its cartography. During the 1990s many cruising vessels in the Caribbean carried a pirated copy of The Capn, courtesy of another cruising buddy boat.

Although this piracy was economically fatal to Nautical Technologies, it created a huge base of Capn users and generated incredible market momentum that continues to this day. Capn users love this application. When Maptech, its current owner, merely hinted at killing the software, the public outcry was vehement—including threats to never again purchase *any* Maptech products. Maptech, aware of the bootlegging problem, granted an "amnesty program" of sorts. Capn users—even without a serial number or replacement CD—could call Maptech and order the upgrade, in effect receiving the complete Capn software and data portfolio set for $149.

Maptech currently insists there is no threat to The Capn's future. In fact, since Maptech picked up The Capn in April 2006 (at version 8.3.14), they've posted several moderate updates, known in the software industry as point releases.

However, as Maptech clearly states, The Capn is tailored to and intended for the commercial market of professional captains. The product ships in a case labeled *Navigation Software for the Professional*. Although its devoted recreational users point out that The Capn has some great features and functions, a new user entering the charting and navigation market would be better served by a package designed for rec-

The Capn

Pros: Highly customizable interface and metrics; transparent raster-over-vector chart display; reads Maptech BSB5 raster charts; multiple GoTo features to locate charts; included User Manual.

Cons: Dated user interface; does not natively support free NOAA ENC vector format; does not display Maptech NavPhotos, topographic maps, or contour charts; limited import and export of waypoints and routes; no weather download.

Coolest Feature: Transparent Raster with Hot Spots

Price: $449.95

Vista Capable: No

Version Tested: 8.3.20a

System Requirements
PC with Intel Pentium (or equivalent)
10 GB available hard disk space
800 x 600 monitor
DVD drive
Windows XP

Maptech
10 Industrial Way
Amesbury, MA 01913
United States
www.maptech.com

If you are having problems getting The Capn to find your GPS, update to the latest version (8.3.20a), which has an improved tool to search for GPS connections and scan a broader range of COM ports.

Raster BSB and Vector DNC: *The Capn's split window capability shows a BSB raster chart (left) and a vector DNC chart (right). Any vector chart object (R4 in this example) may be queried, with important Light List data immediately available.*

reational use on a more contemporary computer platform, such as Maptech's Chart Navigator Pro.

Getting Started

The Capn is sold as a very professional package in a 9 x 12-inch zippered nylon portfolio containing 18 discs and a 64-page comb-bound User Manual. Seventeen DVDs contain complete Maptech cartography for the U.S., including coverage of the Great Lakes and inland waterways. The User Manual is an updated version, recently revised and rewritten by Maptech, who also makes it available through their support web page (ftp://ftp.maptech.com/downloads/Capn_Manual.pdf).

The application easily installs from the DVD installer disc. Once the program is installed, you can load charts by region from the remaining DVDs. As with Maptech's Chart Navigator Pro, The Capn provides more charts and supplemental data than most boaters should load. If you don't need some charting regions or data, watch the installation prompts and only load desired regions.

Today's version of The Capn requires activation by calling a toll-free number or sending an email to a designated technical support email address. When you provide Maptech with your serial number and machine code they issue an unlock code. Although a phone registration process may seem outdated and inconvenient compared to an automated Web registration, it makes sense for some users. Designed for captains who are out on a vessel, possibly in bad conditions after a system crash, it is assumed that an Internet connection may not be available. Maptech makes the process easy. Our registration was complete in less than two minutes with a representative.

Technical support is abundant through Maptech's support web page, which includes technical documents, various downloads, and discussion forums (www.maptech.com/support).

Look and Feel

The Capn uses a traditional Windows interface that integrates large plotter-style buttons for use in rough sea conditions. This is a nuts-and-bolts, no-frills user interface, but one that is relatively easy to learn and use.

Unfortunately, the pioneer in e-charting is starting to show its age. It has a definite Win95 look and feel, most noticeable in a cluttered screen display and ragged fonts. There are only 14 rather non-intuitive shortcuts, most based on function keys such as [F4] for Next Chart Down.

In addition, The Capn occasionally exhibits display artifacts because its monitor refresh was designed for the slow computers of the 1990s. For example, portions of the previous feature occasionally remain on the screen as a new window loads, creating an odd temporary patchwork on your PC display. Newer applications take advantage of graphic display speed and clear the screen instantaneously before re-displaying a new window.

The Capn also frequently violates Windows interface conventions and is inconsistent within its own interface, such as whether to use an ellipsis ("...") indicating upcoming submenu choices. The most troubling instance of a "close-but-not-quite" Windows metaphor is Capn's inability to resize data windows. Since windows of waypoints or routes can't be resized, the information is only legible by scrolling side-to-side or up-and-down with horizontal and vertical elevator bars. At a minimum, users should be able to make windows wider to show complete lines of data.

To its credit, The Capn loads faster than any program we've reviewed, starting up in less than five seconds. It's also one of the most customizable charting and navigation programs we've seen. It includes a proverbial "boat-load" of customization options, such as metrics, colors, ship information, data inputs, Nav Console settings, and even time zones.

Working with Charts

Countless times at boat shows and seminars, we've been asked, "What charts can I use with my Capn software?" There seems to be quite a bit of confusion about the compatibility of The Capn with Maptech and non-Maptech raster and vector chart formats. As it turns out, the confusion is warranted: the answer is a bit complicated.

The Capn does work with a wide assortment of raster charts, including the free NOAA RNCs, SoftCharts in GEO/NOS format, and Maptech's BSB formats. In fact, The Capn is one of the few Maptech software offerings that reads Maptech's own encrypted BSB5 format, which are reproductions of its paper ChartKit pages that include route lines and waypoints. However, The Capn does not read the common international raster format, British Admiralty (ARCS).

The Capn version 8.0 added "support" for vector charts in S-57 and DNC format. Recall that DNCs are Digital Nautical Charts, vector format files produced by the National Geospatial-Intelligence Agency and used by the military.

The confusion arises because users of The Capn are unable to download and use standard vector chart formats—such as standard NOAA ENCs, Army Corps IENCs, international S-57s, or DNCs—straight from NOAA, the Army Corps, or the National Geospatial-Intelligence Agency. In order for these files to be read by The Capn, they must be converted to a proprietary format that Capn reads. In other words, the ENCs and DNCs that are provided on the DVDs are actually *SENCs* and *SDNCs*, vector charts that have been converted to an "S"

(for system) format. Unfortunately, without this Maptech proprietary conversion step, standard vector format charts cannot be read by The Capn.

Although this proprietary format forces Capn users to purchase their vector charts through Maptech, there is a value-add. With traditional ENC and IENC naming conventions, such as those employed on downloaded NOAA chart files, it is difficult to identify a chart by its file name because those file names do not reference the numbers on the equivalent paper chart. As part of Maptech's conversion, they re-name each vector chart to reference its paper chart counterpart. For example, the vector chart file for Mobile Bay (US4AL11M) can easily be found by searching for its paper chart number 11376. The Capn SENC file is labeled 11376-US4AL11M.

When switching between North Up and Course Up, be sure to adjust the settings in *two* places: Options>CAPN Preferences>Charts and by right-clicking on your chart and choosing Rotate Chart To...

SoftChart Color Palette: *Proceeding seaward along a predefined route, The Capn's split screen shows the Merrimack River Entrance (left, using the SoftChart color palette) and a mated 3D bathymetric (right).*

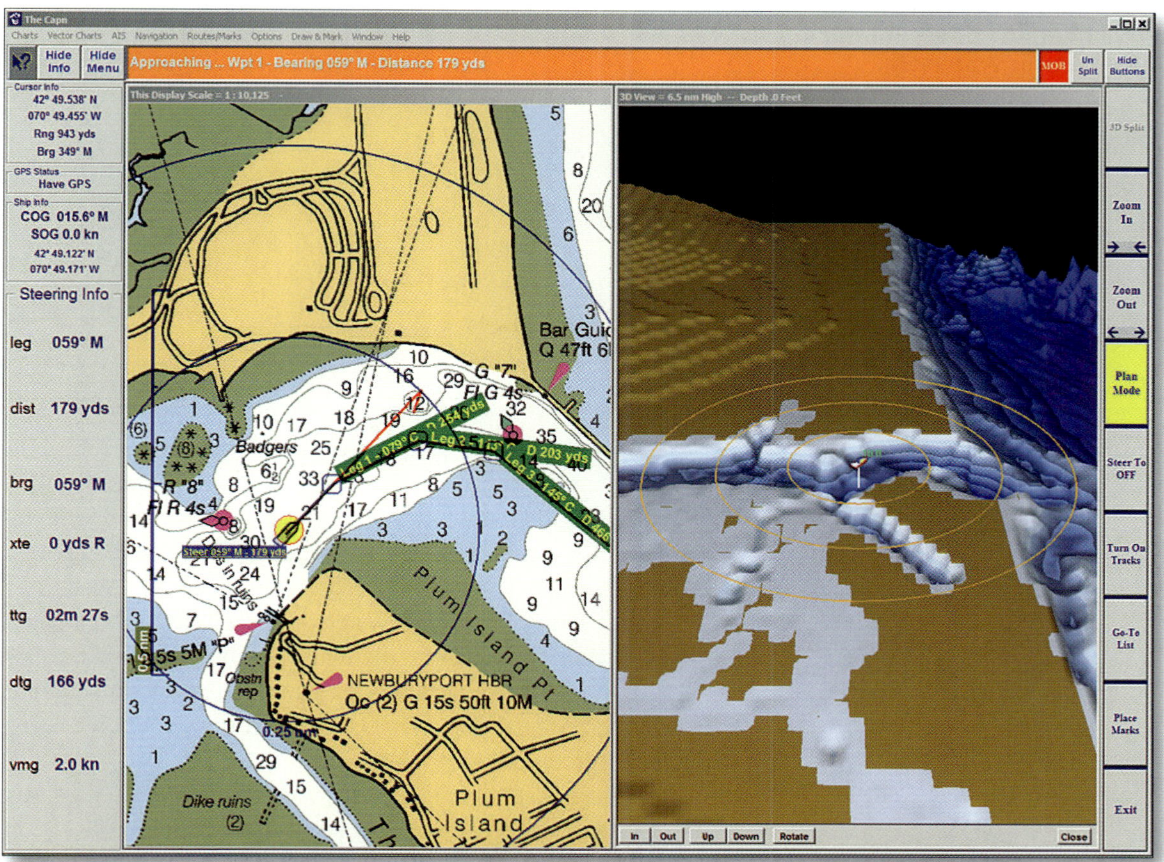

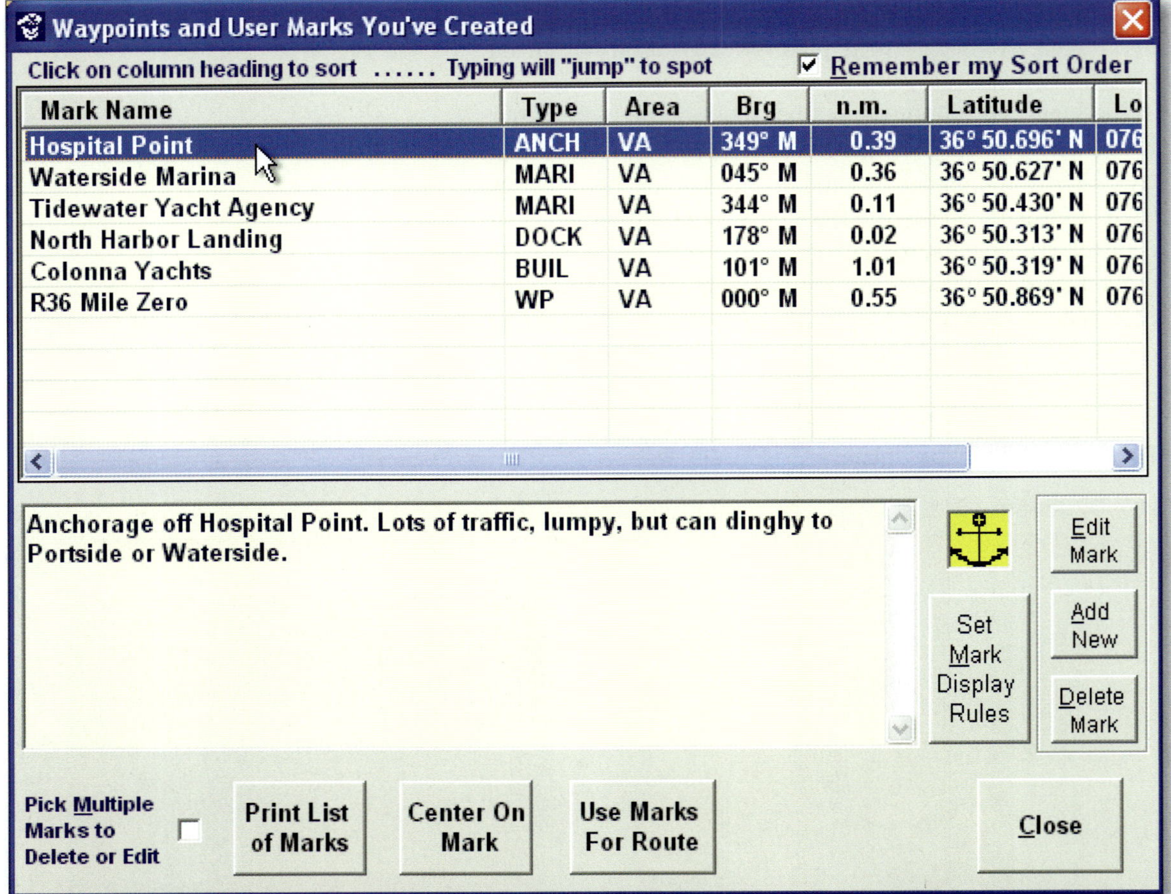

Mark Library: *Waypoints and marks are stored in a library table which can be sorted by column heading. To quickly GoTo a charted location, highlight a way-point or mark and select* Center On Mark *(bottom center).*

Although Capn warns you this option may sacrifice display speed, we didn't notice any negative impact. The new faster computers allow for regular use of this helpful feature.

The Capn also has an option to Use High Visibility Colors. This feature will be familiar to users of SoftCharts, because it is the SoftChart palette. The SoftChart look was always popular for its high legibility in bright light.

Like most advanced charting and navigation applications, The Capn integrates intelligent chart display, which it calls *Smart Chart Logic Technology*. Charts are seamlessly quilted—unless you choose to turn this option off in order to view the chart edges. Capn also automatically selects the best scale chart for the current zoom and location. Charts can be printed, with very nice printouts including customizable scale marks (graticules). Charts can be displayed in a split screen, such as a chart on one side of the screen and 3D bathymetric view on other side.

Panning and scrolling a chart is a bit inconvenient in The Capn, largely because of the lack of a conventional grabber hand. You must pan a chart by zooming out, left-clicking until the new area is reached, then zooming in. Most contemporary applications now use a grabber hand as the convention to pan and scroll.

However, Capn has a very nice GoTo feature to move across charts. To Capn's credit, the program includes distinct commands for "Go To" (pick a waypoint and bring up that chart) versus "Steer To" (pick a waypoint and create a route to it). Many applications combine or confuse these two distinctly different actions.

The GoTo feature makes it very easy to pull up a new chart location. You can flip to any region where you have created a waypoint or route. Alternately, you can flip to a new chart by entering a latitude and longitude position. A very nice option is Go To>General Location, which brings up a window of named chart locations for waters in the U.S. and Bahamas. For example, the Bahamas window lists 1,448 locations. Scrolling and selecting "Marsh Harbor" instantly brings up the best chart for that port, centered on your screen.

The General Location feature would be even stronger if it included a search box. Here again, Capn shows its age by not

As we've noted in other charting and navigation applications, there are some dated assets on the Maptech DVDs. Marketing departments like to shovel the maximum amount of assets onto the discs, even if the data is a bit stale. Maptech, with its front-line access to chart data, does better than most, but we found the Inland Waterway raster charts were dated back to 1998. To be fair, there may not be newer editions for that region, but consumers of bundled cartography should recognize that today's DVDs of charting assets don't necessarily represent today's data.

By using some of The Capn's special features, charts display very crisply. For example, a Use High Resolution Display option is available to display charts with much better rendering.

including this now-standard feature. However, a workaround is to click on any location name (as if you are about to edit it), then start to type the location you're searching for. The window will jump to the name you are typing.

One of The Capn's best features is Allow Transparent Raster, which overlays a transparent raster chart over a vector chart. This feature lets you view a more familiar-looking raster display while querying objects from the underlying vector chart. It's also a nice feature to check if the two charts agree, such as on a shifting shoreline. The option Show Feature Hot Spots helps you see where to click on the raster chart to invoke the underlying vector object. With Hot Spots on, vector chart objects display a red dot so you know where to click on the raster chart. For example, a raster chart may show a green dayboard square near a depth reading. The Hot Spot ensures you don't inadvertently click on the depth reading if you wanted the dayboard information.

The Capn allows you to customize your vector chart displays, including display layers, depth metrics, shaded areas to warn of unsafe depths, and maximum soundings. It also has an option to Favor Vector over Raster, which displays the vector chart whenever one is available, otherwise reverting to a raster chart.

WAYPOINTS AND ROUTES

It is very easy to create waypoints and routes in Capn. You can also create a *mark* to pinpoint an anchorage or special fishing location. Photos can be linked to these marks.

Routes are displayed with course and bearing, although the text is a bit difficult to read because of transparency, overtype, and ragged fonts. The Capn can record a track, but cannot create a route from a track. Unfortunately, it uses basic waypoint and route management, placing data in one collection rather than subfolders defined by the user.

Waypoints and routes are not freely importable or exportable. Version 8.0 added the ability to export routes only to a GPS using GPX format and a GPS transfer utility such as EasyGPS (described in Chapter 8). Capn users can, however, import and export data to other Capn users.

The Capn also includes many alarm features, including visual and audible alerts for waypoint approach or arrival, alarm zone approach or entry, man overboard, anchor watch, cross-track error, and GPS malfunction.

The Capn's Trip Calculator is a nicely designed feature. It not only calculates your trip data, but you can use it in reverse. By beginning at the last waypoint it will calculate departure times or speeds needed to make a restricted bridge opening or to arrive at a location before dark.

Underscoring The Capn's commercial orientation, it will also calculate the area inside a route, in metrics ranging from square meters to nautical square miles. This feature is useful for search and rescue operations, commercial fishing, or lobstermen if they'd like a quick calculation of their trap field area.

Circumvent The Capn's weak search capability by using the edit tool as a surrogate for a search tool. Click on any location name (as if to edit it), but then type the first letters of the location you're searching for. The window will jump to the best match of your typed characters.

Integrated Predictions: *Tide and current data is provided within The Capn. Stations are shown on the chart (T and C) and the data plotted in table form (inset).*

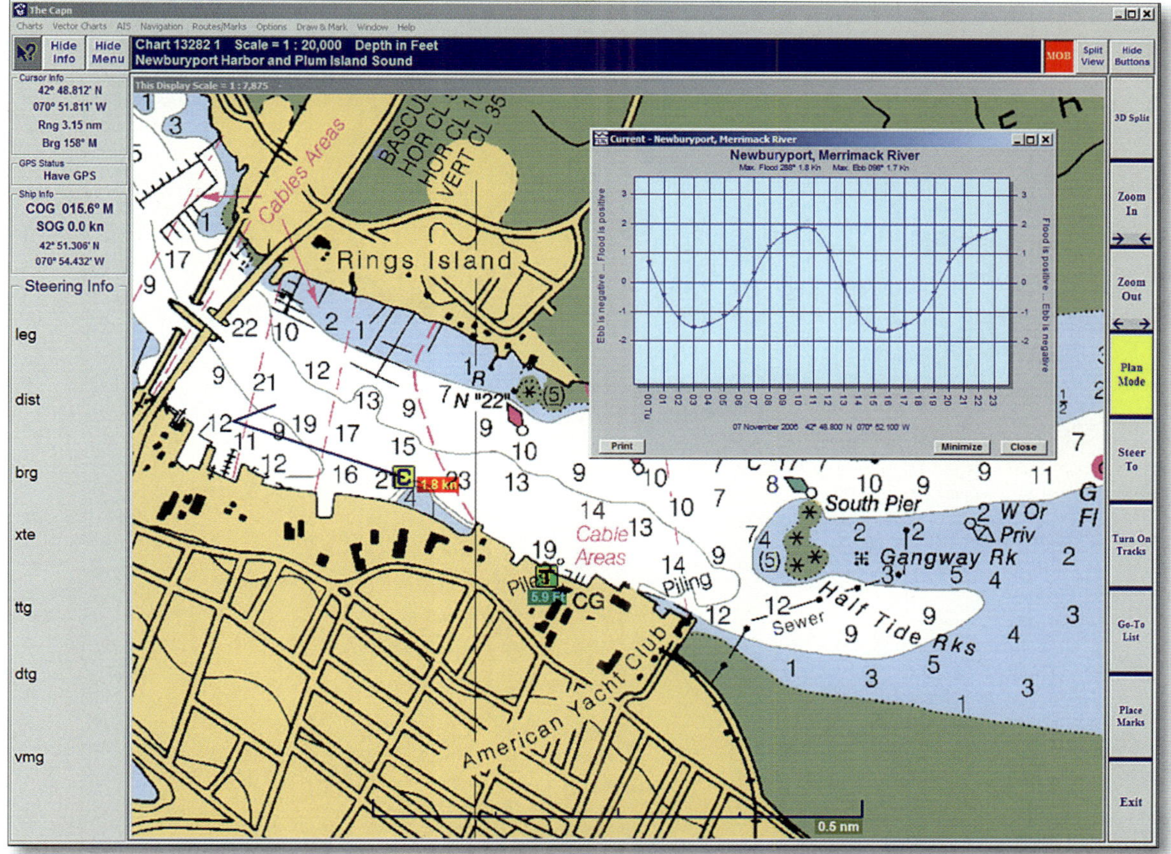

The Capn includes additional marine data, such as tide and current predictions, a nautical almanac, and Coast Pilots. It does not support weather downloads.

Tide and current predictions are displayed in small pop-up windows that open on top of the chart. The interface is a bit dated in the sense that many other programs show tides and currents for the vessel's current time and location. In The Capn, the display is a traditional tide or current graph for a 24-hour period and you must cross-reference a particular time.

However, it is very easy to access present tide and current predictions with the Choose Tide Station or Current Station features. With these features on, all tide or current stations within 30 nautical miles of your vessel are shown on the chart with a T or C. Right-click a station and view that station's tide or current table in a very concise shorthand format, such as "15:38 0.8 ft (Rising)." One easy improvement would be to include the tidal range, such as "15:38 0.8 of 3.2 ft (Rising)." Unfortunately, invoking tide and current predictions severely slowed down The Capn's performance. You cannot feasibly run Capn with tides and currents enabled. Only enable this feature to obtain the information you need, then disable it immediately.

The Capn's nautical almanac, reached by choosing Navigation>Rise and Set Times, brings up a window that calculates the rise and set times for the sun and moon (for that day or month), meridian passage of selected planets, and a list of star positions.

Coast Pilots can be displayed if you choose to load them. Once loaded, choosing the feature Go To>Coast Pilot Page will open the Coast Pilots in a large text window. There is limited search capability, but you can access the index to locate relevant pages.

To compensate for your compass deviation, The Capn includes a Compass Deviation Card. This window lets you fill in your vessel's deviation correction, allowing the software to display accurate course routes. Fill in the deviation for the four cardinal points and The Capn automatically calculates the intermediate values in 15-degree increments.

The Capn interfaces with many external devices, including an autopilot, GPS, compass, depth sounder, and video camera. The Capn has a standard AIS interface, including a position report on a vessel showing a ship's status, course over ground, heading, speed, rate of turn, and MMSI number for Digital Select Calling. Right-clicking on a target provides Full Ship Info with more details on the vessel. To reduce clutter on your chart display, such as in a busy harbor with many ships, the tracks and closest point of approach lines can be turned off.

As we've said, The Capn includes many features not intended for recreation boaters. For example, it can set watches for fishing areas or calculate the length of your vessel and tow. Licenses can be purchased in quantity for fleet manage-

AIS Targeting: All AIS targets are shown in a list (inset). Right-click on any charted target to get Full Ship Info (not shown).

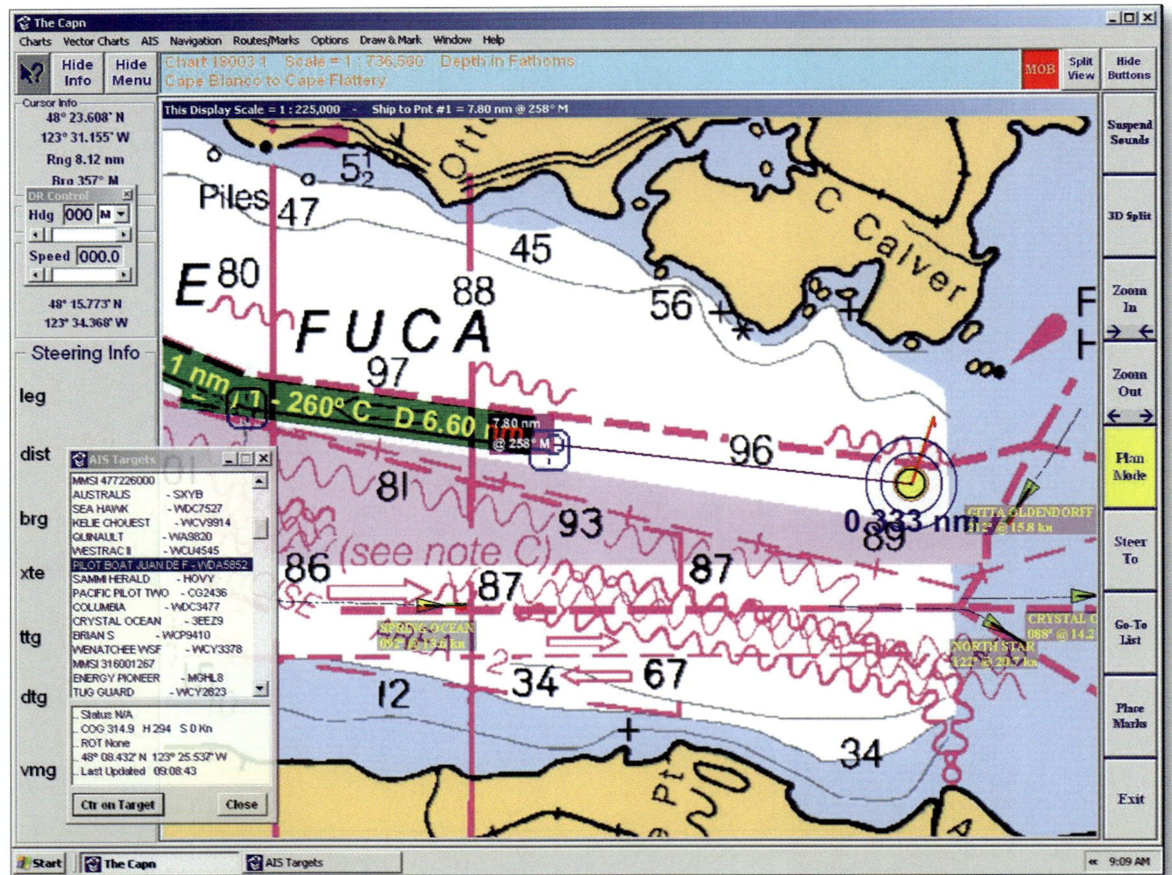

ment. Maptech will also customize your Capn purchase to add special features or functions, or customize your charts with information overlays, specific to a commercial vessel's needs. All these additional features are reminders that The Capn is designed and marketed for the commercial or government user.

Assessment

Devoted users of The Capn can breathe a little easier. It seems there are no longer any assassins lurking in Maptech's corridors. In fact the United States Power Squadrons have recently announced that beginning Spring 2008, The Capn will be taught in their most advanced navigation class. Students not only receive instruction on The Capn, but also take home a personal copy for their own use.

And although The Capn's founder, Dennis Mills, has retired after a stint with the company as product manager, Maptech has responded to consumer demands, recently posting several updates and a revised manual.

The Capn is a jewel—and truly the first e-charting application—but it is showing its age. It remains a powerful, feature-rich program and should improve with access to Maptech's extraordinary resources: technical support, sales channels, development team, and charts and supplemental data. But Maptech must invest in this product or it will eventually die. Its interface and features require a serious dusting-off.

Despite Maptech's stated direction for The Capn as a program geared for government and commercial vessels, today it is used by two groups of boaters: recreational users who learned on the program and remain loyal, and commercial captains. Both groups want a simple stable charting and navigation application.

If you're a recreational boater and already own The Capn, it's still a great program. You should take advantage of the latest updates to the software and cartography that Maptech offers for $149. However, if you don't already own The Capn, or if you want to switch to a more contemporary application specifically geared to recreational users, a better choice is Maptech's Chart Navigator Pro.

The Future of Maptech

At press time, Maptech was for sale—including its paper charts, guide books, Captn. Jack's mail-order website, electronic cartography, and e-charting software such as The Capn and Chart Navigator Pro.

Until recently, Maptech was owned by Gary Comer, the founder of Lands' End clothing. An avid boater, Comer enjoyed the Maptech business venture, perhaps even at the expense of profitability. According to the boaters' coconut telegraph, Maptech was more of a passion of Comer's than a money-making venture.

When Comer passed away in 2006, the family trust inherited ownership of Maptech. Not surprisingly, there is little incentive to maintain the company for sheer boating interest. Maptech is absolutely not closing its doors, but change is certainly on the horizon.

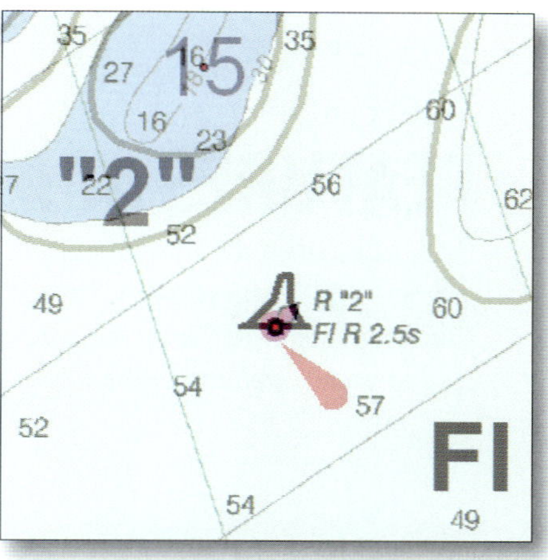

Transparent Raster and Hot Spots: *The Capn allows you to view a transparent raster chart over a vector chart. With Hot Spots turned on (a small red dot appears on vector navigation objects), you can accurately click to query feature information from the vector database.*

The Capn
FEATURES AT A GLANCE

Each full-featured e-charting application has been evaluated according to the following 81 features. Some ratings are binary, "yes" or "no." Others are more subjective, evaluated on a scale from 1 to 5 (5 being "Excellent"). Any other comparison is defined in the Legend.

Out of Box Experience

Packaging	5
Documentation	3
Software install	3
Load charts	4
Load supplemental data	4
GPS hookup	4
Technical support	4

User Interface

GUI metaphor	Win
Screen space management	3
Program responsiveness	3
Chart display speed	3
Chart display quality	2
Customizable GUI	4
Customizable metrics	4
Split windows	Yes
Tabbed interface	No

Basic Features

One-button MOB	Yes
Waypoint creation	3
Route creation	3
Steer-to function	Yes
Track creation	3
Convert track to route	No
Waypoint and route management	3
Chart management	3
Search waypoint and route	4
Range/bearing tool	4
Chart annotation	No
Chart printing	4

Data Exchange and Networking

GPX support for waypoints	No
GPX support for routes	Yes
Waypoint exchange formats	None
Route exchange formats	G
Integrated GPS transfer	No
Multiple monitor support	No
Network application sharing	No

Cartography

Free U.S. rasters BSBs (RNCs)	Yes
Free U.S. vectors S-57s (ENCs, IENCs)	No
SoftCharts	Yes
CHS Canadian rasters	Yes
DNC	No
Seafarer	No
British Admiralty ARCS	No
International S-57s	No
International S-63s	No
Scan and geo-reference paper charts	No
Navionics cards	No
C-Map cards	No
C-Map CD/DVD	No
Satellite geo-referenced photos	Yes
Aerial informational nav photos	No
Topographic maps	No
Bathymetric data	Yes

Instruments and Sensors

GPS	Yes
Autopilot	Yes
AIS receiver	Yes
Wind	Yes
Depth	Yes
Water temperature	Yes
Radar	No
Heading sensor	Yes
Video camera	Yes

Advanced Features

Tides	4
Currents	4
Coast Pilot	4
POIs	No
Streets	No
Google Earth support	No
Advanced search	4
AIS	Yes
Buddy boat	No
Radar (ARPA/MARPA)	No
Radar (display overlay)	No
Fuel calculator	No
Transit calculator	Yes
Celestial calculator/Almanac	Yes
Auto route planning	No
Great Circle Route planning	No
Sail performance	No
GRIB weather integration	No
Weather options	No
Customizable bathymetric recorder	No

Legend

5	Excellent	CM	C-Map
4	Above average	NV	Navionics
3	Average	SC	SailCruiser
2	Below average	MT	Mr. Tides
1	Poor	PP	Plus Pack
CP	Chartplotter	C	CSV
Mac	Macintosh	E	Excel
NT	Non-traditional	G	GPX
Win	Windows	K	KML
N/A	Not applicable		

Chapter 21
Nobeltec VNS MAX Pro

Dave Steckler and Jay Phillips, two former Microsoft engineers, founded a company in 1993 to pursue their dream: creating a PC-based software program that could display a vessel's position on electronic navigation charts.

That company was Nobeltec and the two software products it developed remain at the forefront of today's market, despite increasing competition from more than a dozen companies.

In the late 1990s Nobeltec made two important strategic moves, positioning the company as a leader in both navigation software and vector cartography. On the software front, it released a major new revision called Visual Navigation Suite. On the cartographic front, it inked a deal with the Russian company Transas, giving Nobeltec an exclusive license to the firm's coveted trove of world vector data. This deal made Nobeltec the first company to offer worldwide vector cartography in a mainstream recreational software product.

In 1999, Nobeltec was purchased by Jeppesen, a subsidiary of Boeing and a major producer of paper and digital aviation charts. However, the marine products retain the recognized Nobeltec name. Continuing the strong connection between software and cartography, Boeing recently purchased C-Map, a leading provider of digital maritime data. This purchase led to a complete re-vamping of Nobeltec software products, culminating in the release in Spring 2008 of VNS MAX Pro and Admiral MAX Pro.

Probably no other charting and navigation software has the market presence of Nobeltec. Most marine retailers carry Nobeltec software and its recently-acquired C-Map cartography. What you get is a one-stop shopping system of software and cartography that grows with the needs of a long-distance cruiser. But the decision to use VNS MAX Pro (or Admiral MAX Pro) represents a commitment. Nobeltec software is designed to work with C-Map charts and is not always compatible with other charting systems. Furthermore, VNS software starts on the higher end of the price spectrum ($490) and the cost of a complete system—with additional chart coverage, subscriptions, and add-ons—will likely climb into the thousands of dollars.

Getting Started

VNS MAX Pro can be purchased through nearly any marine retailer, as well as directly from Jeppesen Marine through its extensive online store (www.nobeltec.com/store).

For boaters who owned VNS in version 9 or earlier (the Passport chart versions), Jeppesen Marine offered an extremely unusual and generous upgrade in an attempt to move all Nobeltec users onto MAX Pro software with C-Map charts. For several months, owners of VNS version 9 received a complimentary upgrade to the new VNS MAX Pro, a copy of the Raster Plus Pack, and a comparable MAX Pro chart in exchange for each Passport chart region owned. In other words, in a nearly unprecedented exchange program, Jeppesen Marine basically offered a complimentary one-for-one swap for old software and charts. Boaters who owned VNS version 5 to 8 upgraded to VNS MAX Pro for only $130.

VNS MAX Pro is very professionally packaged in a sturdy boxed set including the software installation disc, four discs of MAX Pro cartography, an eight-page Welcome Guide, and a hardware key (also called a dongle). The discs contain an incredible amount of data, including the assets commonly associated with C-Map MAX Pro cartography such as tides and currents, planning charts, aerial and satellite photos, street and marina information, and 3D views.

Software installation is straightforward, with the USB hardware key included in the boxed set unlocking the software.

Nobeltec VNS MAX Pro

Pros: Convenience of single-vendor software and cartography; very customizable Toolbar and Console; sharp raster chart rendering with CrystalView tool.

Cons: Limited chart formats supported; weak import/export capabilities.

Coolest Feature: Route Wizard

Price: $490

Vista Capable: Yes

Version Tested: 10.0.1.6

System Requirements
PC with Intel Core 2 Duo
1 GB RAM
600 MB available hard disk space
1024 x 768 monitor, 32-bit color
DVD drive
USB port (required for dongle)
Windows XP or Vista

Jeppesen Marine
15160 NW Laidlaw Road
Suite 100
Portland, OR 97229
United States
www.nobeltec.com

Nobeltec VNS MAX Pro and Nobeltec Admiral MAX Pro are nearly unique in their inclusion of Undo and Redo commands. You can undo Delete, Move, Move Text, and Hide for all Nav Objects, including marks, routes, range bearing lines, and boundaries.

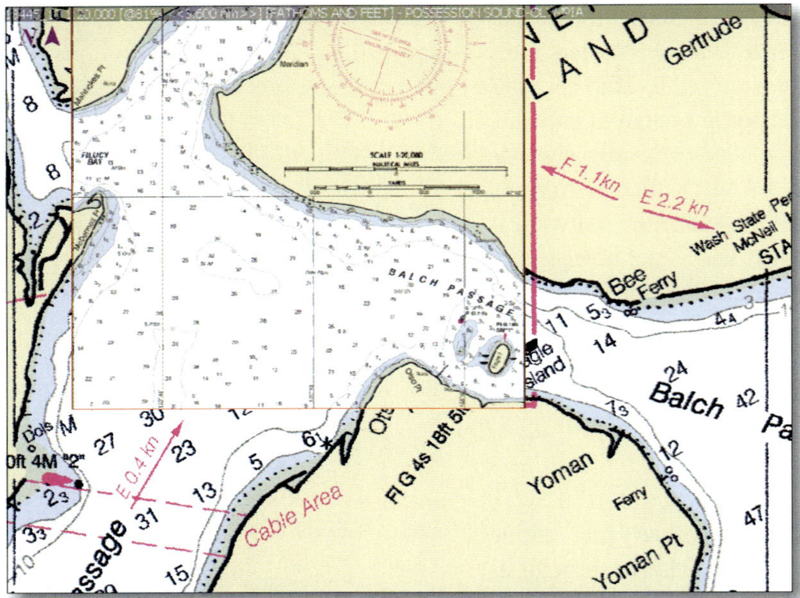

Raster Chart Support: *Raster charts, such as NOAA BSBs and SoftCharts, can be displayed with the addition of the Raster Chart Plus Pack.*

If you're mourning the loss of Passport Deluxe charts, consider their limitations relative to C-Map MAX Pro cartography: the Passport chart library had less coverage (12,000 charts versus C-Map's 30,000), emphasized commercial ports at the expense of light marine ports, could not be automatically updated, and had poor supplemental data. Ready to switch now?

VNS MAX Pro now uses a hardware key, which plugs into a PC's USB port. Connect the hardware key, then install the program and the chart regions you purchased. Additional MAX Pro chart regions can be purchased through dealers or directly from Jeppesen Marine. Prices vary by region, but for the U.S. a Wide region currently costs $339 and a MegaWide region is $499. There are also options for rentals, annual renewals, and upgrades. The purchase of a chart region provides an unlock code, allowing use of that region from the MAX Pro disc included with the software set.

VNS does not include a traditional user's manual, but there are several options for obtaining this information. A 252-page spiral-bound User's Guide can be purchased through the Nobeltec store (www.nobeltec.com/store). Alternatively, it's also available as a PDF document in the application's Help Menu or can be downloaded at www.nobeltec.com/products/prod_suite.asp. Translations to Spanish, French, German, and Italian should be available soon.

Jeppesen Marine provides toll-free phone and email customer support. Additional technical support is available through Nobeltec's extensive website, which includes a searchable Knowledge Base and FAQ page. The Knowledge Base (www.nobeltec.com/support/knowledgebase.asp) is a great concept, but may be in its infancy in terms of topics it covers. The FAQs (www.nobeltec.com/support/support_faq.asp) are nicely grouped into categories, including Installation, Chart Usage, Integrated Devices, and Usage/Miscellaneous. Jeppesen Marine also offers an electronic newsletter, which includes information on updates and technical tips (www.nobeltec.com/company/newsletter_subscribe.asp).

Look and Feel

Like most charting and navigation programs, VNS uses a combination of toolbars, menus, and floating windows. It scores very high marks for its customizable Toolbar and Console, including nearly 100 Toolbar button choices and nearly 50 Console panels. Once your Toolbar and Console are customized, a Save User Interface command lets you maintain different user interfaces for different vessels, helmsmen, or boating conditions such as fishing or long voyages.

VNS also includes many icons to label your information. But as we have recommended for other software packages, we'd love to see an "icon palette" so users could define a subset of frequently used icons.

In Tools>Options>Depth you can customize the color and font size for depth soundings. Customizing sounding text with color adds an extra visual warning, especially when panning over a zoomed-out chart.

However, the customizability of the Toolbar and Console do not extend to the layout of the screen. Screen "real estate" has been one of our important benchmarks, as screen clutter makes navigation more difficult. VNS loses about a third of the screen to the combined area of the Toolbar, NavBar, and Console. Of course, you can choose to turn the NavBar or Console off, but then you forfeit valuable visual information—not a good compromise. In addition, these data windows cannot be resized or repositioned, a standard metaphor in Windows displays.

As expected from a program now on a tenth version, VNS MAX Pro includes quite a few shortcuts or *Hot Keys*. Jeppesen Marine provides a nice two-page list of all VNS and Admiral Hot Keys, downloadable as a PDF at www.nobeltec.com/products/prod_suite.asp. The new MAX Pro version also added a much-appreciated Undo feature.

Working with Charts

The new line of VNS and Admiral software is designed to be used in conjunction with C-Map cartography. In fact, it's unclear why one would use VNS MAX Pro or Admiral MAX Pro without C-Map charts. In addition, MAX Pro's *Quick Sync Updates* automatically downloads recent Notice to Mariner advisories through the C-Map MAX Pro chart update server. A year of chart corrections is included with the purchase of a MAX Pro chart region.

VNS MAX Pro also works with a few other chart vendors and formats, including Maptech (BSB3 through BSB5) and SoftChart (GEO/NOS format). VNS MAX Pro also supports Maptech Photos. However, support of any of these raster chart formats requires the Raster Plus Pack, an additional $50 option.

The transition from VNS version 9 to VNS MAX Pro also entailed major changes to the support of standard S-57 vector charts, which includes free NOAA ENCs and U.S. Army Corps IENCs. Simply put, S-57 vector charts are no longer supported in VNS MAX Pro. Jeppesen Marine cites "concern about the safety associated with importing free data charts (NOAA S57 charts) and in a process that cannot be controlled, monitored and verified." It is assumed vector cartography is obtained through the purchase of C-Map charts.

Panning and scrolling over charts works very well in VNS, with smooth chart quilting for seamless integration. VNS provides two different modes to move over charts: hand panning and cursor mode. The hand panning Toolbar button is a traditional interface, letting you pan a chart by dragging a hand icon. Information about an object is displayed by hovering, which opens a small pop-up window of information. Cursor mode also pans over a chart, but it accesses a wide range of additional features and tools, including lassoing to zoom, interrogating objects, displaying object data in the NavBar, and pan-scrolling in eight compass directions by pointing to the edge of the chart.

Autoscroll Mode is a particularly handy feature when underway. VNS uses the position of your vessel to automatically keep the boat in the chart window. You can choose from one of three Autoscroll Modes, depending on your viewing preference. Look-Ahead keeps the vessel icon at the edge of the chart window, opposite the side where you are headed. Follow the Predictor keeps the predictor centered in the chart window, allowing a user-defined forward projection to be your navigational emphasis. Follow the Boat keeps the vessel centered in the chart window.

We found that VNS did get slightly bogged down with certain chart display tasks. Most noticeably, charts lagged and were sluggish when the Tides and Currents option was turned on. Similarly, the search engine is a bit slow, particularly during a new search. An alternate approach is to use the Locate This feature within the Chart Table, which lets you shortcut directly to a particular known location or object.

If you need to return to a particular chart, scale, or chart location, the Bookmark feature is handy. Typing the function key [F10] flips the display back to your *bookmarked* chart location. We liked this feature so much we wondered why VNS only included a single bookmark capability.

The SplitScreen option lets you display a 3D Navigator window and a vector chart window simultaneously. VNS allows you to navigate in full 3D using its 3D Navigator feature. Charts can be rotated and displayed course-up.

For viewing raster charts, Nobeltec integrates a software feature called *CrystalView*. Unlike Macintosh OS X computers, which have built-in Quartz technology to sharpen screen images, PCs must have these enhanced rendering capabilities added into each software application. VNS includes this technology, with a right-click option or a Toolbar button turning CrystalView on or off for all charts.

Two of VNS's features, Chart Coloring and Shaded Relief make charts more visually interesting and help with readability. Chart Coloring lets you select the chart color scheme you prefer, such as S-57 colors or Explorer chart colors. Shaded Relief displays lightly shaded land or sea features, which helps convey a visual sense of the topography. Nobeltec's 3D bathymetric displays were very impressive and can even be printed at high resolution.

Nobeltec programmers were thinking ahead when they

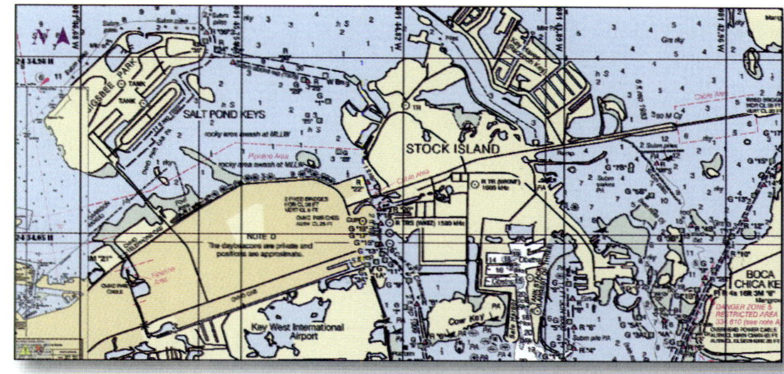
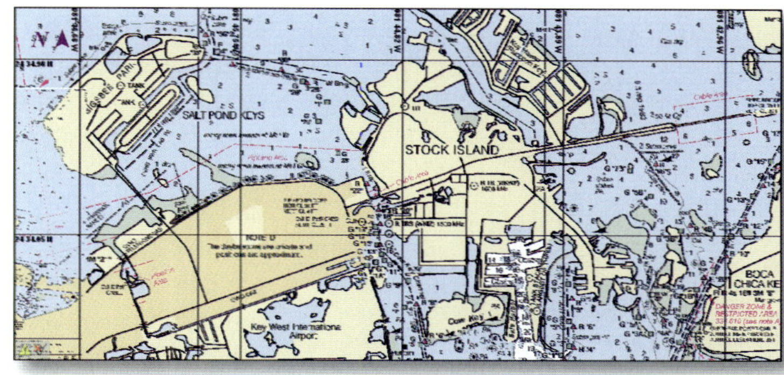

CrystalView: *Unlike the built-in Macintosh OS X Quartz technology, PC programs must improve screen rendering in software. Nobeltec's solution is called CrystalView: turned on (top) and off (bottom).*

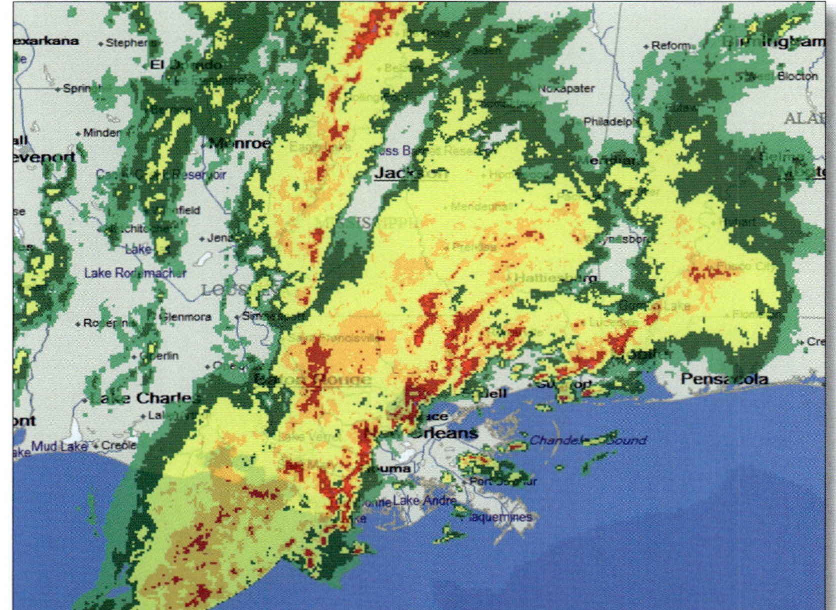

Weather Options: *VNS and Admiral extend marine weather data options beyond simple GRIBs with the Weather Plus Pack, supporting XM WX (below) or Sirius satellite weather.*

integrated a warning screen if you begin to open too many chart windows. We admit we were guilty, opening new chart windows and not closing others. The window alert is a great feature, preventing poor performance—or worse—before a problem arises. In fact, the only time we were able to crash VNS was when we had seven windows, the PlanBook, and the linked Tides & Currents Pro program open while manipulating a 3D Bathy chart. Quite impressive!

Waypoints and Routes

Any application as established as Nobeltec VNS MAX Pro includes extensive waypoint and route features, including unlimited route and waypoint placement, distance and bearing displays, an integrated ETA calculator, and extremely flexible boundaries and alarms.

We had no trouble creating and working with waypoints and routes in VNS. The interface was exceptionally clean. Like Coastal Explorer or Chart Navigator Pro, VNS displays the range and bearing line along the route, just as you would write it in pencil on a paper chart. Furthermore, you can customize the display in VNS to show range and bearing data selectively or on all routes. Routes also can be customized to include arrows for direction of travel. The Instant Waypoint feature creates a quick simple route from the boat's current position to a destination mark, keeping you vigilant on your cross-track error.

VNS MAX Pro also includes a Route Wizard feature that uses the C-Map vector charts and its tide and current data to automatically create a route from a user-selected origin to a destination. You can set parameters such as depth, distance from land, or minimum and maximum route leg lengths.

However, importing and exporting waypoints and routes is not as impressive. Nobeltec applications do not support the three main standard formats for exchanging data: tab-delimited, CSV, or GPX. Instead, Jeppesen Marine has chosen a standalone approach, only supporting its own system, ironically called Open Navigation Format, or ONF. In other words, users can exchange information between Nobeltec software installations but cannot import or export navigational assets.

This is a real problem for boaters with existing waypoint data who are considering switching to Nobeltec. Likewise, a flotilla or club cruise captain is unable to share waypoint files with other non-Nobeltec participants. When we spoke with Jeppesen Marine about this limitation, they acknowledged that using a chartplotter as an envoy was the only way to import existing waypoints (for us, that's more than 3,000 waypoints!)—a tricky and cumbersome workaround.

The flip side of that equation is a Nobeltec system is designed to work well with other Nobeltec users, external sensors, and gear—and to be extensible. You may begin by connecting a GPS or autopilot, and eventually grow to integrate advanced features such as radar overlays, multiple monitors, or video cameras. Both VNS and Admiral support nearly all NMEA-compatible devices, including depth sounder, fluxgate compass, wind and speed indicator, radar, DSC-equipped radio, bathymetric recorder, and AIS receiver.

Additional Features

Nobeltec VNS MAX Pro includes the capability to read GRIB weather files, but you must manually retrieve the .GRB file yourself, typically through an email query. (See GRIB Weather in Chapter 7.) The GRIB overlay worked smoothly and displayed the weather information on the chart, but VNS doesn't use this data to its fullest. Not all data options available in the GRIB download are viewable through VNS. More robust weather reporting and display requires the Weather Plus Pack ($300), which supports either Sirius Marine Weather or XM satellite weather.

VNS has several options to display tide and current data,

including an overlay on a C-Map chart or a separate tidal data window. As we noted above, the overlay option hampered performance. In addition, tides are only displayed for that day and cannot be advanced or retarded. We preferred the more information-rich and flexible window display of the stand-alone linked application, *Tides & Currents Pro.*

Sailing and Bathy Recorder Plus Packs are also available for VNS MAX Pro. These software plug-ins add specialized features to the basic VNS MAX Pro application. The Sailing Plus Pack ($300) integrates laylines, wind arrows, and polar diagrams. The new MAX Pro version adds Windvantage Weather Route, which uses the Route Wizard to make calculations based on the boat's performance polars and a GRIB weather file. The Bathy Recorder Plus Pack ($800) enables you to record sea floor topographic information using your sounder/depth finder and incorporate that data into your VNS 3D display.

Jeppesen Marine also sells specific hardware accessories designed to supplement Nobeltec software. For example, its InSight Radar2 Black Box integrates an existing radar system with Nobeltec software, overlaying radar images on your electronic chart display. Their sunlight-readable Wireless Display lets you view your VNS display on a small portable handheld device.

ASSESSMENT

Overall, Nobeltec VNS is a solid and very extensible package, integrating an established software application, proprietary vector cartography, and supplemental software applications and hardware devices. It's a package designed to grow with a serious long-distance boater while staying within the Nobeltec family. In fact, an initial choice of Nobeltec marks a long-term decision to stay within Nobeltec. VNS and Admiral are not "welcome the outsiders" packages, as evidenced by their inability to read many common chart formats or to easily import and export data.

With the release of the MAX Pro versions, it doesn't make sense to use Nobeltec software without C-Map cartography. The power of Nobeltec is the synthesis of their software, cartography, and probably even hardware. But this is a charting

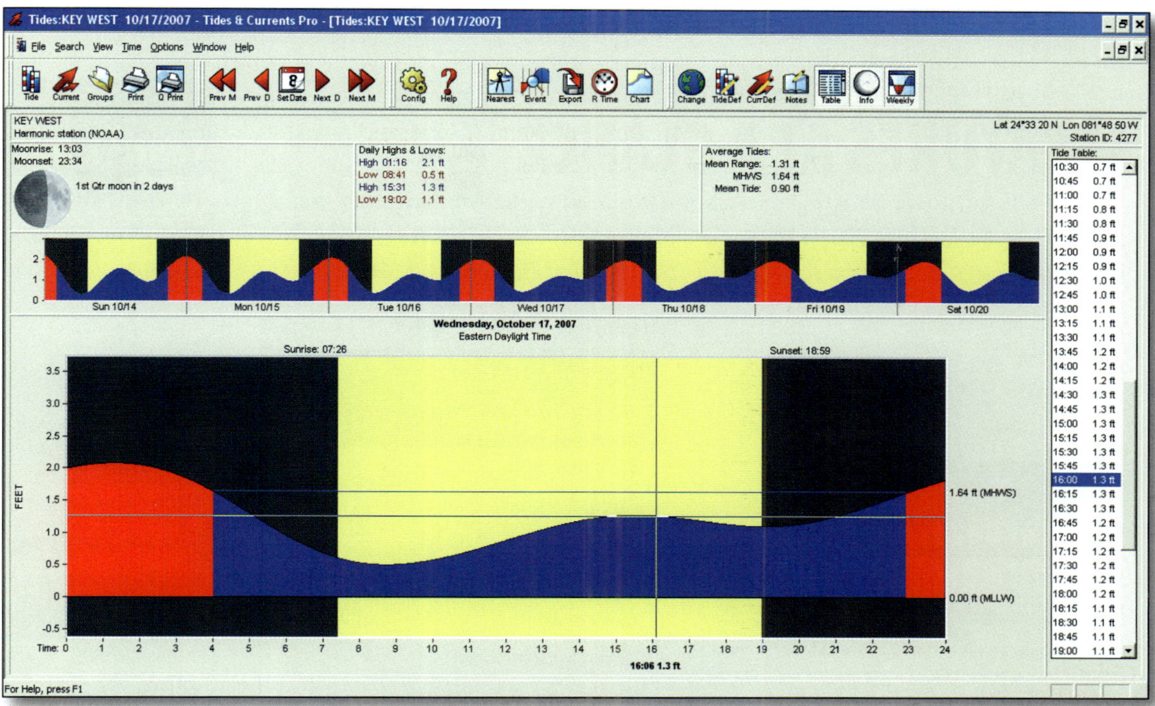

and navigation package that comes with a price. Although free NOAA raster charts remain an option with the purchase of the Raster Plus Pack software plug-in, free S-57 vector charts are no longer supported. A commitment to VNS MAX Pro is a commitment to C-Map cartography which, while not free, is widely recognized for top-notch marine data.

Tide & Current Options: *VNS MAX Pro's tidal information can be overlaid directly on a chart, summarized through a tabbed window, or viewed more richly with a graphical display using the separate Tides & Currents Pro program (above).*

Nobeltec VNS MAX Pro
FEATURES AT A GLANCE

Each full-featured e-charting application has been evaluated according to the following 81 features. Some ratings are binary, "yes" or "no." Others are more subjective, evaluated on a scale from 1 to 5 (5 being "Excellent"). Any other comparison is defined in the Legend.

Out of Box Experience

Packaging	4
Documentation	3
Software install	3
Load charts	3
Load supplemental data	3
GPS hookup	4
Technical support	3

User Interface

GUI metaphor	Win
Screen space management	3
Program responsiveness	3
Chart display speed	3
Chart display quality	3
Customizable GUI	5
Customizable metrics	4
Split windows	Yes
Tabbed interface	No

Basic Features

One-button MOB	Yes
Waypoint creation	3
Route creation	4
Steer-to function	Yes
Track creation	3
Convert track to route	Yes
Waypoint and route management	3
Chart management	4
Search waypoint and route	2
Range/bearing tool	3
Chart annotation	4
Chart printing	3

Data Exchange and Networking

GPX support for waypoints	No
GPX support for routes	No
Waypoint exchange formats	None
Route exchange formats	None
Integrated GPS transfer	Yes
Multiple monitor support	No
Network application sharing	No

Cartography

Free U.S. rasters BSBs (RNCs)	Yes
Free U.S. vectors S-57s (ENCs, IENCs)	No
SoftCharts	Yes
CHS Canadian rasters	Yes
DNC	No
Seafarer	No
British Admiralty ARCS	No
International S-57s	No
International S-63s	No
Scan and geo-reference paper charts	No
Navionics cards	No
C-Map cards	No
C-Map CD/DVD	Yes
Satellite geo-referenced photos	Yes
Aerial informational nav photos	No
Topographic maps	No
Bathymetric data	Yes

Instruments and Sensors

GPS	Yes
Autopilot	Yes
AIS receiver	Yes
Wind	Yes
Depth	Yes
Water temperature	Yes
Radar	Yes
Heading sensor	Yes
Video camera	Yes

Advanced Features

Tides	4
Currents	4
Coast Pilot	2
POIs	2
Streets	2
Google Earth support	No
Advanced search	3
AIS	Yes
Buddy boat	No
Radar (ARPA/MARPA)	No
Radar (display overlay)	No
Fuel calculator	Yes
Transit calculator	Yes
Celestial calculator/Almanac	Yes
Auto route planning	Yes
Great Circle Route planning	Yes
Sail performance	PP
GRIB weather integration	2
Weather options	PP
Customizable bathymetric recorder	PP

Legend

5	Excellent	CM	C-Map
4	Above average	NV	Navionics
3	Average	SC	SailCruiser
2	Below average	MT	Mr. Tides
1	Poor	PP	Plus Pack
CP	Chartplotter	C	CSV
Mac	Macintosh	E	Excel
NT	Non-traditional	G	GPX
Win	Windows	K	KML
N/A	Not applicable		

Chapter 22
Chart Navigator Pro

Maptech is an established charting and navigation company, with a product portfolio that includes printed paper chartkits, cruising guides, digital charts on CD, and charting and navigation software. In fact, Maptech has so many offerings—including many with overlapping names and features—that it is sometimes difficult to keep it all straight.

Here we keep it simple, reviewing only Chart Navigator Pro, which Maptech markets as its full-featured flagship navigation software package for recreational boaters. As we discussed in Chapter 19, Chart Navigator Pro is a branded version of Coastal Explorer, developed by Rose Point Navigation Systems.

Rose Point benefits from this arrangement because Maptech's stronger marketing presence and distribution reach allow them to sell more retail copies. Maptech adds value to Chart Navigator Pro by including 15 DVDs full of resources—including extensive chart coverage and extras such as 3D contour maps, shoreline topographic maps, aerial photos, and satellite images—and then charges $100 more.

The result is Chart Navigator Pro ($499), an application that takes Rose Point's extraordinarily stable program and marries it with Maptech's extensive charting resources to produce a full-featured package that will work well both for serious weekend boaters and long-distance cruisers.

Navigating Maptech

Start looking at Maptech products and it's easy to get confused. With more combinations than a Starbuck's, it even takes a bit of time to read the menu. We think we've sorted it out, so let's give it a try.

A good first step is to sort out the differences between the *Lites* and the *Pros*. Maptech has two applications in its Offshore Navigator line: Offshore Navigator ($249.95) and Offshore Navigator Lite ($99.95).

Offshore Navigator is the navigation application included with many of Maptech's past and current Digital Charts offerings. This includes products branded as Chart Navigator, Marine Navigator, Digital ChartKit, and Digital ChartKit Pro.

Offshore Navigator Lite is the e-charting application included on the Companion CDs that accompany their paper ChartKits and Waterproof Chartbooks. In the past, Offshore Navigator Lite has also been included with other chart CDs, usually for waters outside of the U.S. It is very similar to Offshore Navigator, but does not include features such as tracks, instant and named waypoints, autopilot support, alarm zones or a ship log.

Maptech has four applications with *Chart Navigator* in the title: Chart Navigator Viewer, Chart Navigator Light, Chart Navigator ($249.95), and Chart Navigator Pro ($499.95).

Chart Navigator Viewer is the free software application for planning with and printing Maptech BSB charts. It looks similar to Offshore Navigator and Offshore Navigator Lite but does not integrate with a GPS nor have a "vessel" with which to navigate. It is often used to demo Maptech charts and is available as a free download (http://www.maptech.com/support/emailed.doc.cfm?docat=123.24).

Chart Navigator Light is the navigation software included with Chartbook Companion CDs for the Bahamas and Caribbean. It has many of the same features of Chart Navigator Pro—but without the DVDs of U.S. charts.

Chart Navigator (without Pro or Light) features Maptech's Offshore Navigator software and includes DVDs of raster, photo, and topographic charts (but does not include vector or 3D contour charts).

Finally, Chart Navigator Pro is Maptech's full-featured

Chart Navigator Pro

Pros: Easy set-up and user interface; excellent creation, management and transfer of waypoints and routes; excellent search capabilities; integrated guidebook information; extensive chart and data libraries.

Cons: Less-than-perfect image quality and chart rendering for raster charts.

Coolest Feature: 15 DVDs of Supplemental Data

Price: $499.95

Vista Capable: Yes

Version Tested: 1.1.61

System Requirements
PC with Intel Pentium (or equivalent)
512 MB RAM
1 GB available hard disk space
800 x 600 monitor
DVD drive
Windows 2000, XP, or Vista

Maptech
10 Industrial Way
Amesbury, MA 01913
United States
www.maptech.com

When it's time to update your charts and supplemental data, Maptech sells an update through its website for $99.95. The package includes updated NOAA raster charts (RNCs), NOAA vector charts (ENCs), Army Corps of Engineers vector charts (IENCs), navigation photos, and an updated marina facilities database.

product for recreational boaters, and the one reviewed in this chapter.

Maptech also recently purchased The Capn. Although The Capn was one of the first and most popular PC charting packages for recreational boaters, it is now marketed primarily to the commercial market of workboat, merchant, and government professionals. It is reviewed in Chapter 20.

Confused yet? We still are. But thankfully, Maptech provides a summary comparison of all these offerings on its website (www.maptech.com/water/chartNavigatorPro/index.cfm?infopg=compare).

GETTING STARTED

Because Chart Navigator Pro and Coastal Explorer were born of the same parents, much of what we highlighted in our review of Coastal Explorer in Chapter 19 also applies here. Good features are good features. But there are some substantial differences, and we have tried to point those out.

Chart Navigator Pro is sold in a boxed set that includes an installation DVD for the software, a 20-page Getting Started Guide, and 15 DVDs of additional regional digital charts and supplemental data. Boaters who already use Maptech's

chart CDs, chartkits, or guides will find these sets familiar. The DVDs even include coverage for Alaska, Hawaii, Puerto Rico, and the U.S. Virgin Islands.

In addition, all U.S. raster charts from the National Oceanic and Atmospheric Administration in Maptech's BSB3 file format are included, as well as all available S-57 ENC and IENC vector charts from NOAA and the U.S. Army Corps of Engineers. The data libraries also include Maptech's satellite and aerial navigation photos, shoreline topographic maps, and contour 3D charts for the entire U.S. coastline.

In some of the other software chapters, we detailed a long list of additional items to purchase. This is not the case with Chart Navigator Pro. When they say, "Everything you need in one box," they're not exaggerating. Because all the digital charts and additional databases are contained in the stack of DVDs, you truly are all set for U.S. cruising.

However, if you are cruising internationally, including Canada, the Bahamas, or Mexico, you must purchase your own digital charts. Maptech and other chart distributors sell international digital charts on CD for most regions of the world.

You don't need to be a computer geek to install Chart Navi-

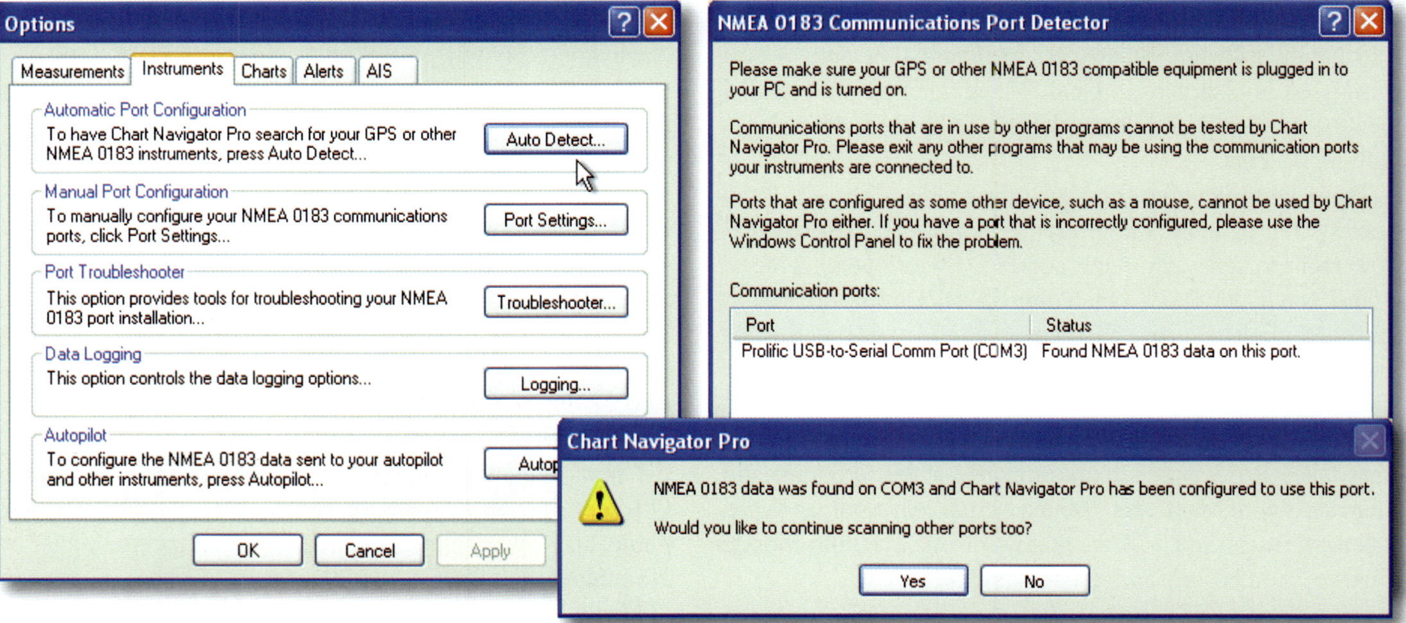

gator Pro. Installation was trouble-free. The software and data loaded immediately. A GPS wizard scanned the ports, found the GPS, and automatically assigned a communication port.

Maptech's infrastructure is part of the value equation if you choose to purchase Chart Navigator Pro versus Coastal Explorer. Maptech has a strong presence in marine retail outlets, boat shows, and on the Internet. For example, Maptech maintains an extensive website for Chart Navigator Pro, including technical documents covering chart and instrument questions; downloads such as software updates and BSB chart registration; and a discussion forum with about 70 topics.

To supplement the brief Getting Started guide included with the software package, additional manuals are available on Maptech's website and through the Help menu on the application itself. Unfortunately, the website download for the User Manual was a disappointing 14-page primer. To create a manual with more detail, we recommend you open and print each of the eight chapters in the Help menu.

That said, Chart Navigator Pro's best customer support is the exceptionally stable code produced by Rose Point. This is an application that is easy to use and—as far as we could tell—impossible to crash.

LOOK AND FEEL

Chart Navigator Pro has one of the cleanest user interface of the e-charting choices. We particularly liked the use of *display modes*, which kept the screen uncluttered by only displaying necessary tools and information. For example, Planning Mode displays planning tools such as creating routes or getting information from the guide books. Cruising Mode switches to a chartplotter-like display, with windows for instrument and position data.

In addition, Planning Mode uses small Windows-style menus suitable for work, home, or in the slip. Cruising Mode displays big easy-to-read buttons and windows with virtual analog-style instruments suitable for viewing while underway. Even more impressive, the application senses whether you are stationary or transiting, changing to the most appropriate display automatically. At any time you can manually toggle between the two displays by pressing the [F12] key.

Like Coastal Explorer, information in Chart Navigator Pro is a mouse-click away. For example, most software applications show a data window with bearing and distance to your next waypoint, squeezing one more window of data on an already-too-small laptop screen. Chart Navigator Pro displays your magnetic bearing and distance directly along your course line, just as you would record it in pencil on a chart.

Like any sophisticated and mature program, Chart Navigator Pro includes many keyboard shortcuts, summarized in a table at the end of the Getting Started Guide. More importantly, it uses standard Windows shortcuts, a particularly nice design feature of an application only used seasonally or on weekends. The application also allows multi-level undos.

WORKING WITH CHARTS

Although Chart Navigator Pro ships with the complete United States Maptech chart library, it is also compatible with nearly

If you have trouble seeing your route on a cluttered chart area, choose **Highlight Objects** in the toolbar. The chart will dim to highlight or emphasize the route.

Looking Ahead: *The Check Route for Obstacles feature warns mariners of wrecks, obstructions, restricted and caution areas, ferry routes, and many other hazards to navigation.*

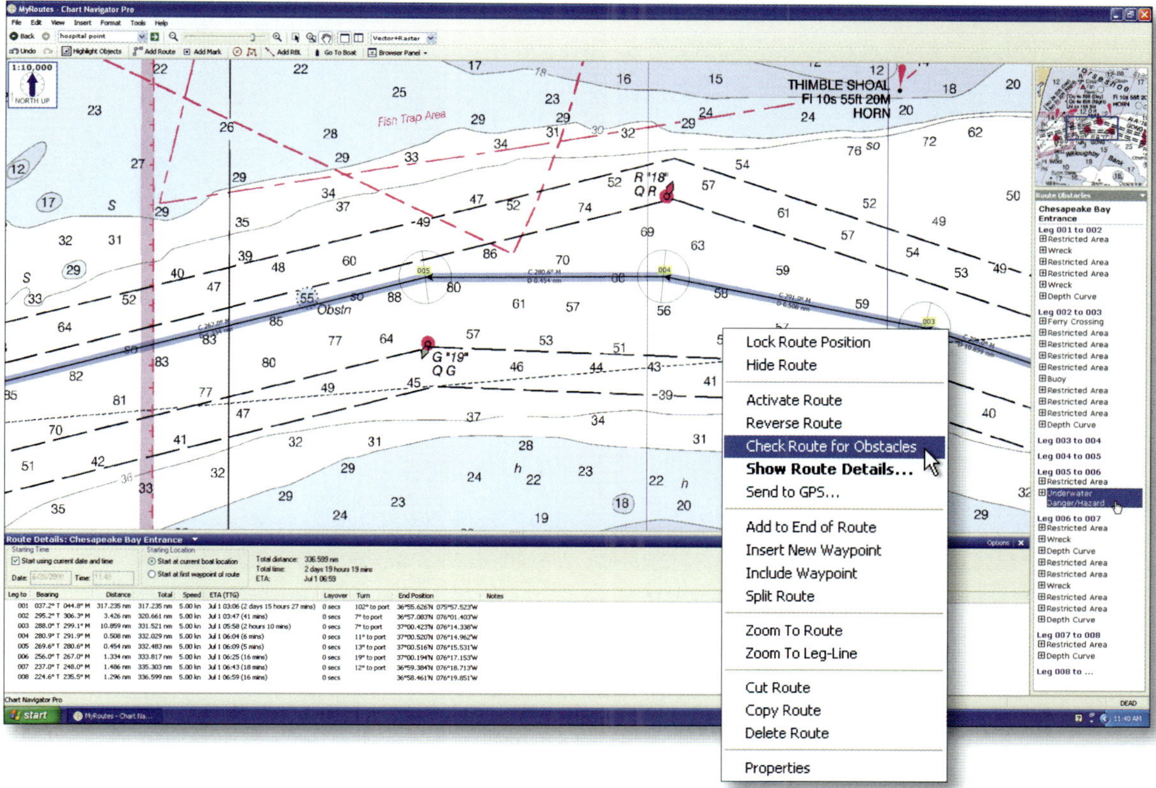

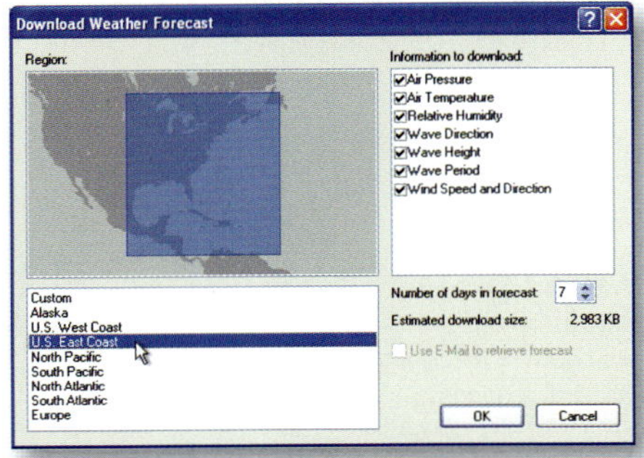

Coastal Explorer Weather: *Begin weather downloads by choosing the region (top left), weather data type (top upper right), and forecast period (top lower right). Then view downloaded weather data such as wind, pressure, temperature, precipitation, and wave height (below).*

all major chart formats. This means you can use the program with digital charts you already own, such as SoftChart GEO/NOS formatted charts; international charts from other sources such as NDI DigitalOcean; or the free charts provided by NOAA or the U.S. Army Corps of Engineers. The only exceptions are that Chart Navigator Pro does not display DNCs (an alternative vector format produced by the U.S. National Geospatial-Intelligence Agency), British Admiralty ARCS, or NV-Verlag BSBs.

We had no trouble panning and zooming over a chart. Chart Navigator Pro's chart quilting seamlessly moved across chart files, even when using charts of different scales. We particularly liked the split screen option, allowing you to display two or three chart or aerial/satellite panels simultaneously.

Like Coastal Explorer, Chart Navigator Pro's search capability is exceptional because it searches the entire database. Many charting and navigation applications only search your waypoints. For example, we typed "Seward"—a location in Alaska where we don't have any waypoint marks—and the chart for Seward Bay, Alaska popped up on our screen. A window of similarly-named locations also opens so you can correct the search result with a single click if need be.

WAYPOINTS AND ROUTES

The creation of waypoints and routes is very sophisticated in Chart Navigator Pro. It has all the expected features: unlimited waypoints and routes, boundary circles and areas with alarms, chart annotations, range and bearing lines, and so on. But Chart Navigator Pro also includes some exceptional features that enhance usability and safety.

For example, to make a route you simply click along, creating a string of waypoints. This is certainly convenient—and a lot less cumbersome than dragging and dropping waypoints into a route folder. But even better, as you approach the edge of the chart, the cursor flips to a white arrow. Continue to nudge the edge and the chart automatically scrolls so you can continue creating your route seamlessly.

A welcome safety feature is the Check Route for Obstacles function. When you create a route, Chart Navigator Pro reads the data from the vector database and alerts you if your proposed route encounters obstacles such as a navigation aid or wreck. Even more impressive, this feature works even if you are viewing a raster chart (remember, raster charts are simply images, with no attendant data)—as long as the vector chart for the area is installed.

After you've created a route, an option called Fork Route lets you use this route to build additional routes. For example, if you have two or three common routes that split or "fork" after you exit your home marina, the Fork Route feature lets you create a route, then build a new route, departing at any waypoint in a new direction.

ADDITIONAL FEATURES

Chart Navigator Pro is designed to work with nearly any NMEA 0183 device that has a PC interface. For example, you can connect a GPS, autopilot, depth sounder, water temperature sensor, knot/log, wind speed and direction indicator, rudder angle indicator, AIS receiver, or radar.

With the inclusion of 15 DVDs of data, you have plenty of information at your fingertips. In addition to the chart library, Chart Navigator Pro ships with a library of aerial and satellite navigation photos, 3D bathymetric contour charts, topographic maps, and tide and current prediction tables. The integrated Coast Pilot and Light List information means you can access data by clicking directly on a chart rather than opening a PDF document. The built-in *Gazetteer* lets you click on a harbor to display a window with a cruising-guide-synopsis of that location.

Chart Navigator Pro obtains up-to-date weather information through a server maintained by Rose Point Navigation. Obtaining weather data is exceptionally easy. From the Weather Browser window, click Download, and select your choices for location and data. You can choose to download all the weather information for your location—GRIB files are not very large—then toggle to display data such as sea temperature, wind speed, or air temperature. The weather data is stored on your hard drive so you can view it underway without an Internet connection.

It may not seem as sexy as flying over 3D hilltops, but Chart Navigator Pro's *documents file* is a very cool feature. All waypoints, routes, tracks, and event marks—called "Navigation Objects"—are packaged and saved in a special user documents file. If it reminds you of the standard computer metaphor, you're correct. Remember, the code for Chart Navigator Pro was written by former Microsoft engineers over at Rose Point.

Why is a documents file cool? It means you can send all of your navigation information to other users of Coastal Navigator Pro (or Coastal Explorer) simply by selecting Send To. It lets you share your navigation information with other boaters as a seamless import/export package. What a great feature for a flotilla captain, who could develop a rich float plan with waypoints, routes, marinas, and attractions, then distribute this information as one package by email or CD.

Assessment

Like Coastal Explorer, Chart Navigator Pro is an excellent choice for a wide range of recreational boaters. Weekend or short-season boaters will appreciate Chart Navigator Pro's easy-to-use interface. Its intuitive and standard design lets you minimize time re-familiarizing yourself with the software every time you use your boat. At the same time, with its ability to read international charts and connect to AIS receivers and MARPA-equipped radars, it is full-featured enough for long-distance cruisers.

Chart Navigator Pro begins with this well-designed software, then harnesses Maptech's extensive resources, ranging from complete raster chart coverage of the U.S. to extras such as 3D contour maps and navigational photos. Contrary to many false marketing promises, this package truly does include everything you need in one box.

The Future of Maptech

At press time, Maptech was for sale—including its paper charts, guide books, Captn. Jack's mail-order website, electronic cartography, and e-charting software such as The Capn and Chart Navigator Pro.

Until recently, Maptech was owned by Gary Comer, the founder of Lands' End clothing. An avid boater, Comer enjoyed the Maptech business venture, perhaps even at the expense of profitability. According to the boaters' coconut telegraph, Maptech was more of a passion of Comer's than a money-making venture.

When Comer passed away in 2006, the family trust inherited ownership of Maptech. Not surprisingly, there is little incentive to maintain the company for sheer boating interest. Maptech is absolutely not closing its doors, but change is certainly on the horizon.

When your chart type is set to photo charts, don't zoom out too far or the program may bog down trying to load the files. A 1:100,000 scale image requires six to twelve photos to fill the screen! The more you zoom out, the more files need to be loaded.

Chart Navigator Pro
FEATURES AT A GLANCE

Each full-featured e-charting application has been evaluated according to the following 81 features. Some ratings are binary, "yes" or "no." Others are more subjective, evaluated on a scale from 1 to 5 (5 being "Excellent"). Any other comparison is defined in the Legend.

Out of Box Experience

Packaging	3
Documentation	3
Software install	4
Load charts	4
Load supplemental data	4
GPS hookup	5
Technical support	4

User Interface

GUI metaphor	Win
Screen space management	3
Program responsiveness	4
Chart display speed	3
Chart display quality	2
Customizable GUI	3
Customizable metrics	3
Split windows	Yes
Tabbed interface	No

Basic Features

One-button MOB	Yes
Waypoint creation	4
Route creation	5
Steer-to function	Yes
Track creation	3
Convert track to route	Yes
Waypoint and route management	4
Chart management	3
Search waypoint and route	5
Range/bearing tool	4
Chart annotation	4
Chart printing	3

Data Exchange and Networking

GPX support for waypoints	Yes
GPX support for routes	Yes
Waypoint exchange formats	G
Route exchange formats	G
Integrated GPS transfer	Yes
Multiple monitor support	No
Network application sharing	No

Cartography

Free U.S. rasters BSBs (RNCs)	Yes
Free U.S. vectors S-57s (ENCs, IENCs)	Yes
SoftCharts	Yes
CHS Canadian rasters	Yes
DNC	No
Seafarer	No
British Admiralty ARCS	No
International S-57s	No
International S-63s	No
Scan and geo-reference paper charts	No
Navionics cards	No
C-Map cards	No
C-Map CD/DVD	No
Satellite geo-referenced photos	Yes
Aerial informational nav photos	Yes
Topographic maps	Yes
Bathymetric data	Yes

Instruments and Sensors

GPS	Yes
Autopilot	Yes
AIS receiver	Yes
Wind	Yes
Depth	Yes
Water temperature	Yes
Radar	Yes
Heading sensor	Yes
Video camera	No

Advanced Features

Tides	5
Currents	5
Coast Pilot	5
POIs	5
Streets	No
Google Earth support	No
Advanced search	5
AIS	Yes
Buddy boat	No
Radar (ARPA/MARPA)	Yes
Radar (display overlay)	Yes
Fuel calculator	No
Transit calculator	No
Celestial calculator/Almanac	No
Auto route planning	Yes
Great Circle Route planning	No
Sail performance	No
GRIB weather integration	5
Weather options	No
Customizable bathymetric recorder	No

Legend

5	Excellent	CM	C-Map
4	Above average	NV	Navionics
3	Average	SC	SailCruiser
2	Below average	MT	Mr. Tides
1	Poor	PP	Plus Pack
CP	Chartplotter	C	CSV
Mac	Macintosh	E	Excel
NT	Non-traditional	G	GPX
Win	Windows	K	KML
N/A	Not applicable		

Chapter 23
RayTech RNS

Although the Raymarine brand and company has only existed a short time, boaters are long familiar with the name of its founding company, Raytheon. Following a management buyout in 2001, and a subsequent re-branding of all products, Raymarine has become an established leader in electronic equipment for recreational and light commercial boaters, including radars, chartplotters, fishfinders, autopilots and more.

Less well known are Raymarine's two navigation software products, RayTech Planner and RayTech RNS, both intended to supplement their hardware product line.

RayTech Planner is well named: it is a planner, not a full-featured charting and navigation application. The software is essentially a subset of the licensed full-featured RayTech RNS package, distributed either as a free download through Raymarine's website or as a disc bundled with a Raymarine/Navionics USB card reader. Designed as an "at-home" tool, RayTech Planner cannot connect to a GPS or other instruments. It is suited for viewing charts and creating and exporting waypoints and routes to a Raymarine C-, E-, or G-Series chartplotter using a CompactFlash memory card.

In contrast, RayTech RNS is a full-featured charting and navigation application. Version 6.0 was the recipient of a 2006 Best in Show Award by the National Marine Electronics Association (NMEA). In many ways, RayTech RNS is very different than other PC-based e-charting applications. Most noticeably, RayTech RNS uses an interface that mimics a chartplotter, rather than a PC interface. The network application is also different. With RayTech RNS, your PC becomes another *node* in a networked system of (presumably Raymarine) devices, rather than the center of a collection of attached devices. These differences, and others we discuss below, are consistent with Raymarine's emphasis on providing an integrated system of marine instruments.

The result is that, for those who make heavy use of Raymarine electronics, the RayTech software package may make a great deal of sense. It is not cheap, but the cost is relatively small in the context of a larger Raymarine network. But for a boater who is not already using Raymarine hardware, other charting and navigation packages may be better suited.

Getting Started

RayTech RNS is available as a boxed set directly from Raymarine's website or for retail purchase from the hundreds of authorized Raymarine dealers.

This very professional set includes two discs, a User's Guide, and two hardware items. The first disc, a CD-ROM, includes the software application, SoftChart Coastal Planning Charts, C-Map NT+ Worldwide Planning Charts, and Navionics Worldwide Tides and Currents data. The planning charts (which are not suitable for navigation) and the tides/currents data install automatically with the software. The second disc, in DVD format, includes a complete set of NOAA U.S. Raster Navigational Charts.

The hardware items are a Navionics Multi Card Reader, which connects directly to a PC and provides a means to display Navionics cartography, and a GPS 9-pin serial data cable to optionally wire your vessel's GPS to your PC.

Given that Raymarine includes a Navionics Multi Card Reader with its software, it's safe to assume that Raymarine expects users will purchase cartography from Navionics. Note that the chart data contained on a Navionics Compact-Flash chart card is encrypted to prevent piracy and the Navionics reader has special circuitry that decrypts the charts for RayTech RNS. You can also read older C-Map NT or NT+ charts with RayTech RNS, but only with a C-Map USB C-

RayTech RNS

Pros: Excellent intra-network capabilities; convenient Ethernet cable hook-up to E- and G-Series devices.

Cons: Additional Raymarine hardware required for optimal use; chartplotter interface reduces chart display area; standard S-57 vector charts not supported.

Coolest Feature: DataTrak

Price: $699

Vista Capable: No (Yes, v 6.1)

Version Tested: 6.0.6275

System Requirements
PC with Intel Pentium IV
256 MB RAM
1024 x 768 monitor, 16-bit color video
NVIDIA GeForce graphics card recommended
CD or DVD drive (DVD drive required for charts)
Windows 2000 or XP

Raymarine plc
Anchorage Park
Portsmouth, Hampshire PO3 5TD
United Kingdom
www.raymarine.com

Raymarine, Inc.
21 Manchester Street
Merrimack, NH 03054
United States
www.raymarine.com

Do you already own or are you considering the purchase of RayTech software? Be sure to read *What's Next?*, at the end of this chapter, for an overview of the upcoming RayTech RNS 6.1 release.

Startup Wizard: *RayTech RNS employs a hybrid interface, part chartplotter and part PC Windows.*

To use third-party GRIB files with RayTech, begin by placing the downloaded GRIB file into the C:\program files\raymarine\raymarine raytech navigator\grib folder. All GRIB files must be in standard NOAA GRIB format and have a .grb file extension. Lastly, configure the GRIB file in File>Layers>Advanced Raymarine>Weather File.

Card Reader (not included).

The package also includes free U.S. raster charts from NOAA, but Raymarine's DVD included chart editions dated back to December 2005. If you rely on raster charts, we suggest installing the latest BSB chart files, which are available as a free download from NOAA or as an entire 1000-plus chart catalog for under $50 from many NOAA-certified chart distributors.

Installing RayTech RNS was straightforward. Chart installation was also trouble-free. It took about 15 minutes to install the 907 files from a Maptech Region 65 DVD (Block Island, Rhode Island to the Canadian Border), which includes BSB, photo, and topographic files.

Connecting a USB GPS sensor was our only installation stumble. Although there is a Startup Wizard, RayTech RNS doesn't include a Port Wizard. Instead one must check Window's Device Manager for an occupied port that suggests a GPS, then select that communication port in RayTech RNS. (See COM Ports and Your GPS in Chapter 5 for more details.) The lack of a Port Wizard is another reminder that the RayTech RNS target configuration is to connect a PC to an integrated marine electronics system rather than a standalone, non-Raymarine device.

Raymarine's customer support for RayTech RNS focuses on two resources: a User's Guide and a detailed "solutions database" on the company website.

The packaged set includes a 208-page wire-bound User's Guide. Unfortunately, although the manual looks meaty, it's thin on software features and functionality, and heavy on connections and wiring. In addition, much of the manual is dedicated to networked features (such as radar and sonar) that come with their own documentation. Additionally, the manual could use an index; it was often difficult to locate topics using only an expanded table of contents.

For detailed information on troubleshooting, system specifications, chart compatibility, and network hook-up, Raymarine maintains one of the best FAQs we've seen (http://raymarine.custhelp.com). Their 37-page database lets you search by product, category, phrase, or search text. An impressive 54 topics were listed for RayTech RNS version 6.0. Even better, you can click a button to be emailed automatically when that topic is updated.

Although their website drives you to the extensive knowledge base, you also can send email requests or call technical support at the number provided in the User's Guide. The online protocol is to register for a "Raymarine Insider Account," which gives you access to "Raymarine Insider Only" features and benefits. You then log in to contact the customer support team directly.

LOOK AND FEEL

Most of the charting and navigation packages strive for Windows interface compliance. In fact, occasionally we dinged an application if its interface deviated from this standard. RayTech RNS takes a completely different approach, designing a hybrid interface that partially mimics a chartplotter display.

This choice is partly demographic-driven (they note older, less PC-savvy users as customers). But it also follows the strong networked design of a Raymarine system, where the PC is intended as another multi-functional display node amid a collection of marine instruments that center on a chartplotter display.

In Raymarine's way of thinking, the PC is not the central component of your marine electronics, but simply a (portable) unit for planning at home and then conveniently transferring this information to your boat's network via card or by plugging the laptop into the network while aboard. For example, icons are shaped and shaded to look like the physical buttons on other Raymarine displays. Data is keyed in and entered as if your screen is a touch-sensitive chartplotter.

For boaters who are more comfortable with chartplotters than computers, RayTech RNS will be a welcome interface. However, boaters who grew up on PCs may find this interface a bit clumsy and outdated. Personally, we started out not liking the chartplotter look, but grew to like it and understand its purpose within the Raymarine gestalt. However, given an independent choice between the two formats, we still prefer the flexibility and efficiency of a contemporary computer user interface.

One downside of a chartplotter interface is the loss of chart display area. The large SoftKeys across the bottom, the Databoxes on the left side, and the Pathfinder panel on the right side all create wide margins that compromise available chart real estate. Although you can choose which buttons, tools, or data to include, wide gray margins remain. You'll wish for a more streamlined interface and/or a larger display screen.

RayTech's interface is also not as customizable as a typical Windows interface. Although you can choose which options to display, it is inconvenient or impossible to customize elements such as icons, fonts, or screen placement. In fact, we couldn't even customize weather data to display Fahrenheit rather than Celsius. (Raymarine, a British company, acknowledges this limitation and has it slated for improvement.)

Conveniently, Databoxes can be quickly moved and resized, which is useful for many chart layouts. For example, to efficiently display long, narrow strip charts of a coastline, you can arrange data horizontally across the top, leaving a large landscape area for the chart.

We've already mentioned several Raymarine hardware devices—in particular the C-, E-, and G-Series displays—that are designed to optimize the power of your PC running RayTech software. We mention one more here: Raymarine's proprietary specialized keyboard. One solution to RNS's limited chart screen space is to use the PC in *RNS mode* with Raymarine's handheld keyboard. This $500 accessory clears the screen of many data entry buttons, reclaiming valuable chart viewing area.

Working with Charts

Raymarine, like Humminbird and Northstar, was one of the early adopters of Navionics Platinum cartography, designing its hardware devices to handle the higher demands of multidimensional data. These types of files—including 3D displays, aerial and satellite navigation photos, and bathymetrics—re-

quire new hardware designs with more memory and faster processors. Raymarine's hardware (such as its E- and G-Series multifunction displays), and its RayTech RNS software, support these advanced Navionics features.

However, in terms of traditional chart files, RayTech RNS is largely a raster chart or proprietary vector chart program. It does not support standard format S-57 vector charts—which includes NOAA ENCs, U.S. Army Corps IENCs, and standard international vector charts. Raymarine hopes to include support for S-57 vector charts in a future software release. In the meantime, vector charts must be purchased from proprietary vendors such as Navionics or C-Map.

To summarize, RayTech RNS 6.0 is compatible with the following cartography: Navionics HotMaps, Silver, Gold, and Platinum charts (version 6.1 will support Platinum+); BSB raster charts, which include NOAA RNCs and equivalent products from Maptech, NDI, and SoftChart; Maptech topographic charts; Maptech aerial photos (which RayTech RNS converts to their own format); and the older C-Map NT and NT+ cartography.

In addition to S-57 vector charts, RayTech RNS is not currently compatible with British Admiralty ARCS, NV-Verlag, nor the new C-Map MAX or MAX Pro cartography.

We tested RayTech RNS using both Navionics and Maptech cartography. Charts displayed promptly and the program was very responsive. A nice touch for chart viewing customization is the freedom to turn quilting off when displaying raster charts. This displays the raster chart in its entirety, letting you read any chart notes or visually check chart editions and correction dates.

RayTech RNS was not as smooth as other applications when it came to panning and scrolling over a chart. Sadly, there is no grabber hand in their chartplotter-style display. In general, one cannot pan over a chart to move the dis-

For detailed instructions on transferring waypoints and routes from RayTech RNS to Raymarine C- and E-Series MFDs, visit http://raymarine.custhelp.com/cgi-bin/raymarine.cfg/php/enduser/std_adp.php?p_faqid=1247.

Remote Keyboard: *The optional Raymarine keyboard removes many RNS buttons and toolbars from your PC screen, recovering more space for the chart display.*

play. Panning is only enabled when you are creating a route. Otherwise, you must click on a point on the chart, which re-centers the display to that point. Panning over a large chart area requires zooming out, clicking on the new geography, then zooming back in to the working scale.

RayTech RNS has two screen configurations to provide the maximum view of your information: a split window display and a somewhat unique tabbed window display.

In the split window environment, several displays can be viewed at one time. For example, you can show four windows simultaneously, such as a chart display, a perspective view, your radar image, and sonar data. Any of the split win-dows can be quickly enlarged to a full window presentation by clicking on an icon. The same action collapses it back to the split window format. Although a common metaphor, Ray-Tech's split window functionality is the cleanest implementa-tion we've seen.

RayTech RNS also incorporates a *tabbed* window func-tion. This display metaphor is reminiscent of Microsoft Excel, which navigates through multiple open "worksheets" using tabs at the bottom of the screen. This is a great utility, letting you tab through full-sized windows of important information rather than using small split windows.

In addition, RayTech RNS utilizes a feature called chart layers, letting you superimpose one type of cartography over one or more others. For example, a vector chart, raster chart, topographic chart, aerial photo, and radar image can all be overlaid. Optional subscription data, such as sea temperature or plankton, can also be added. The presentation of any layer (except Navionics and C-Map vector charts) can then be ad-justed using a transparency control.

RayTech RNS also includes many improvements on prac-tical bread-and-butter features. For example, their ruler tool to measure distance and bearing has multiple rulers and al-lows you to pick up and move both the origin and destination points. Their latitude/longitude grid was the most sensible we've seen so far, using clean whole number values (such as W 070 10.0') rather than forcing a display grid (with values such as W 070 09.334'). These small details give RayTech RNS the feel of a professional, competent marine package.

Unfortunately, the chartplotter-like interface of RayTech RNS affects some features that would be considered standard on a PC application. Most notably, just like a chartplotter, RayTech RNS does not have a printing feature. If you want to print your charts you must purchase a screen capture soft-ware package such as SnagIt for $39.95 (www.SnagIt.com).

Waypoints and Routes

Working with waypoints and routes was a mixed experience, stemming in part from RayTech's chartplotter interface and hurdles with import and export of data outside the Rayma-rine collection.

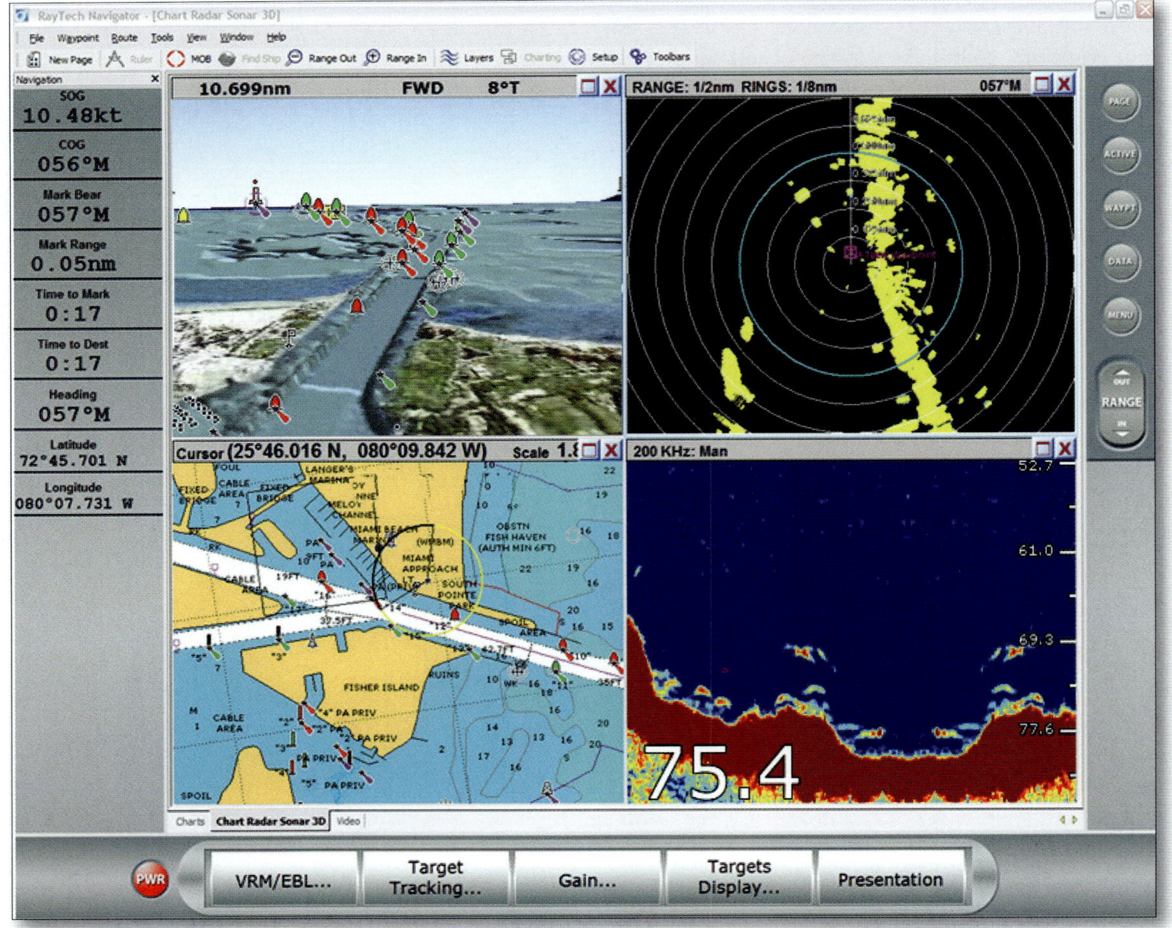

Multifunction Display: A four-way split window showing (clockwise from upper left) a perspective view, radar display, sonar display, and Navionics vector chart.

Waypoints and routes are easy to create, but the process is not as user-friendly as some applications using a PC interface approach. Yet the creation and management of waypoints is very robust in RayTech RNS. The only inconvenience we encountered was the inability to have a waypoint in more than one folder or to transfer more than one waypoint at a time to a new folder. For example, if you want to transfer a collection of Coconut Grove waypoints from your Intracoastal Waterway folder to your Florida Keys folder, you must move each waypoint individually.

As one would expect from a chartplotter interface, there is no mechanism to scroll or flip from the waypoint or routes list to the appropriate chart. To go to a waypoint's chart location, you must search for the appropriate chart by going back to the planning chart overview display.

RayTech's strongest waypoint and route feature is its exportability to other Raymarine devices. If you use a Raymarine chartplotter, the data seamlessly transfers using their SeaTalk network over an Ethernet cable, which plugs into your PC's Ethernet port.

However, the extent of information you can "push" up to your chartplotter depends on the capabilities of that chartplotter. We've already mentioned the new hardware needs for multidimensional data such as 3D perspectives, bathymetrics, or aerials. Similarly, the software and hardware constraints of your chartplotter dictate the number of waypoints and/or routes you can export. For example, a C-Series display is limited to 1000 waypoints and 100 routes; the E-Series to 1200 waypoints and 150 routes. If you cruise extensively, you will have to work around this limitation by grouping waypoints into subsets and uploading each set as you need it.

Importing waypoints and routes to RayTech RNS, such as from another vendor's software application or device, is trickier. Although in theory Raymarine can accommodate Microsoft Excel spreadsheets for data entry and ASCII text files for import/export, the process needs a helper application such as GPSBabel (www.gpsbabel.org). (See GPS Helper Applications in Chapter 8 for more details.)

RayTech RNS also includes several advanced sailing navigation and tactical tools as part of its basic package. The po-

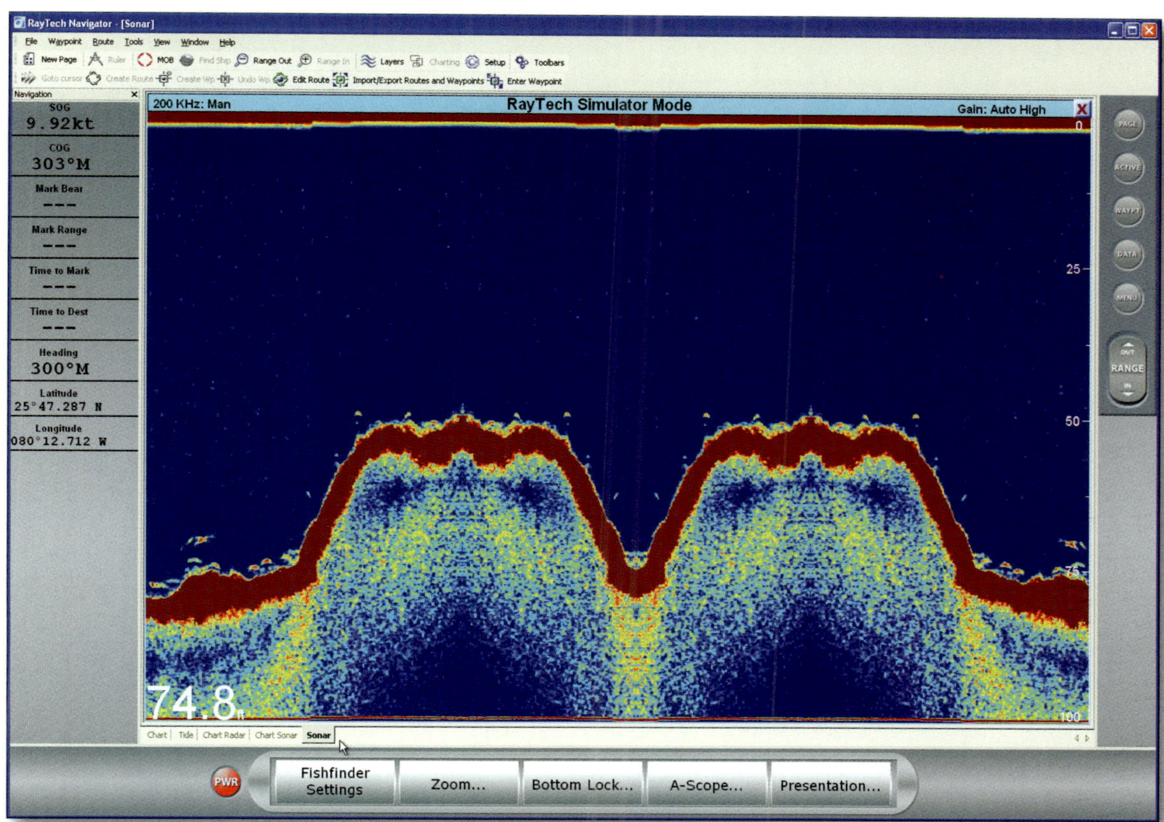

lars and pre-start features are great free inclusions for club racers. These are comparable to the *basic* features in Nobeltec or MaxSea sailing packs, each requiring an additional purchase.

RayTech's tools also include DataTrak which, when connected to your boat's instruments, can record and display collected data as a time-based graph. This is useful for looking at trends, averages, real-time data, and comparing instrument non-agreement. You can create graphs to display course and speed over ground, sea temperature, depth, apparent and true wind, and other data useful for route optimization, advanced weather routing, creating sailing polars, or race pre-start calculations.

Tabbed Interface: *Similar to Microsoft Excel worksheets, RayTech RNS accesses multiple pages in a tabbed interface, such as the sonar page selected here (see cursor).*

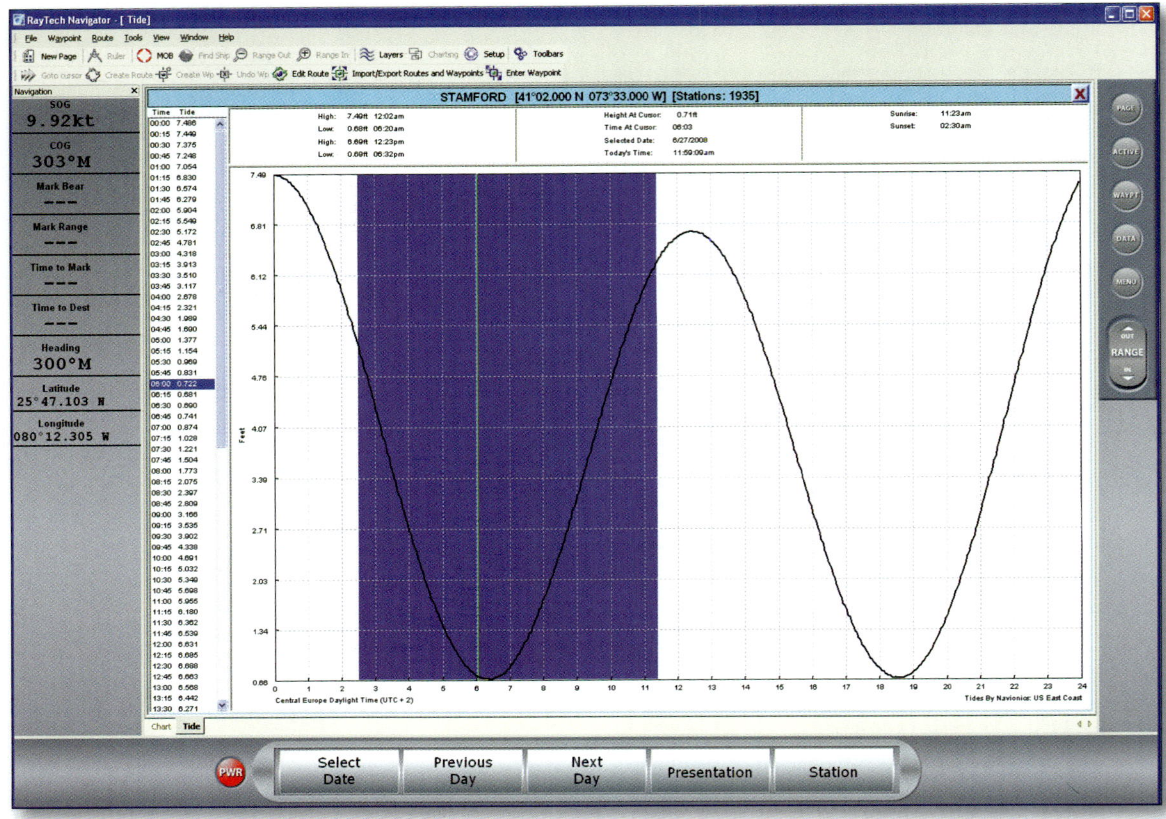

Navionics Tides: *World Tide and Current data is provided by Navionics. In any chart view, right-click* Open Nearest Tides *or create a new tide chart page with* File>Open New Page>Tides.

ADDITIONAL FEATURES

RayTech RNS is designed to be part of a fully-networked Raymarine electronics installation. This is an important distinction. With most PC-based charting and navigation applications, you plug devices into the PC, such as a GPS or autopilot. In other words, you tell the PC which external sensors are now part of the system. With RayTech RNS, you plug the laptop into an existing network of sensor data. In this application, you tell an existing Raymarine network that a PC with RayTech RNS is now part of the system.

This difference creates an important distinction between what RayTech RNS does by itself and what it does as a repeater node on the vessel network. For example, the engine panel feature is an excellent illustration of how your Ray-Tech-loaded PC is an extension or node on your existing system. With your laptop connected to a Raymarine E-Series chartplotter, which is connected in turn to an NMEA 2000 engine interface, you can display engine instrument data on your PC screen. The software is not generating any information. Rather, it is repeating the data sent by the engine via the E-Series network.

RayTech RNS is designed to really shine when used in conjunction with a Raymarine E- or G-Series multifunction display. Your PC running RayTech RNS connects to these chartplotters with a simple, single Ethernet cable between the PC and a Raymarine SeaTalk network switch. Unfortunately, E- and G-Series units are a bit of an investment, costing several thousand dollars depending on the model and screen size. RayTech RNS also integrates with Raymarine's C-Series, but these units are more limited—think alphanumeric data sharing only. For example, they cannot accommodate multidimensional image sharing.

Obviously, nearly any NMEA or Raymarine device can be part of this extensive network. RayTech RNS 6.0 does not support Automatic Identification System (AIS) tracking. Raymarine intends to include support for AIS in their next release. However, collision avoidance is currently supported through a MARPA-equipped radar.

Because your PC is configured as an additional display, it is important to understand the extent to which data can be shared or repeated across the network. For example, relatively "smaller bandwidth" images, such as radar, vessel data, or charts, are repeated to both PC and chartplotter.

Multidimensional data, such as 3D perspectives, bathymetrics, or satellite photos, can be repeated only to the more capable Raymarine E- or G-Series models. Video, a notoriously high bandwidth data stream, can only be transmitted directly to your PC—there is simply too little bandwidth to repeat this data to a chartplotter display. However, as a standalone device, RayTech RNS can display up to four video inputs on your computer screen.

RayTech supports many additional data features such as street and map data and satellite, topographic, and aerial images. These databases are not included with the software. It's assumed the consumer will purchase upgraded Navionics Platinum cartography for these additional features.

The software includes built-in worldwide tides and currents data provided by Navionics. (RayTech Planner also includes tides and currents data, but only for the U.S.) This data, part of a co-marketing agreement between Raymarine and Navionics, is top quality. The data is extensive and displayed in an easy-to-use format.

Weather data is obtained by downloading GRIB files. We were particularly impressed with RayTech's animated weather forecasts. This feature lets you pull down a seven-day GRIB and fast-forward the display minute-by-minute. This smooth replay lets you follow isobars as they expand or contract, or wind barbs as they clock or back.

Like many other marine software vendors, Raymarine also offers subscription services. Two fishing data features, sea surface temperature and ocean plankton data overlays, are available by subscription through Raymarine's website.

Assessment

RayTech RNS is a bit different than the other PC applications. It functions less as a free-standing PC application that connects to external devices, and more as an additional portable node in an existing, extensible network of Raymarine devices. This creates important distinctions in both its user interface and how it is intended to be used.

In its intended configuration, your PC is not the center of a marine network, but is an additional mobile or mirrored display. Your PC becomes another display down below or at the helm, as well as a convenient and mobile way to do waypoint and route planning at home.

Because of this, RayTech RNS is best suited as part of a Raymarine networked package—ideally including a powerful E- or G-Series multifunction display. If you want a fully integrated system, or have recently purchased a new Raymarine chartplotter, RayTech is a logical addition. The added display and flexibility to your system is a small relative cost and gives you all the convenience of PC charting and navigation with an incredibly easy Ethernet or CompactFlash connection.

What's Next?

At press time, RayTech RNS 6.1 was on the horizon but not yet available for testing. Version 6.1 is designed to work with Raymarine's new G-Series multifunction displays (the successor to its C- and E-Series). It also supports Raymarine's new HD digital radars and HD digital sounders. AIS target tracking, not supported in RayTech RNS 6.0, will be included, as well as support for Navionics Platinum+ charts. Also expect greater Vista compatibility.

To determine which RayTech features can be used in conjunction with your existing onboard Raymarine electronics, visit: http://raymarine.custhelp.com/cgi-bin/raymarine.cfg/php/enduser/std_adp.php?p_faqid=1316.

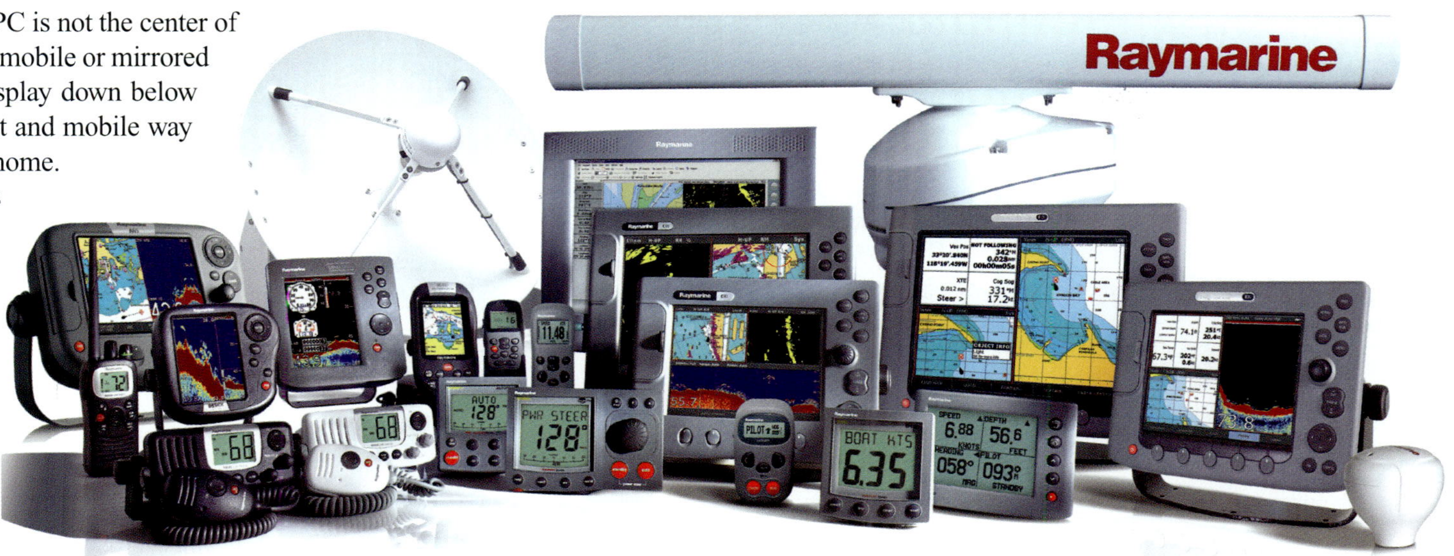

Family Affair: *Unlike many standalone PC charting and navigation applications, RayTech RNS is designed to be part of an existing extensive network of sensor data.*

RayTech RNS
FEATURES AT A GLANCE

Each full-featured e-charting application has been evaluated according to the following 81 features. Some ratings are binary, "yes" or "no." Others are more subjective, evaluated on a scale from 1 to 5 (5 being "Excellent"). Any other comparison is defined in the Legend.

Out of Box Experience

Packaging	3
Documentation	4
Software install	3
Load charts	3
Load supplemental data	3
GPS hookup	2
Technical support	4

User Interface

GUI metaphor	CP
Screen space management	2
Program responsiveness	4
Chart display speed	3
Chart display quality	3
Customizable GUI	2
Customizable metrics	2
Split windows	Yes
Tabbed interface	Yes

Basic Features

One-button MOB	Yes
Waypoint creation	3
Route creation	3
Steer-to function	Yes
Track creation	3
Convert track to route	No
Waypoint and route management	4
Chart management	2
Search waypoint and route	2
Range/bearing tool	4
Chart annotation	No
Chart printing	No

Data Exchange and Networking

GPX support for waypoints	No
GPX support for routes	No
Waypoint exchange formats	EC
Route exchange formats	EC
Integrated GPS transfer	No
Multiple monitor support	Yes
Network application sharing	Yes

Cartography

Free U.S. rasters BSBs (RNCs)	Yes
Free U.S. vectors S-57s (ENCs, IENCs)	No
SoftCharts	Yes
CHS Canadian rasters	Yes
DNC	No
Seafarer	No
British Admiralty ARCS	No
International S-57s	No
International S-63s	No
Scan and geo-reference paper charts	No
Navionics cards	Yes
C-Map cards	No
C-Map CD/DVD	No
Satellite geo-referenced photos	Yes
Aerial informational nav photos	NV
Topographic maps	Yes
Bathymetric data	NV

Instruments and Sensors

GPS	Yes
Autopilot	Yes
AIS receiver	No
Wind	Yes
Depth	Yes
Water temperature	Yes
Radar	Yes
Heading sensor	Yes
Video camera	Yes

Advanced Features

Tides	4
Currents	4
Coast Pilot	No
POIs	NV
Streets	NV
Google Earth support	No
Advanced search	No
AIS	No
Buddy boat	No
Radar (ARPA/MARPA)	Yes
Radar (display overlay)	Yes
Fuel calculator	No
Transit calculator	No
Celestial calculator/Almanac	No
Auto route planning	No
Great Circle Route planning	No
Sail performance	4
GRIB weather integration	4
Weather options	No
Customizable bathymetric recorder	No

Legend

5	Excellent	CM	C-Map
4	Above average	NV	Navionics
3	Average	SC	SailCruiser
2	Below average	MT	Mr. Tides
1	Poor	PP	Plus Pack
CP	Chartplotter	C	CSV
Mac	Macintosh	E	Excel
NT	Non-traditional	G	GPX
Win	Windows	K	KML
N/A	Not applicable		

Chapter 24

Furuno MaxSea Explorer

Since 1986, every winner of the Volvo Around The World/Whitbread, BOC, and Vendee Globe Races has used MaxSea software. Beginning as one of the first e-charting applications, MaxSea is now an established player in electronic charting, forming a technical alliance in 2004 with Furuno to combine software and hardware strengths. MaxSea is currently the leading navigational software in Europe—available in English, Spanish, Portuguese, Danish, Italian, Norwegian, Swedish, Dutch, French, Icelandic, and German—with more than 40 percent marketshare in the fishing industry and 30 percent in the yachting industry. Over 25,000 copies have been sold worldwide.

MaxSea is a modular application, meaning it consists of a base software package with optional add-on modules for specialized features. Understanding these modules is important because they govern which features are included or can be added. The basic software comes in five levels: Planner (free), Navigator+ ($500), Commander ($1,000), Explorer ($3,000), and Professional ($5,000).

MaxSea Planner is simply planning software, available free on DVD from Furuno dealers. It does not connect to a GPS, but you can create and transfer routes to your Furuno chartplotter using an SD memory card. It is a "MaxSea Lite" sample, a planning tool or trial before upgrading to a full version of MaxSea.

Navigator+ is MaxSea's base package developed for the U.S. market, including a full set of NOAA raster charts. Although it does not include any add-on modules, it connects to a GPS and autopilot and can download and overlay weather data. Performance Sailing, Weather Routing, and AIS modules can be added.

Commander adds features such as events recording, data layer editing, and the option to add additional modules beyond Navigator+, such as depth profile window or advanced AIS display.

Explorer is a recently-created offering, designed to bundle the best features of Commander with the PBG (Personal Bathymetric Generator) 2D/3D Module and MaxShell Windows Desktop Management Software.

Professional is the same software as Commander, but with nearly all modules included. It also integrates a fleet tracking system that works via e-mail, designed for shipping, fishing, or charter companies.

Furuno also offers NavNet-Commander and NavNet-Explorer, which are the same as the basic Commander and Explorer packages but allow direct connectivity and interaction with Furuno NavNet 1 and vx2 systems.

Clearly, MaxSea has no shortage of e-charting options at an assortment of price points. There are entry-level packages that compare in price and functionality to Nobeltec VNS MAX Pro or Maptech's Chart Navigator Pro. There are also options that cost thousands of dollars with all the modules, and are designed for the professional. Which package you choose depends on which modules you currently need or may want in the future. But note that not all software choices are compatible with every module. For this chapter, we review MaxSea Explorer.

Indeed, navigating the modules alone can be daunting. For example, MaxSea provides two modules for integration with ARPA devices. The ARPA Module ($250) allows the display of ARPA and AIS targets, but does not include the ARPA Center or AIS Messaging. The Mobile Target Tracking Module ($250)—MaxSea's most popular—displays targets, includes Target Centers and AIS Messaging and allows target tracks to be saved in a specified data layer.

The Personal Bathymetric Generator that comes with Ex-

Furuno MaxSea Explorer

Pros: Overall speed and responsiveness; superior image quality; extensive and varied documentation; modular software options; organization by layers; supports British Admiralty Charts (ARCS); GPX import/export of waypoints.

Cons: Does not support NOAA free vector charts; mark names limited to eight characters; no undo command.

Coolest Feature: Sail and Weather Routing

Price: $500 (Navigator+); $1000 (Commander); $3,000 (Explorer); $5,000 (Professional)

Vista Capable: Yes

Version Tested: 12.6.3

System Requirements
PC with Pentium processor
512 MB RAM
40 GB available hard disk space
DirectX9 compatible video card
Windows 2000, XP, or Vista

Furuno Electric Co., Ltd.
9-52 Ashihara-cho
Nishinomiya City, Hyogo. 662-8580
Japan
www.furuno.co.jp

Furuno USA, Inc.
4400 N.W. Pacific Rim Boulevard
Camas, Washington 98607
United States
www.furunousa.com

Do you already own or are you considering the purchase of MaxSea software? Be sure to read *What's New: MaxSea Time Zero*, at the end of this chapter, for an overview of the upcoming major changes to MaxSea.

MaxSea offers its software in many versions at varying price points and compatible (or bundled) with different add-on modules.

Planner

Free planning software for transferring way-points and routes to a Furuno chartplotter.

Navigator+

The base package without any modules. The Performance Sailing, Weather Routing, or ARPA Modules can be added.

Commander

Like Navigator+, but adds features such as advanced AIS display. In addition to the above modules, supports the 2D/3D Module.

Explorer

A bundle that includes the Commander software and the 2D/3D and MaxShell Modules.

Professional

A bundle that includes the Commander software, a fleet tracking system, and all add-on modules.

The easiest way to enter MapMedia Chart and Bathymetric Data codes is to first close MaxSea. Next, connect the computer to the Internet and turn off any firewall. Open **Configuration Codes Internet** from the **MaxSea Utilities** located under Windows Programs. The new permits will be downloaded and MaxSea will then open.

plorer allows you to continuously update your 3D charts with data from your sounder. A Ground Discrimination Module ($250) combines with third-party hardware to record bottom hardness and bottom type, which is a popular feature among fisherman.

The Sailing Performance Module ($250) adds tools such as upwind and downwind lay lines displayed on the chart, which are continuously corrected for changes in true wind direction and current. MaxSea's Weather Routing Module ($250) uses a boat's theoretical performance data, tide and current information, and GRIB weather files to calculate an optimal sailing route.

The NavNet Module ($500) upgrades MaxSea Commander, Explorer, or Professional to the NavNet version, allowing for direct data exchange with a Furuno NavNet system. The MaxShell Windows Desktop Manager Software ($250), included with Explorer and Professional upon request, is a module that replaces the Windows interface. It is designed for use on professionally crewed vessels, separating MaxSea from the PC's other functions, while giving direct access to MaxSea utilities.

Getting Started

MaxSea Explorer comes in a boxed set that includes an installation CD, a Quick Reference Card, a Getting Started Guide, and more than 1000 NOAA raster charts on DVD. Optional MapMedia discs, such as regional bathymetric charts, are also available. Any add-on modules are marked on the box and included with the software.

Licenses are transferred using a software key, obtained by registering over the Internet, or via a hardware key (also called a dongle). Furuno recommends the flexibility and safety of the hardware key option. A hardware key allows the software to be installed on any number of computers for backup or planning purposes and allows for re-start without contacting Furuno in the event of a catastrophic computer crash.

At first, MaxSea's installation is a bit intimidating. This is an extensive program, involving a lot of documentation in a mix of French, English, and "Frenglish." Surprisingly, it was easy to learn, thanks in part to detailed documentation in a va-

riety of formats. MaxSea includes an 84-page Getting Started Guide, a 34-page Installation Manual, and a 530-page Operator's Manual (installed in the My Documents folder). Fifteen tutorials are loaded into My Documents>My MaxSea Tutorials and are also available to view directly on the Web (http://comen.maxsea.fr/MaxSea/Products/RollingDemos/default.aspx). The robust HTML Help is accessible through the program from the Help Menu or by pressing [F1]. It's better than a typical help menu, with a table of contents, index, search, glossary, and even a way to save favorites for tasks that are complex or seldom used.

The good news about the reams and megabytes of documentation is that the diversity of formats accommodates different learning styles. You can get started simply by reading the Installation Manual and viewing the Tutorials. Or, if you are a meticulous document-reader, you can go through all the written material at home over a winter. We recommend beginning with the excellent tutorial movies, then reading the Getting Started Guide, then filling in any detailed questions using the Help Menu, Operator Manual, or Installation Manual.

Chart installation was straightforward. Explorer automatically recognized our C-Map charts on discs and C-Map cartridges read from a USB Multimedia Reader. It also recognized our NOAA BSB and Maptech BSB charts. The first installation of Maptech's 53 BSB charts from Region 78 "Flamingo to St. Mark's, Florida" was a bit slow. According to Furuno, Explorer verifies the entire chart catalog upon the initial load. Once raster charts were loaded they opened promptly.

Although there is no Wizard, the GPS installation was also easy. Very clear instructions guide you through the process screen-by-screen. The only possible confusion arises when a USB GPS is in use, requiring you to download the device's USB-to-serial driver from the manufacturer. Furthermore, since your device is no longer reading as a USB device, choose the button marked Serial Port (COM) rather than USB.

Look and Feel

MaxSea's interface centers on a Tool Palette and a Chart Palette. The icons of MaxSea's Tool Palette are designed to emulate the tools mariners use with paper charts. Though it is a computer-based tool, MaxSea emphasizes traditional navigation tools and concepts. Recall that MaxSea is the most popular navigation program in Europe—a part of the world that often requires recreational boaters be licensed and demonstrate a basic knowledge of navigation. MaxSea assumes users are familiar with basic navigation and are inclined to think in terms of traditional navigational reasoning. For example, it is very easy in MaxSea to quickly use traditional triangulation (three crossed bearings) to confirm your electronic or dead reckoning position.

The *Tool Palette* consists of a collection of icons which change the functionality of the cursor and let you perform various actions. For example, the pencil tool creates marks, the finger tool highlights more information, and the delete tool removes objects. The Tool Palette can be customized to include only the tools you typically use.

One of the toolbar icons looks like a yellow Pacman. This is the delete tool. The Pacman icon is appropriate: it is ruthlessly efficient, deleting in a single click without warning or confirmation. This tool is especially dangerous since MaxSea has no "undo" option. We found ourselves accidently deleting objects until we learned to be more cautious.

The downside of a Tool Palette is the amount of mousing associated with each function. For example, one must enable the pencil tool to create a waypoint, then return to enable the "move position" tool to move it. But there are some shortcuts and workarounds. Instead of enabling the finger tool to request object information, you can also simply hover the cursor over the object to display the information link.

To really cut down on your clicking, important windows such as the Waypoint Center or Route Center can be saved with a tab at the bottom of the screen. This optional tabbed interface is similar to Microsoft Excel and provides much faster access to windows you need to open repeatedly.

The *Chart Palette*, across the top of the screen, works a bit differently than the Tool Palette. This row of buttons is for tasks such as changing the orientation of the chart (north-up, head-up, or course-up), displaying different chart or object layers, or switching to a 2D or 3D display. Unlike a tool, which stays enabled to perform multiple cursor-based actions,

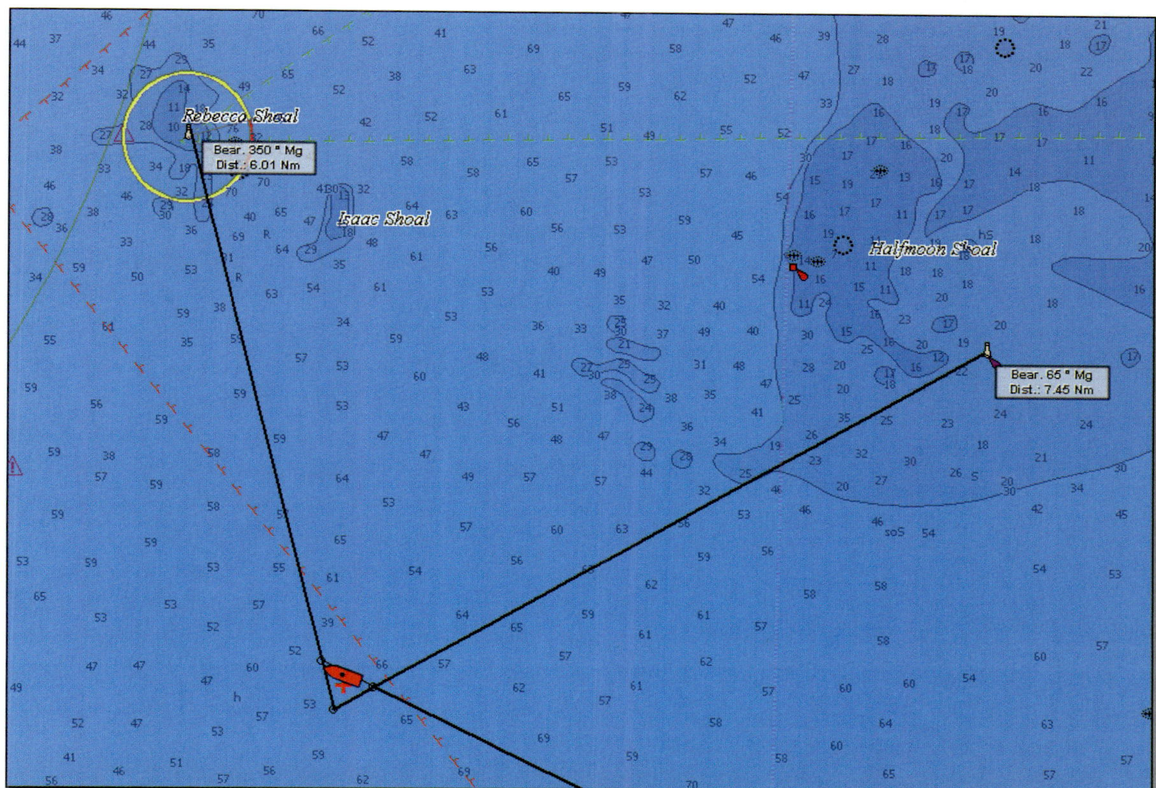

Plotting Multiple Bearings: *MaxSea's roots lie in traditional navigation. Three Dry Tortugas lights— bearing 350°M, 65°M, and 120°M—confirm your DR or electronic position.*

a button initiates a single action. Like the Tool Palette, the Chart Palette can be customized; you can add or omit buttons, display a button's name, or display large icons for easier viewing underway.

MaxSea also uses traditional menus along the top of the screen, adding 54 keyboard shortcuts for the power user. Data boxes are shown as windows along the right edge of the display. The Cursor Data window shows information such as the latitude and longitude of the cursor position, its distance and bearing from your vessel, and the depth at that point. The Nav Data window shows the latitude and longitude of your vessel position, speed, course, and heading. Both windows can be customized for the data, level of transparency, and font size.

MaxSea's interface is highly customizable visually and operationally. Features such as line width, shading, icons, objects, and color can all by user-defined. More importantly, the menu Utilities>Option provides options for accessing and cus-

MaxSea's tools can be accessed through simple contextual menus, available with a right-click of the mouse.

tomizing specific windows, such as GPS/Track, Fuel Consumption, or Radar.

WORKING WITH CHARTS

MaxSea reads charts in several formats, including both raster and vector. It is one of the few navigation applications that reads British Admiralty Raster Charts (ARCS). In raster format, MaxSea can display NOAA BSBs, Maptech BSBs, Canadian Hydrographic Office charts such as NDI/Digital-Ocean, and ARCS. It can also read SoftCharts and MapMedia's new charts (MaxSea's cartography company is providing raster charts for Europe, Africa, and Pacific territories).

However, MaxSea does not support S-57 vector charts, which includes free NOAA ENCs and Army Corps IENCs for U.S. coastal and inland waters. These charts were supported in version 12.0 but were dropped in version 12.5 "for internal reasons," relating to strategic decisions with MapMedia, MaxSea's cartography division. MapMedia sells digitized charts and maps in raster format and is moving toward vector (www.mapmedia.com).

In vector format, MaxSea can display C-Map NT+ and C-Map MAX charts. Although MaxSea can read C-Map charts from CD or cartridge, Furuno recommends CD-ROM. Reading charts from an external device, such as C-Map's USB 1.0 multimedia reader, is too slow for the program. We can attest to this.

With the 2D/3D Module, MaxSea can also display contour and bathymetric charts. These bathymetric charts are purchased separately from MaxSea.

Like all upper-end navigation applications, MaxSea displays charts seamlessly and with split windows. Charts are panned with a grabber hand and zoomed with the mouse scroll wheel. Panning has a bit of a different look: new charts beyond the "seam" are not dragged along, but are left as black, only appearing when the panning click is released. This behavior is disconcerting at first, because it suggests a missing chart section. But the idea makes sense: rather than dragging along all those pixels the display waits until you have stopped panning, then displays a fully rendered chart.

Most notably, MaxSea has stunning image quality. The display is incredibly vivid and sharp, with excellent line quality for raster, vector, and satellite images. While charts typically use 256 colors, charts produced by MaxSea's cartographic company MapMedia use 16-bit color (that's 65,536 colors). MaxMedia claims to be the only company to produce 16-bit color charts.

Although MaxSea originated as a Mac program, it is now firmly in the Windows camp. In fact, MaxSea's display architecture uses the same Windows technology that drives a Microsoft Xbox for interactive gaming. The technology is DirectX, a graphics acceleration add-in to the Windows operating system that is used by game and CAD programs. In the MaxSea setting, DirectX lets you quickly render high-quality 3D graphics that can be rotated in perspective.

WAYPOINTS AND ROUTES

Waypoints, routes, and marks are easy to create in MaxSea. The Insert Waypoint tool creates a waypoint on the chart location, which can be dragged and dropped to adjust its position. Unfortunately, mark names are limited to an eight-character

Image Quality: MaxSea has a stunning display. Here a split window shows off excellent screen rendering of a satellite photo (left) and NOAA raster chart (right).

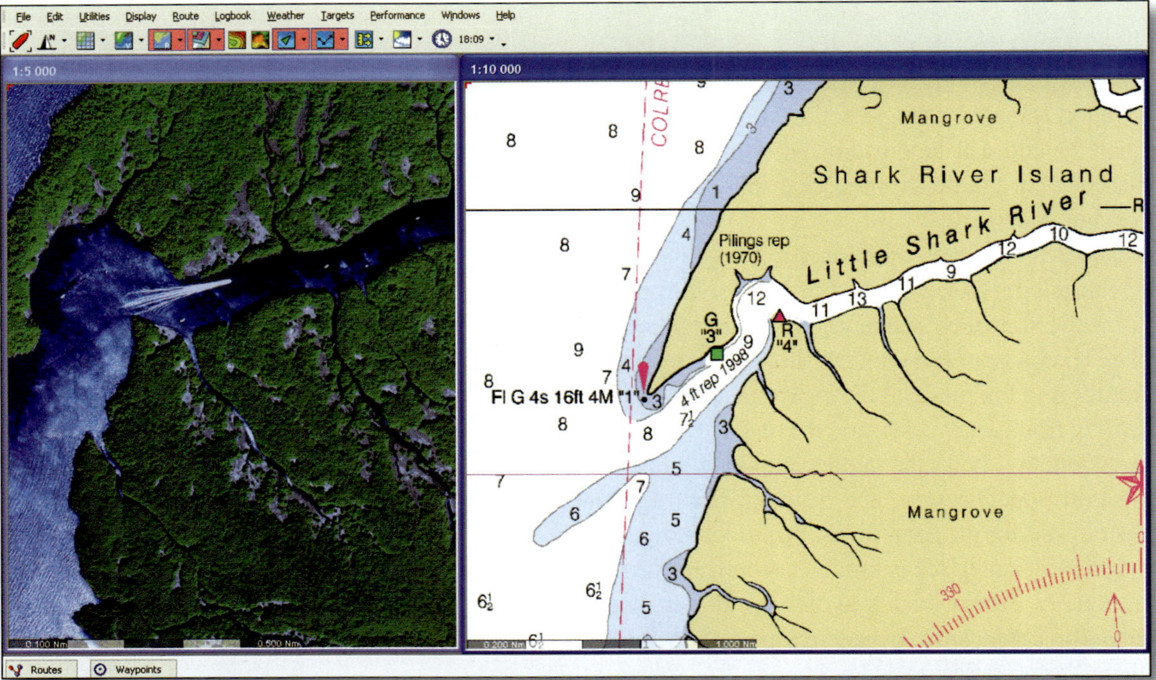

field. You can circumvent this restriction by choosing Text instead of Point to set a mark, then choosing an icon to associate with the text. When the mark is entered on the chart it will have an associated long text field for a more complete name.

Charts can be annotated with a palette-full of options, including marks, lines, areas, circles, text, color, and/or shading patterns. All objects can be saved on a layer or multiple layers to be displayed or hidden, letting you customize and organize your charts with fishing areas, danger zones, or anchorage notes.

The Go To tool (really a SteerTo) quickly creates one-stop routes, a commonly-used navigation trick to monitor one's cross-track error. Selecting the Go To tool and clicking on a point ahead on the chart drops a waypoint, creates and activates that route, and opens a Steering Data window with information such as your estimated arrival time. It automatically transmits that information to your autopilot, if it is connected. Any active route also displays cross-track error boundaries with dotted lines (red for port and green for starboard). You can customize this from the default setting of 0.2 NM.

Longer routes with multiple waypoints are also easy to create. Drawing and clicking across a chart creates a temporary route with QuickWaypoints. These points are linked to the route and won't be saved in the waypoint database unless instructed. MaxSea's routes include many nice features and details. For example, routes can show distance and bearing (if turned on in Route>Catalog). You can calculate departure or arrival times or estimate fuel use. Routes can be activated at any leg along the route by selecting the Go To tool and clicking on a route leg. An active route can be quickly cancelled by double-clicking on the Go To tool.

Routes are also very flexible in MaxSea. For example, you can customize routes to exclude certain areas when planning a route, automatically avoiding areas such as exclusion or danger zones. You can be running an active route while displaying a planning route. In other words, you can plan and create a route for future use while actively navigating a previously-created route.

Waypoints and routes are managed through the Waypoint Center and Route Center, windows that appear across the bottom of the chart, or as tabs for easier access. The Waypoint Center includes detailed information such as waypoint names, comments, and creation dates. You can customize these data columns. The Route Center is similar, summarizing information about each route. The Routing Center (different from the Route Center) shows the information for the active route, such as leg number, course, speed, and direction.

These windows are linked to the actual chart, letting you bring up charts from waypoints or routes. By highlighting a waypoint in the Waypoint Center and clicking on the Center icon, a chart opens with that waypoint in the center of the display. Any waypoint you touch on the chart is highlighted in the Waypoint Center; and selecting a waypoint in the Waypoint Center highlights that mark on the chart.

All waypoints, marks, objects, and tracks are created on layers. Every object you create—points, lines, text, or anything else—is assigned to a layer. Each layer can be displayed and edited separately, or they can be combined. But you must plan your layer construction so you can hide or display each layer intelligently. Obviously you don't want everything on one layer, nor do you want every single object on its own layer. If you think about how to layer your personal navigation information you can create extremely data-rich charts without cluttering your display.

Fortunately, it's easy to change an object's layer. For example, a waypoint can be assigned to another layer simply by pulling down the new layer's name in the Waypoint Center. The buttons in the Chart Palette let you toggle between displayed layers. Once you have created your information in layers, you can turn on and off individual layers to only show those within your chosen groupings.

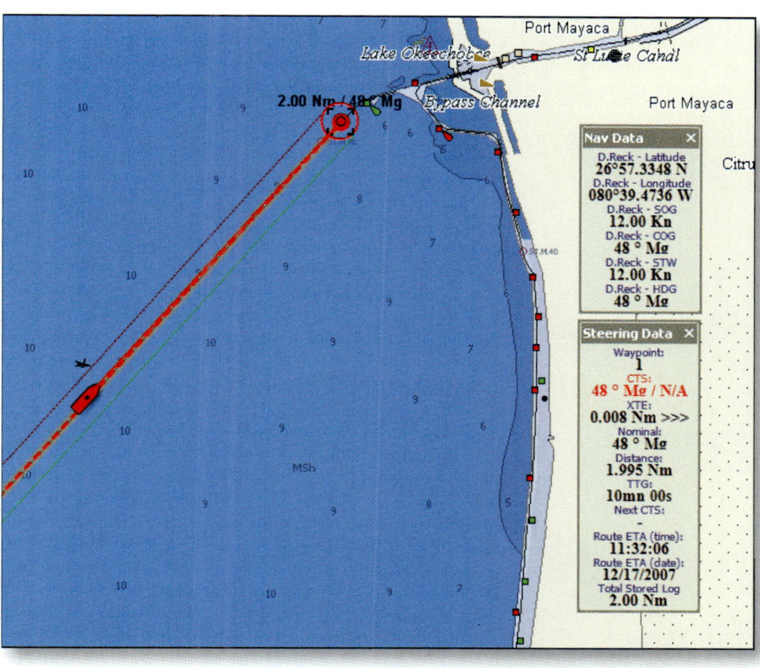

Straight and Narrow: *Crossing Lake Okeechobee, a Go To Waypoint can be set to monitor cross-track error (XTE). MaxSea adds many nice touches, such as red (port) and green (starboard) dotted lines to visually indicate user-chosen XTE boundaries (in this case 0.10 NM).*

Always select the Hand Tool after finishing with another MaxSea tool, such as the Waypoint, Erase Object, or Zoom tool. This avoids any unintentional actions such as erasures or creating unwanted waypoints, routes, or objects.

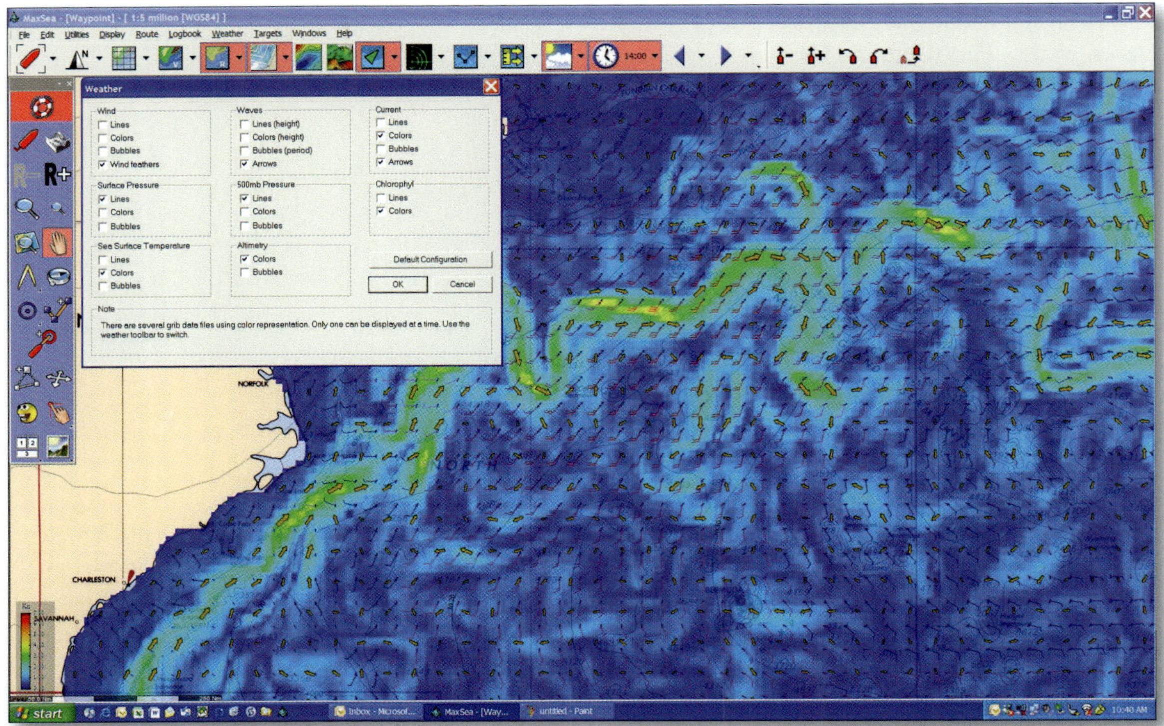

MaxSea Chopper: *MaxSea maintains its own weather server and integrates excellent observation and routing tools.*

Tide and current predictions are displayed with a single click. The data can be shown as a chart overlay or as complete tide or current tables for a chosen station. Weather data is also integrated, using a utility called *MaxSea Chopper*. This free service requests a GRIB weather file by email from the MaxSea server. You can choose different sources of weather data, weather from many regions of the world, the type of data (such as wind, pressure, waves, and sea surface temperature), and the forecast period. An area of weather data can be specified by simply dragging a rectangle over the weather download map. The data can be sent either as a single email, or as part of an excellent free subscription option that delivers a new weather file based on your saved parameters each day.

Weather is displayed as an overlay on the chart, with a pull-down menu to select the particular data values shown. For fastest access to your weather layers, you can drag the button off the Chart Palette to create a tiny weather toolbar, letting you single-click to display or hide weather information as you need it. The display can even be animated and saved to a file as a movie for future reference.

Some of MaxSea's additional features require purchase of optional add-on modules. For example, displaying contour and bathymetric data requires the 2D/3D Module and additional bathymetric charts. We loaded MaxSea's East Coast 04 Cape Charles to Cape May disc, which covers Hatteras Canyon to just north of Hudson Canyon. We've already mentioned the stunning displays of this data, where perspectives of the sea bottom rotate instantly, showing your vessel on a transparent sea surface. However, these extras are specifically geared to fishermen. They not only cost extra, but the single region we loaded was 1.2 GB, included 50 files of 3D data, and took nearly 90 minutes to load. These are not extras to add unless you need them.

If you are a deep-sea fisherman (or treasure-hunter), you can go a step further with the Personal Bathymetric Generator (PBG). The PBG updates the bottom contour in real time using your echo sounder or fish finder. Your collected data is stored in a PBG database separate from the database provided on disc—the two databases seamlessly blend for the actual display.

MaxSea also records tracks, customized to record either all your instrument data or only your position by points at a specified time or distance. Tracks can be saved to a file, deleted, or copied onto a CD or USB drive. If your display becomes cluttered with too many tracks or layers, they can be made inactive, which temporarily removes them from the menu pulldown choices. These files can even be encrypted, accessible only with a password—a useful feature for fisherman to hide their hot spots or for philanderers to hide something else.

Because it supports GPX technology, MaxSea's ability to import and export is excellent.

ADDITIONAL FEATURES

Although MaxSea supplements its features with add-on modules, many important features—such as tides, currents, and weather—are integrated into all packages. This extends from Navigator+ at the lower end all the way up to Professional at the high end.

For sailors, MaxSea provides a Sailing Performance Module to visually display and calculate the relationships between true wind angles, true wind speeds, and boat speed. Performance efficiency is shown graphically and numerically, letting you evaluate your speed relative to your boat's theoretical potential.

The Weather Routing Module is a significant build on the Sailing Performance Module, calculating optimal routes based on weather, sailboat characteristics, and currents. This is an impressive algorithm. You begin by choosing a polar file that corresponds to your vessel—or you can set your own custom specifications. Then the route is initialized by setting start times and position. You can limit your route to sailing conditions below a particular wind speed, or choose "less than 100 knots" to let the algorithm calculate the most efficient direct route. You can pre-set the algorithm to avoid certain areas, such as night sailing through regions reported with pods of right whales. The algorithm then computes the most efficient and safest route between points based on the *isochrons*, which are lines representing potential positions given equal time. This route is saved and can be invoked as an active route.

ASSESSMENT

At first glance, MaxSea Explorer may appear to have the same features as many other full-featured packages: waypoints, routes, annotated charts, tide and current prediction, weather, and so on. If so, then what's the difference? What is there to justify a $1000-plus price tag?

The distinction is not simply the presence of particular features, but that Explorer includes all of them—and each with excellent implementation. MaxSea's marine software does well on the details of each of these features, such as visually showing cross-track error on all routes, integrating waypoint management and layers, or providing GPX data transfer.

We evaluated Explorer, which is definitely one of Max-

Point A to Point B: *MaxSea's Weather Routing uses a weather forecast, tidal current, and theoretical speed of the sailboat to calculate an optimal route. Step 1 is to choose a polar file to use and/or modify (left). Step 2 inputs a start time, then graph, calculation, and routing parameters such as isochron intervals, sailing efficiency expectations, current data, maximum wind tolerance, and route preferences (center). A detailed plot is then calculated showing an original (straight-line) route and a route optimized for meteorological factors, currents, boat characteristics, and your navigation preferences and style (right).*

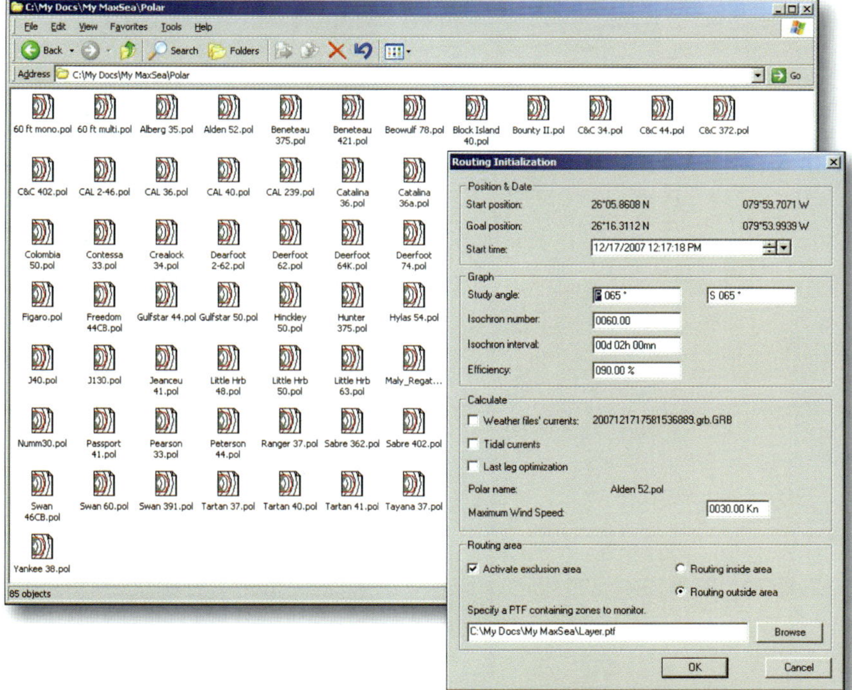

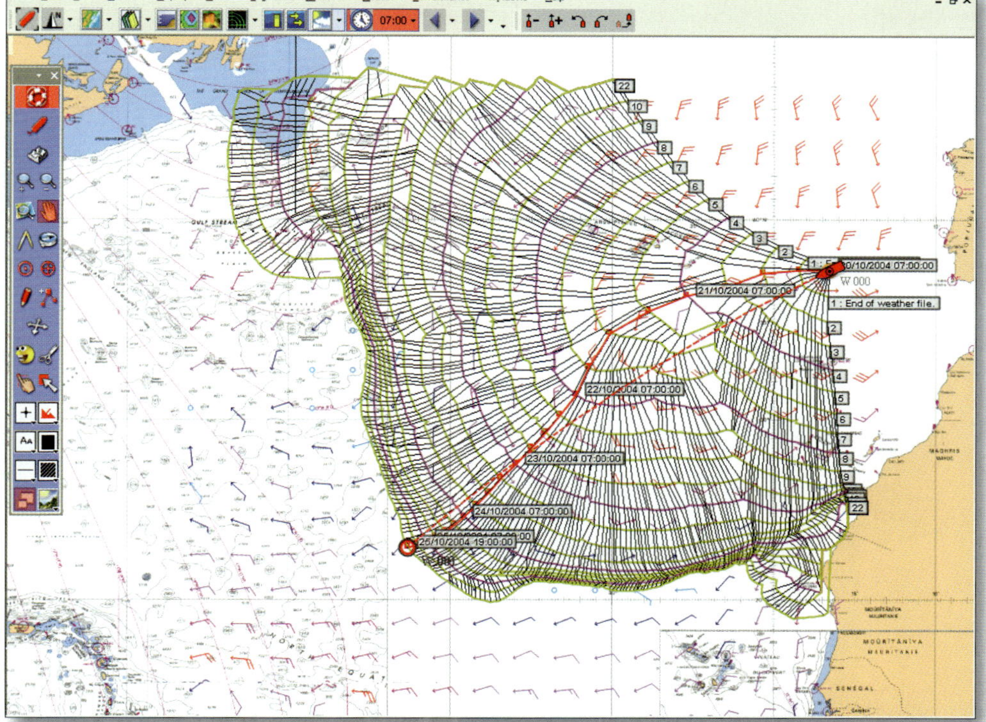

Compare the forthcoming features of MaxSea Navigator Time Zero and MaxSea Explorer Time Zero at: www.maxsea.fr/timezero/Produits/SolutionsComparées/tabid/87/language/en-US/Default.aspx.

Sea's more expensive packages. But it is a mistake to think of MaxSea software as expensive products that are only for professional navigators or racers.

To do so would be to discount MaxSea's lower-priced option, Navigator+. You get the same framework as Explorer or Commander—including the same interface and basic features—but without the specialized modules. If you're looking in the $500 range, Navigator+ is worthy of comparison to Maptech's Chart Navigator Pro or Nobeltec VNS MAX Pro. You may never grow into the full package, but we bet you'll add the Sailing Performance Module or the Weather Routing Module to next year's holiday wish list.

WHAT'S NEW: MAXSEA TIME ZERO

At press time, MaxSea was preparing for the U.S. launch of MaxSea Time Zero, a new version of MaxSea with incredibly fast chart redraw—as in *zero wait time*. Using video game flash technology, Time Zero displays realistic 3D charts with instantaneous response from the software. Recently launched in France, their advertisements quip, "Maybe we made it a bit too realistic," while showing a person with their head in a waste paper basket as they plan their boating routes on their home computer.

The beta version we tested certainly is fast and realistic. With Time Zero, you can zoom from 3000 miles to 500 yards in literally a few seconds. The chart never leaves the screen nor lags as you zoom, even while it automatically chooses the chart file with the scale that shows the most detail. You can work in full 3D, rotating and zooming charts as if you are on a helicopter circling and banking over a harbor.

Three fundamental changes enable Time Zero's speed increase. First, MaxSea has incorporated a new cartographic engine comparable to video game technology. Second, Time Zero uses its own MapMedia charts in the .mm3d format. Third, Time Zero incorporates a feature MaxSea calls *photofusion*, where charts and aerial photos are morphed to create a more realistic chart display.

In addition to software innovations, MaxSea Time Zero depends on its MapMedia charts. These cartographic collections, bundled by region with the software, drive the 3D ren-

MaxSea Time Zero: *A technique called* photofusion *morphs topographic and aerial images with nautical charts, yielding incredibly realistic displays (left). Weather routing will also get a face-lift in the new version (right).*

Information Route

Start at Boat Location — Date of Departure: Now 11:49:54 11/06/2008 — Speed: Actual 005,0 N

Leg To...	Nom	SOG	COG	Distance	Total	ETA	TTG	TTG Total	STW	TWD	TWA	TWS	AWA	AWS
8		18,7 N	2,3 °V	37,48 mn	237,8 mn	03:51:16 12/06/2008	2h00m	14h00m	18,7 N	214,0 °V	T148,3°	28,2 N	T109,5°	15,7 N
9		17,6 N	1,1 °V	35,25 mn	273,0 mn	05:51:16 12/06/2008	2h00m	16h00m	17,6 N	213,0 °V	T148,1°	26,6 N	T109,5°	14,9 N
10		16,9 N	1,6 °V	33,83 mn	306,9 mn	07:51:16 12/06/2008	2h00m	18h00m	16,9 N	214,0 °V	T147,6°	25,8 N	T109,4°	14,7 N
11		16,8 N	357,0 °V	33,60 mn	340,5 mn	09:51:16 12/06/2008	2h00m	20h00m	16,8 N	210,0 °V	T146,9°	25,3 N	T107,7°	14,5 N
12		12,4 N	24,2 °V	24,74 mn	365,2 mn	11:51:16 12/06/2008	2h00m	22h00m	12,4 N	185,0 °V	B160,7°	21,7 N	B138,6°	10,8 N
13		11,7 N	25,9 °V	23,45 mn	388,7 mn	13:51:16 12/06/2008	2h00m	1j00h	11,7 N	173,0 °V	B147,0°	17,6 N	B107,6°	10,1 N
14		10,1 N	23,9 °V	20,26 mn	408,9 mn	15:51:16 12/06/2008	2h00m	1j02h	10,1 N	164,0 °V	B140,0°	13,4 N	B90,9°	8,6 N
15		10,2 N	17,5 °V	20,45 mn	429,4 mn	17:51:16 12/06/2008	2h00m	1j04h	10,2 N	144,0 °V	B126,4°	12,4 N	B74,0°	10,4 N
16		10,4 N	5,1 °V	20,89 mn	450,3 mn	19:51:16 12/06/2008	2h00m	1j06h	10,4 N	122,0 °V	B116,8°	12,1 N	B65,2°	11,9 N
17		9,9 N	4,6 °V	19,77 mn	470,0 mn	21:51:16 12/06/2008	2h00m	1j08h	9,9 N	300,0 °V	T64,7°	11,8 N	T35,5°	18,4 N

dering. Each collection includes raster and/or vector charts, 3D data, and satellite photos, fueling the seamless display of realistic 3D charts.

Although most Americans are less familiar with MapMedia charts than sources such as C-Map or Navionics, they are established charts in Europe. In fact, MapMedia's vector charts are based on Navionics data. Coverage includes the U.S. (including Hawaii), Canada, Mexico, Central America, Europe, Australia and New Zealand, and the Caribbean.

Chart display speed may be the big splash with Time Zero, but this new software series incorporates other changes, including revisions of the graphic user interface, product line, and module offerings.

Time Zero has a redesigned graphical user interface that, while less cluttered, will still be familiar to MaxSea users. The Tool Palette and Chart Palette remain the mainstay of the interface, but drop-down menus have been replaced, leaving more screen area for chart display.

Time Zero also initiates a streamlining of MaxSea's product line. Instead of several software choices (Navigator+, Commander, Explorer, Professional), there will be two: Navigator Time Zero and Explorer Time Zero. Navigator Time Zero will include the software and one MapMedia chart region. Explorer Time Zero will include the software, one MapMedia chart region, and the AIS and NAVnet modules.

The Time Zero product line also entails changes to the optional software modules. At launch, three modules will be available: Routing, AIS, and NAVnet 3D. The Routing Module combines the former Sailing Performance Module and Weather Routing Module. The AIS Module combines the former AIS Module and Mobile Target Tracking Module. According to Furuno, the MaxShell Windows Desktop Manager Software has been dropped for now and the Ground Discrimination Module will be developed later.

At press time the French version of MaxSea Time Zero was at version 1.3.0. However, U.S. prices were still to be determined. Current MaxSea users can likely expect an upgrade package that transitions them to Time Zero. However, older versions of MaxSea (what Furuno now calls *MaxSea Classic*) will still be available and supported during the tran-

sition. Expect MaxSea Time Zero to be available in the U.S. by late 2008.

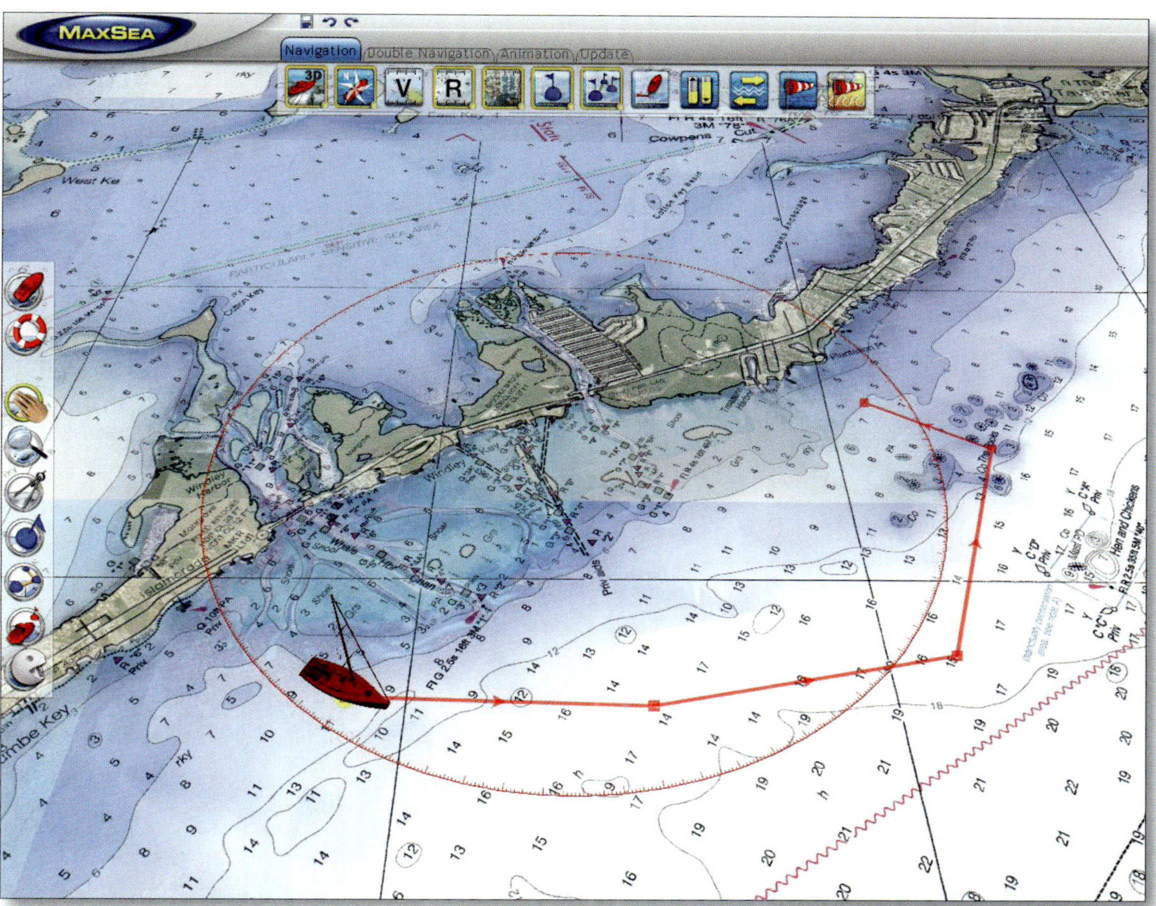

Real-Time Viewing: *Time Zero can continuously zoom and pan a small-scale chart of the Florida Straits to a large-scale Hawk Channel close-up in seconds.*

FEATURES AT A GLANCE

Each full-featured e-charting application has been evaluated according to the following 81 features. Some ratings are binary, "yes" or "no." Others are more subjective, evaluated on a scale from 1 to 5 (5 being "Excellent"). Any other comparison is defined in the Legend.

Out of Box Experience

Packaging	4
Documentation	5
Software install	3
Load charts	3
Load supplemental data	3
GPS hookup	3
Technical support	4

User Interface

GUI metaphor	Win
Screen space management	4
Program responsiveness	5
Chart display speed	4
Chart display quality	4
Customizable GUI	4
Customizable metrics	4
Split windows	Yes
Tabbed interface	Yes

Basic Features

One-button MOB	Yes
Waypoint creation	3
Route creation	4
Steer-to function	Yes
Track creation	3
Convert track to route	Yes
Waypoint and route management	5
Chart management	3
Search waypoint and route	4
Range/bearing tool	4
Chart annotation	4
Chart printing	4

Data Exchange and Networking

GPX support for waypoints	Yes
GPX support for routes	Yes
Waypoint exchange formats	G
Route exchange formats	G
Integrated GPS transfer	Yes
Multiple monitor support	Yes
Network application sharing	Yes

Cartography

Free U.S. rasters BSBs (RNCs)	Yes
Free U.S. vectors S-57s (ENCs, IENCs)	No
SoftCharts	Yes
CHS Canadian rasters	Yes
DNC	No
Seafarer	Yes
British Admiralty ARCS	Yes
International S-57s	No
International S-63s	No
Scan and geo-reference paper charts	No
Navionics cards	No
C-Map cards	Yes
C-Map CD/DVD	Yes
Satellite geo-referenced photos	Yes
Aerial informational nav photos	No
Topographic maps	No
Bathymetric data	Yes

Instruments and Sensors

GPS	Yes
Autopilot	Yes
AIS receiver	Yes
Wind	Yes
Depth	Yes
Water temperature	Yes
Radar	Yes
Heading sensor	Yes
Video camera	No

Advanced Features

Tides	4
Currents	4
Coast Pilot	2
POIs	2
Streets	2
Google Earth support	No
Advanced search	4
AIS	Yes
Buddy boat	No
Radar (ARPA/MARPA)	Yes
Radar (display overlay)	Yes
Fuel calculator	Yes
Transit calculator	Yes
Celestial calculator/Almanac	No
Auto route planning	Yes
Great Circle Route planning	Yes
Sail performance	5
GRIB weather integration	4
Weather options	4
Customizable bathymetric recorder	Yes

Legend

5	Excellent	CM	C-Map
4	Above average	NV	Navionics
3	Average	SC	SailCruiser
2	Below average	MT	Mr. Tides
1	Poor	PP	Plus Pack
CP	Chartplotter	C	CSV
Mac	Macintosh	E	Excel
NT	Non-traditional	G	GPX
Win	Windows	K	KML
N/A	Not applicable		

Chapter 25
Nobeltec Admiral MAX Pro

Jeppesen Marine, the maker of Nobeltec software, released a major update to its Admiral and Visual Navigation Suite (VNS) products in spring of 2008—and it was no ordinary upgrade!

The new versions, branded Admiral MAX Pro and VNS MAX Pro, do not support the company's well-established Passport charts. Instead, they transition to C-Map MAX Pro charts, premium cartography produced by C-Map, which was recently acquired by Jeppesen.

Admiral users on version 9 or prior needed to update both their software and their chart format. Jeppesen offered an extremely unusual and generous upgrade in an attempt to move all Nobeltec users onto MAX Pro software with C-Map charts. For a limited time, owners of Admiral version 9 received a complimentary upgrade to the new Admiral MAX Pro, a copy of the Raster Plus Pack, and a comparable MAX Pro chart in exchange for each Passport chart region owned. In other words, in a nearly unprecedented exchange program, Jeppesen offered a complimentary one-for-one swap for old software and charts. Boaters who owned Admiral version 6.5 to 8.x upgraded to Admiral MAX Pro for only $280.

Admiral MAX Pro is the same base program as VNS MAX Pro (reviewed in Chapter 21), but adds specialized features geared to larger vessels, most notably Nobeltec's GlassBridge Network, multi-monitor support, and certain Plus Pack options such as Tender Tracker. Admiral also adds a customiz-able NavView window of data, enhanced AIS target tracking, ARPA radar support, and OCENS WeatherNet, a subscription weather service.

These rarified features—and the matching $1,200 price tag—are likely to attract large-boat owners who have special requirements, such as multiple redundant navigation stations or the ability to track a fleet of tenders. But most recreational boaters will almost certainly be better served by a more economical package (such as Nobeltec VNS MAX Pro) at less than half the price.

GETTING STARTED

Admiral MAX Pro is very professionally packaged in a sturdy boxed set including the software installation disc, four discs of MAX Pro cartography, an eight-page Welcome Guide, and a hardware key (also called a dongle). The discs contain an incredible amount of data, including the assets commonly associated with C-Map MAX Pro cartography such as tides and currents, planning charts, aerial and satellite photos, street and marina information, and 3D views.

Nobeltec is a modular charting and navigation application. You begin with a boxed set, to which you add optional add-ons called *Plus Packs*. Five Plus Packs are available for Admiral MAX Pro: Weather, Sailing, Bathy Recorder, Tender Tracker, and the new Raster Plus Pack.

Admiral includes OCENS WeatherNet pre-installed, which means it is available should you choose to subscribe for an annual license fee and per usage charges. You can also opt for the Weather Plus Pack ($300), which provides streamed XM or Sirius satellite weather with a subscription and appropriate hardware devices.

The Sailing Plus Pack ($300) integrates lay lines, wind arrows, and polar diagrams. This module requires a GPS and a wind sensor that outputs apparent wind speed and direction. Admiral has added a folder with 50 default polars, ranging from cruising class boats such as Valiant and Tayana to racers such as JBoats, Farr, and Santa Cruz. You can choose to use these polars directly, or modify them for your particular vessel. You can also purchase your Performance Package ($275 member price) from U.S. Sailing, which includes polar dia-

Nobeltec Admiral MAX Pro

Pros: Customizable Toolbar and Console; CrystalView Tool; GlassBridge within-vessel networking; Route Wizard; modular applications and options; ETA Calculator with Estimated Fuel Use; Night Vision and Twilight Vision slider.

Cons: No S-57 support for free NOAA and U.S. Army Corps vector charts; weak import/export capabilities.

Coolest Feature: GlassBridge Network

Price: $1,200

Vista Capable: Yes

Version Tested: 10.0.1.6

System Requirements
PC with Intel Core 2 Duo
1 GB RAM
600 MB available hard disk space
1024 x 768 monitor, 32-bit color
DVD drive
USB port (required for dongle)
Windows XP or Vista

Jeppesen Marine
15160 NW Laidlaw Road
Suite 100
Portland, OR 97229
United States
www.nobeltec.com

Tablet Option: *The Wireless Nobeltec Display (WND) is a private-labeled Panasonic Toughbook. You can mirror the Admiral display of any 802.11b-enabled PC within 300 feet.*

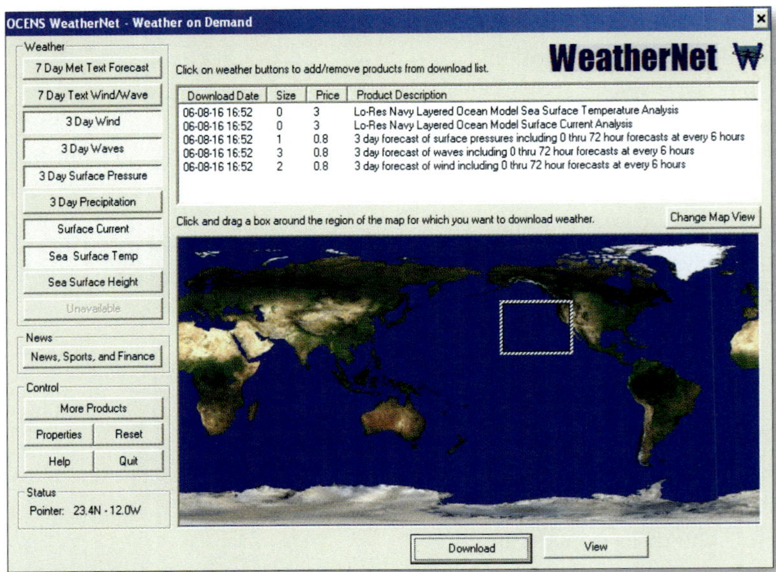

Weather On Demand: *OCENS WeatherNet ships with Admiral. For an additional subscription fee you receive downloadable weather data in GRIB format.*

Nobeltec Plus Packs

Weather
Integrates OCENS WeatherNet, XM WX, or Sirius Marine Weather subscriptions and hardware.

Sailing
Adds an optimal route feature using wind speed, direction, and your boat's polars.

Bathy Recorder
Records sea floor topography using your depth sounder and integrates the data into a 3D bathymetric chart display.

Tender Tracker
Monitors affiliate crew and watercraft for safety or theft.

Raster
Allows the software to display raster format BSB and GEO/NOS charts.

grams and target boat speeds for hundreds of sailboats. Using the apparent wind data, GPS position, and your boat's polars, Admiral calculates true wind speed and direction and creates an optimal route.

The Bathy Recorder Plus Pack ($800) enables you to record sea floor topographic information using your depth sounder, and then incorporate the data into a 3D display. This module is designed to allow fishermen or divers to create their own customized, highly-accurate sea floor charts.

The Tender Tracker Plus Pack ($1,500) uses a Seetrac Tender Tracking System to monitor affiliate watercraft, or even people if they carry the small wireless Seetrac Tender Unit. It tracks the location of up to 99 units, reporting back to your PC display. Note that you must purchase the Seetrac hardware separately, at about $7500 for the first unit and $4000 for additional units. Tenders can be tracked up to five nautical miles away, including boundary alarms to ensure children stay within range or to alert you of a stolen tender. Data transmissions are encrypted for additional security.

The Raster Plus Pack ($50), a new plug-in for Admiral MAX Pro, allows the software to read raster format charts. No charts are included, but with the addition of this Plus Pack you can display NOAA raster (BSB format) and SoftChart (GEO/NOS format) charts.

Installation of the Admiral base program and Plus Packs is straightforward. A Setup Wizard and a GPS/Port Setup Wizard guide you through the process. Like VNS MAX Pro, the software, your computer(s), and your purchased C-Map chart regions are all keyed to the dongle included in the boxed set.

Admiral MAX Pro does not include a traditional user's manual, but there are several options for obtaining this information. A 252-page spiral-bound User's Guide can be purchased through the Nobeltec store (www.nobeltec.com/store). Alternatively, it's also available as a PDF document in the application's Help Menu or can be downloaded at www.nobeltec.com/products/prod_suite.asp. Translations to Spanish, French, German, and Italian should be available soon.

Jeppesen Marine provides toll-free phone and email customer support. Additional technical support is available through Nobeltec's extensive website, which includes a searchable Knowledge Base and FAQ page. The Knowledge Base (www.nobeltec.com/support/knowledgebase.asp) is a great concept, but may be in its infancy in terms of topics it covers. The FAQs (www.nobeltec.com/support/support_faq.asp) are nicely grouped into categories, including Installation, Chart Usage, Integrated Devices, and Usage/Miscellaneous. Jeppesen Marine also offers an electronic newsletter, which includes information on updates and technical tips (www.nobeltec.com/company/newsletter_subscribe.asp).

LOOK AND FEEL

Since they are the same base program, with the exception of some additional tabs and icons, Admiral and VNS have the same interface. In fact, the two packages are so similar that Nobeltec only shows Admiral—not VNS—at boat shows (note, both VNS and Admiral cannot be installed on the same computer), pointing out the small button and icon differences if the customer is interested in VNS.

Like most charting and navigation programs, Admiral uses a combination of toolbars, menus, and floating windows. Like VNS, Admiral also has a highly customizable Toolbar and Console, including scores of Toolbar button and Console panel choices. The Night or Twilight screen intensity modes are also very customizable, letting you set the shade of red (Night Vision) and/or the shade of gray (Twilight Vision) using a slider bar rather than choosing from a few pre-set factory selections.

Admiral loses about a third of the screen to the combined area of the optional Toolbar, NavBar, and Console—data windows that cannot be resized or repositioned. It is designed with many shortcuts or *Hot Keys*, summarized in a nice two-page PDF at www.nobeltec.com/products/pdf/HotKeys.pdf. The Admiral MAX Pro upgrade also adds a handy Undo command.

Since Admiral has more features than VNS, there are more choices under Tools>Options. Admiral has an active GlassBridge tab (which is missing in VNS). It also adds tabs for any installed Plus Packs. Admiral's Enhanced AIS filtering shows up within the Targets tab, letting you filter AIS data to display only particular types of vessels, such as cargo ships, tankers, passenger ships, or even law enforcement vessels. Of course data in is only as good as data out: for this feature to work properly the vessel must broadcast their class or type, in addition to their MMSI number.

The most important difference between the VNS and Admiral interface is the NavView button on the toolbar. NavView is an alternate interface with larger buttons designed for touch screens or rough water. It includes an InfoBar at the bottom of the screen with navigation information, a trip odometer, and a display of GPS signal strength. Within NavView, a Nav-Info panel displays the most important navigation data, such as Course Over Ground (COG), Speed Over Ground (SOG), and GPS data in a large-font display. This information also can be displayed in a time series format, which Jeppesen Marine calls a "Strip Chart."

WORKING WITH CHARTS

Admiral MAX Pro software is designed to be used in conjunction with C-Map MAX Pro cartography. Passport Deluxe charts are soon to be part of a "sunset plan," with Jeppesen currently considering the appropriate time frame in which to completely discontinue their production. As a result of Boeing's acquisition of C-Map, when you think Nobeltec you should think C-Map charts. In fact, it's unclear why one would use VNS MAX Pro or Admiral MAX Pro without C-Map charts. In addition, MAX Pro's *Quick Sync Updates* automatically downloads recent Notice to Mariner advisories through the C-Map MAX Pro chart update server. A year of chart corrections is included with the purchase of a MAX Pro chart region.

Admiral MAX Pro also works with a few other chart vendors and formats, including Maptech (BSB3 and BSB4),

SoftChart (GEO/NOS format), and Maptech Photos. However, the support of any of these raster chart formats requires the Raster Plus Pack, an additional $50 option.

The transition from Admiral versions 8 and 9 to Admiral MAX Pro also entailed major changes to the support of standard S-57 vector charts, which includes free NOAA ENCs and U.S. Army Corps IENCs. S-57 vector charts are no longer supported in Admiral MAX Pro. It is assumed that vector cartography is obtained through the purchase of C-Map charts rather than with free government cartography.

The good news is raster charts display very well in Admiral. Like VNS, Admiral integrates a software feature called *CrystalView* to sharpen screen images. Charts are easily panned and scrolled, with smooth chart quilting for seam-

Route Optimization: *Sailing Plus Pack polars and GRIB weather data establish a route and target level for sail performance.*

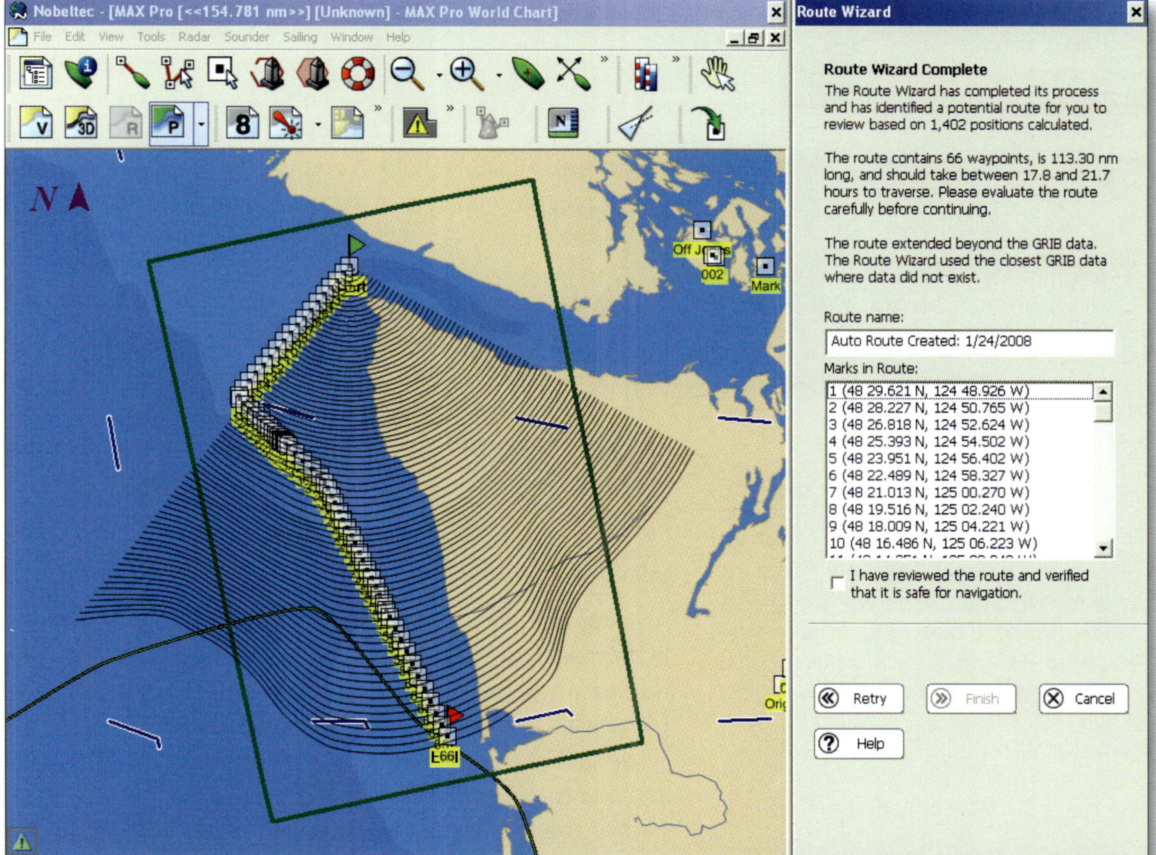

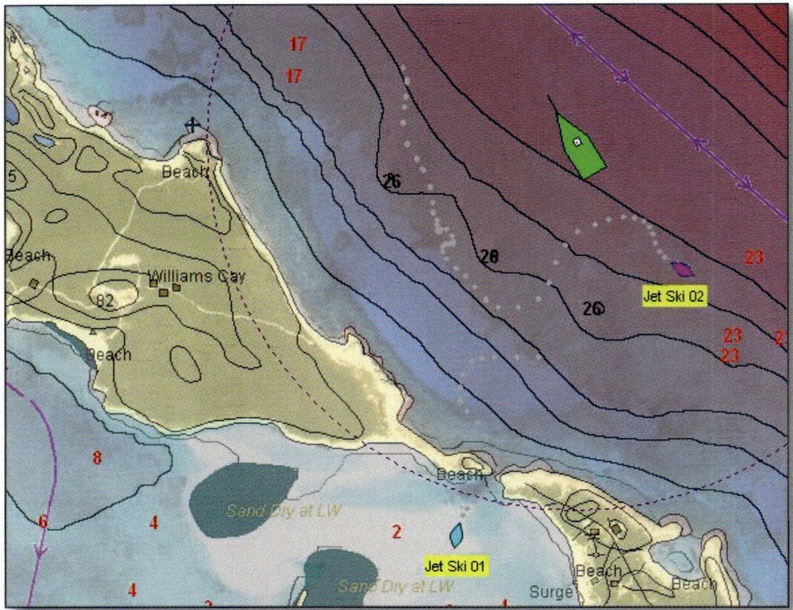

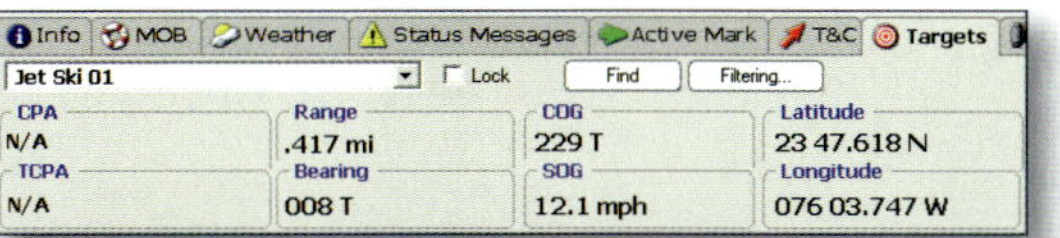

Mind the Assets: *The Tender Tracker Plus Pack is available only with Admiral and a Seetrac Tender Tracking System. The Seetrac Base Unit can track up to 99 tenders, dinghies, jet skis, or even passengers carrying Seetrac Tender Units (STUs).*

A new radar feature will be especially welcome on "big boat" Admiral installs. *Sector blanking* stops radar transmission for specific rotation angles—a useful remedy for dual radar installations, mast shadows, and building echoes.

less integration. Various Autoscroll Modes automatically keep your vessel in the chart window, and you can customize the vessel position relative to the underlying chart.

Admiral adds a nice feature for larger vessels, called GPS Position Onboard. With a GPS now providing accuracy to within three meters, the physical location of the GPS on a large boat is a factor. This feature lets you customize where the GPS is physically located onboard your vessel and shows your vessel icon scaled to its correct size. A custom-sized vessel icon is more than a cosmetic touch. It helps a user gauge the scale of the chart relative to their boat, particularly when trying to sense the breadth of a narrow swash or privately-maintained channel.

With the purchase of a Nobeltec InSight Radar2 Black Box ($3,000) and any NMEA 0183 heading sensor, your existing radar output can be overlaid on a chart. This device converts analog radar data into digital format, allowing it to be transmitted and displayed on your PC. Combined with a split screen display, this feature really helps with interpreting the radar targets.

WAYPOINTS AND ROUTES

Any application as sophisticated as Admiral MAX Pro includes extensive waypoint and route features, including unlimited waypoints and routes, a Route Wizard, an integrated ETA calculator, and a Great Circle route builder.

We had no trouble creating and working with waypoints and routes in Admiral. The interface was exceptionally clean. Routes can be customized to show range, bearing, and direction. The Instant Waypoint feature creates a quick, simple route from the boat's current position to a destination mark for monitoring cross-track error (XTE).

The Route Wizard automatically creates a route when given parameters for depth, distance from land, or minimum and maximum route leg lengths. It requires detailed vector data or 3D bathymetric data, available on the C-Map MAX Pro charts. The Route Wizard is an advanced feature, but is very easy to use when following the field-by-field prompts.

The ETA calculator uses a similar interface, letting the user fill in a collection of fields to obtain estimates for best departure times, transit times, required speed, and fuel consumption. In order to use all the features, the tide and current data for your geography must be loaded. The calculator is easy to use and provides a slew of data to help plan your transit.

For longer passages, Admiral integrates a Great Circle route builder. Simply select Edit>New>Great Circle Route and choose an origin and destination waypoint. A route is automatically created with a series of waypoints along the shortest Great Circle path, which can be named and edited like any other route. For celestial navigation, Admiral includes a built-in celestial sight reduction tool called Star Navigator. This separate program, available under Tools>Other Tools>Star Navigator, has its own online help.

Admiral can also associate any NMEA input—such as water temperature or depth—with a corresponding track. This additional feature is available standard with Admiral MAX Pro. It is a nice feature for sport fishing.

Like VNS, Admiral does not support the three main standard formats for exchanging data: tab-delimited, CSV, or GPX. Instead, Jeppesen Marine has chosen a standalone approach using their own Open Navigation Format (ONF). This format allows for the exchange of information between Nobeltec software, but any import or export of waypoints requires using a chartplotter as an envoy.

ADDITIONAL FEATURES

Admiral has an exceptional system for the exchange of data within a Nobeltec system, called Nobeltec's GlassBridge Network. With this feature, your vessel can have multiple PCs running Admiral, with full data sharing between displays.

For example, one PC can receive AIS data and another receive wind, speed, and depth—and both copies of Admiral will have both sets of data. Charts can be shared over the network, allowing them to be installed on a single computer. However, each computer on the GlassBridge network must have its own copy of the Admiral software—and that's $1,200 a copy. One can use multiple PCs or a single PC with up to four monitors.

If you are concerned about too many devices sending the same data to your PC, the Port Priorities feature resolves conflict by prioritizing multiple sensors. For example, a GPS and LORAN may both send slightly different position coordinates, which would cause your vessel icon to reposition with each set of instructions from each device. With devices added to Port Priorities, data from the highest-priority device is used first, with the software automatically switching to the next prioritized device if the primary device stops sending data.

Admiral also has a wireless option with its Wireless Nobeltec Display ($2,500). This handheld device works with any wireless-enabled computer and provides portable access to your PC screen. It does not require its own copy of Admiral. Rather, it mirrors the display on your PC wirelessly.

Assessment

Admiral MAX Pro is clearly designed for a larger and perhaps professionally-captained vessel. The additional features over Nobeltec VNS MAX Pro, such as GlassBridge Network, multi-monitor support, and Tender Tracker, are of value primarily to larger or longer-distance yachts. Only under rare circumstances would a typical coastal 40-foot sailboat, trawler, or powerboat require multiple networked computers or the ability to track a $2,000-dollar dinghy.

Jeppesen Marine knows this and understands that this demographic won't be bothered by the additional costs inherent with Admiral software. Modular packages are a bit like purchasing a Harley-Davidson motorcycle: the cycle engine and frame are just the beginning, then comes the chrome and options. Harnessing the power of Admiral—and it is a robust and full-featured package—begins with the purchase of proprietary cartography, then includes add-on modules, hardware devices, and annual subscriptions.

The typical recreational boater, concerned about the value of their purchase, will not be happy with these significant additional costs. Yet without these extra purchases, which are necessary to fully engage its features, Admiral at $1,200 holds little advantage for the money over VNS at $490.

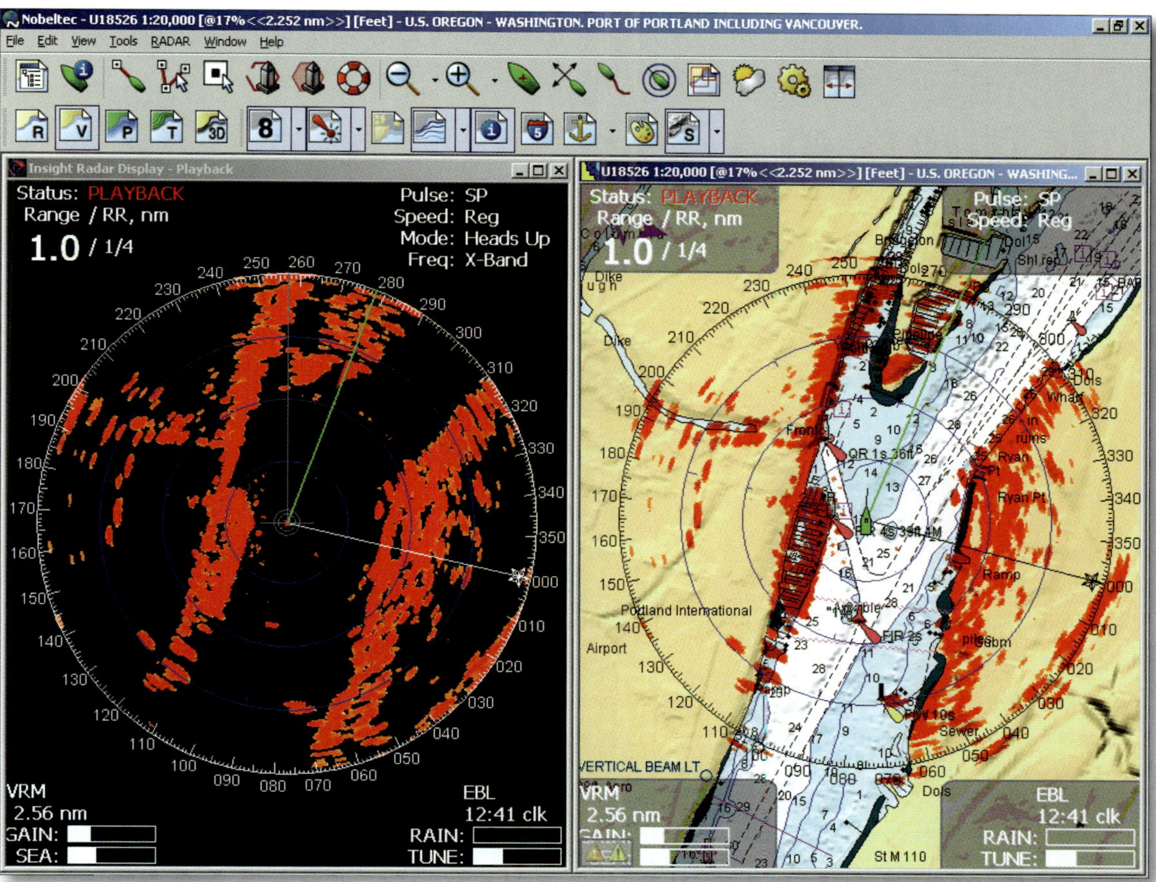

Radar Overlay: *Digital InSight radar data is sent to your computer via Nobeltec's Black Box (analog-to-digital converter). Admiral's NavView can display the raw radar display (left) or, with a heading sensor, can overlay the data on a chart (right).*

FEATURES AT A GLANCE

Each full-featured e-charting application has been evaluated according to the following 81 features. Some ratings are binary, "yes" or "no." Others are more subjective, evaluated on a scale from 1 to 5 (5 being "Excellent"). Any other comparison is defined in the Legend.

Out of Box Experience

Packaging	4
Documentation	3
Software install	3
Load charts	3
Load supplemental data	3
GPS hookup	4
Technical support	3

User Interface

GUI metaphor	Win
Screen space management	3
Program responsiveness	3
Chart display speed	3
Chart display quality	3
Customizable GUI	5
Customizable metrics	4
Split windows	Yes
Tabbed interface	No

Basic Features

One-button MOB	Yes
Waypoint creation	3
Route creation	3
Steer-to function	Yes
Track creation	4
Convert track to route	Yes
Waypoint and route management	3
Chart management	4
Search waypoint and route	2
Range/bearing tool	3
Chart annotation	4
Chart printing	3

Data Exchange and Networking

GPX support for waypoints	No
GPX support for routes	No
Waypoint exchange formats	None
Route exchange formats	None
Integrated GPS transfer	Yes
Multiple monitor support	Yes
Network application sharing	Yes

Cartography

Free U.S. rasters BSBs (RNCs)	Yes
Free U.S. vectors S-57s (ENCs, IENCs)	No
SoftCharts	Yes
CHS Canadian rasters	Yes
DNC	No
Seafarer	No
British Admiralty ARCS	No
International S-57s	No
International S-63s	No
Scan and geo-reference paper charts	No
Navionics cards	No
C-Map cards	No
C-Map CD/DVD	Yes
Satellite geo-referenced photos	Yes
Aerial informational nav photos	No
Topographic maps	No
Bathymetric data	Yes

Instruments and Sensors

GPS	Yes
Autopilot	Yes
AIS receiver	Yes
Wind	Yes
Depth	Yes
Water temperature	Yes
Radar	Yes
Heading sensor	Yes
Video camera	Yes

Advanced Features

Tides	4
Currents	4
Coast Pilot	2
POIs	2
Streets	2
Google Earth support	No
Advanced search	3
AIS	Yes
Buddy boat	No
Radar (ARPA/MARPA)	Yes
Radar (display overlay)	Yes
Fuel calculator	Yes
Transit calculator	Yes
Celestial calculator/Almanac	Yes
Auto route planning	Yes
Great Circle Route planning	Yes
Sail performance	4
GRIB weather integration	2
Weather options	4
Customizable bathymetric recorder	Yes

Legend

5	Excellent	CM	C-Map
4	Above average	NV	Navionics
3	Average	SC	SailCruiser
2	Below average	MT	Mr. Tides
1	Poor	PP	Plus Pack
CP	Chartplotter	C	CSV
Mac	Macintosh	E	Excel
NT	Non-traditional	G	GPX
Win	Windows	K	KML
N/A	Not applicable		

Chapter 26
E-Charting on a Mac

Until recently, members of the "1% club" who owned a Macintosh computer had an incredibly short list of electronic charting options. The Mac's weak market-share and Apple's seemingly suicidal product lurches made the platform downright frightening for software developers.

Many companies and individuals attempted to develop e-charting packages for the Mac, but stumbled through the mid-1980s and 1990s with Apple's uncertain future and frequent operating system changes.

Today, with Apple's near 20% laptop marketshare, and OS X now solidly in place, choices for Mac boaters have expanded significantly. MacENC and NavimaQ are the two most established packages, and the two reviewed in this chapter. MacENC is the successor to GPSNavX and currently the only Mac e-charting application that supports both raster and vector chart formats. GPSy and MacGPS Pro, both raster-only packages, are two newcomers. PassagePlus is an unusual Mac charting application in that it reads UK Admiralty and New Zealand raster charts, filling the niche of New Zealand boaters who use OS X. At $70 (with the entire set of 150 New Zealand Mariner charts for an additional $50), PassagePlus is a great option for boaters cruising or chartering in Australia or New Zealand.

Although some Macintosh packages are not as full-featured as the long-established PC products, Mac e-charting has come a long way. Coastal cruisers and weekend sailors will find more than enough functionality—affordably priced and on the user-friendly Mac platform.

NavimaQ

NavimaQ (pronounced "Nav-eh-mac") was originally written in the mid-1980s by husband-and-wife software team Larry and Barb Bauer. When Apple launched OS X in 2001, requiring a complete re-write of all applications, the Bauers decided enough was enough. In 2004, it was purchased by Barco Software, which began the arduous task of making NavimaQ OS X compatible. NavimaQ is now universal binary, meaning it runs natively on older PowerPC Macs as well as the newer Intel Macs.

Getting Started

NavimaQ is available online through its website (www.barcosoftware.com) or through select retailers such as Landfall Navigation and Celestaire. The boxed version sells for $195 retail or $99.95 direct from Barco, an unfortunate discrepancy blamed on retailer margin requirements. Alternatively, you can opt for a software-only download from Barco for $75.

Do not expect an Apple out-of-the-box experience with the $99.95 purchase. The boxed version clearly wasn't conceived by Apple's packaging design team in Cupertino. The slim-case contains an install CD and a photocopied 32-page Getting Started guide.

An HTML manual, accessed through Help on the main menu, is your most important resource when working with NavimaQ. This detailed manual, with 20 hot-linked chapters, is much more helpful than the pamphlet included with the CD. We suggest immediately printing the Help chapters to create a manual.

NavimaQ includes unlimited email support for the life of the product and telephone support for the first 30 days to help with installation. You can purchase a support contract for extended and priority email and telephone support. Barco Software also has an online community site (www.barcosoft.net) with FAQs and forums for discussion.

In order to get up and running, you need raster charts in BSB/KAP or GEO/NOS (formerly SoftChart) format. In our

See Parallels Desktop in Action discussion and examples at end of chapter.

You can't add waypoints, logpoints, or chartmarks in NavimaQ until you've created a "home" for these navigational assets. Choose **File>New>Select File Type** and then the radio button for **Waypoints**, **Logpoints**, or **Chartmarks**. When a window opens, save and name the new file by choosing **File>Save As**.

To convert an old NavimaQ 3.0.3 (OS 9) waypoint, chartpoint, or logpoint file to the current version, you'll need to run the Convert2NavimaQ application, which can be found in the Applications folder.

When selecting the location of the charts in the Chart Open window, you are creating a directory, not selecting a file. Always pick the top level folder in which you want NavimaQ to search for charts.

tests, NavimaQ crashed with Maptech's encrypted BSB4 and BSB5 charts, but had no trouble opening standard NOAA BSB3 files. Barco Software is aware of this problem and expects to have it resolved soon.

We also recommend keeping a copy of NOAA's *Catalog of Charts & Publications* handy. This brochure displays all the chart regions and their numbers so you can easily choose the chart files you wish to view and, more importantly, identify adjacent charts as you move through a region.

Like most Macintosh set-ups, installing chart files is very straightforward: simply copy your chart files or folders onto your hard drive. You can place chart files in any folder or organize them in any way as long as they are on your hard drive. Choose File>Open>Open Chart and select Show Charts On to browse your files and select your chart folder. It takes a few minutes exploring the Open Chart window, but it will eventually become clear where to click to see a list of your chart files. If you click on Show Preview, a side drawer opens with a preview of the chart. It's a neat feature that is useful if you are familiar with the region, but it's not always big enough to help you identify the chart.

Note that NavimaQ only allows you to browse in one folder at a time. If you keep your charts organized by type (BSB or GEO) or by region (Chesapeake Bay or Alaska), this isn't much of a hindrance. In fact, limiting to one folder at a time may prevent the average boater from having too many charts loaded. Because the application is only pointing to one folder at a time, it stays organized and your laptop remains very responsive. For this reason, NavimaQ works well with CDs or DVDs of charts that are pre-organized into regional folders.

Look and Feel

Like many e-charting applications, NavimaQ is designed to operate with three "views" always open: a Chart window showing the chart display, an Overview window, and a Coordinates window.

Obviously, the Chart window is the most prominent. Unfortunately, NavimaQ does not yet take advantage of OS X's Quartz technology, so charts viewed at less than 100 percent appear soft and line quality suffers (see Quartz Rendering). Barco is aware of this issue and states that adopting Quartz technol-

ogy is a high priority effort.

The Overview window displays the larger area from which the chart display is chosen. Dragging the rectangle outline on the Overview window pans the main chart display accordingly. However, the window has a quirky tendency to expand and contract, depending on the scale of the displayed chart. For example, a Chart window of St. Lucie Inlet results in a tiny, barely-legible, one-inch-by-one-inch Overview window.

NavimaQ's graphical user interface is largely the standard Mac metaphor, with a main menu bar along the top, an icon-based toolbar, and floating windows. Unfortunately NavimaQ sometimes ignores standard Macintosh conventions. Certain shortcuts, window names, and tab positions are at odds with OS X standards and occasionally even behave inconsistently. For example, to a Mac user, a title bar placed on the left side of a window signifies a pull-out drawer. NavimaQ does not abide by this convention. Similarly, NavimaQ does not allow you to float your cursor over an icon and display its function. Instead, it describes groups of icons.

On the positive side, NavimaQ allows reasonable control over the user interface. NavimaQ also lets you customize your favorite screen display—setting window sizes, placements, and column widths—saving them for your next start-up.

Working with Charts

NavimaQ uses the standard metaphor for working with chart files. You can pan and scroll to view the chart. The charts displayed very promptly, though the program performs best when working with a single chart.

Although NavimaQ automatically rolls up the next chart while navigating, it is more difficult to move between charts while planning. The chart display is exactly like a paper chart on your screen, complete with the printed border and abrupt edge. Charts are not "stitched" together smoothly as they are in most other e-charting programs.

In addition, the application lacks a chart outline feature, which helps you choose the next chart. Panning the chart doesn't indicate which chart to use at what scale or when it's best to switch to a new chart. You must actively decide which chart you want, using a reference such as NOAA's *Catalog of Charts & Publications*.

NavimaQ has a somewhat unique and useful feature called the Spyglass tool (only TIKI Navigator has a similar function). Clicking on this toolbar icon simulates a glass bubble chart magnifier (but without the distortion!). This feature—perfect for us middle-aged boaters—is almost worth the price of the download. You can interactively drag the Spyglass icon over the digital chart to magnify that area of the chart, letting you read those tiny navigation aid or depth notations.

Waypoints and Routes

NavimaQ uses three different points of reference on charts: waypoints, logpoints, and chartmarks.

Waypoints are navigational points and are used to create a route. Logpoints are analogous to tracks, showing your path over time. Chartmarks refer to non-navigable geographic marks, such as anchorages or marinas. Chartmarks also provide a way to annotate chandleries, laundromats, and dinghy docks for quick and easy visual reference. It's handy to have a way to mark these secondary locations without bloating your waypoint list.

It was very easy to create a waypoint or chartmark with NavimaQ using the icon-based toolbar. However, we experienced difficulty trying to save these points. Unlike most programs, NavimaQ does not create a default repository for your marks. You must first create a file manually.

Additional Features

NavimaQ integrates with GPS receivers and autopilots, but currently does not accept data from wind, compass, depth, AIS, weather, or radar instruments. Waypoint data exchange is compatible with some Garmin and Magellan handheld GPS units.

NavimaQ incorporates tide data through XTide, a separate program that produces tide predictions (www.xtide.com). When Barco Software acquired NavimaQ it also acquired WeathermaQ, a weather fax software package. However, Barco has discontinued this product.

Assessment

After hearing stories of NavimaQ's instability and bugs, we were pleasantly surprised with the improvements made since the early OS 9 days, and particularly with Version 3. It is clear that much work has been put into the product.

We found NavimaQ to be relatively stable and bug-free, and the program has some useful features, such as the Spyglass tool and chartmarks. On the other hand, crashes caused by BSB4 and BSB5 charts would give headaches to the average user, as might some of the departures from Apple's standard interface. The company is aware of the program's shortcomings and much seems to be riding on future releases and future products (see below).

The bottom line is that NavimaQ is a work in progress. For now, it is attractive as a budget-conscious option (assuming you take advantage of the $75 download) for coastal and weekend boaters who wish to use navigation software while maintaining their allegiance to the Mac.

What's Next?

More importantly, Barco Software has been working on Sea-Farer Pro, a new application which is intended to run on Mac OS X (Leopard) and Linux. At press time, a working copy was not yet available for testing. Pre-announced improvements over NavimaQ include chart quilting, better waypoint and route management, and GRIB weather.

MacENC

Many boaters are probably not familiar with MacENC, a relatively new Macintosh e-charting application that evolved from GPSNavX, a program that was specifically designed for OS X and worked only with raster format charts.

Despite the "ENC" in its name—which implies vector charts—MacENC displays both raster and vector formats.

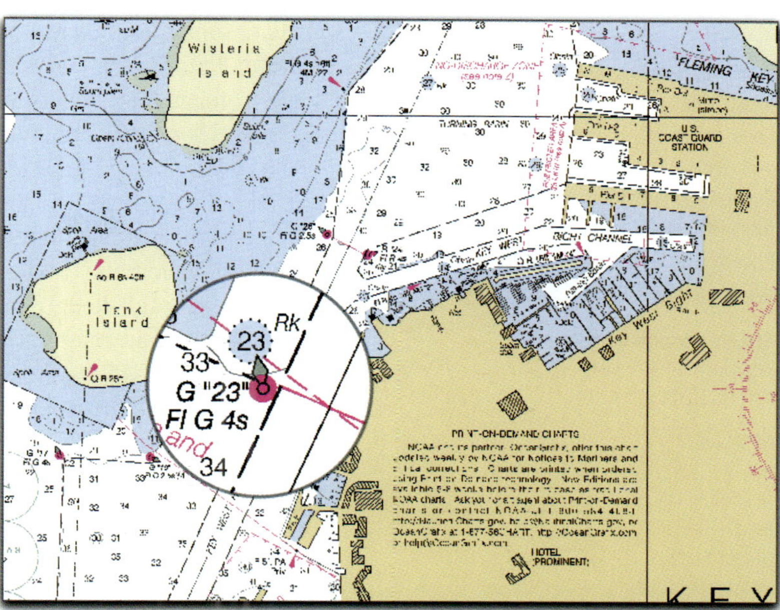

Spyglass Tool: *This unique NavimaQ tool works like a bubble-glass magnifier, enhancing often difficult-to-read chart features.*

To quickly convert a NavimaQ logpoint (track history) into a route or list of waypoints, make a copy of the logpoint file and then change the extension. For example, make a copy of **mylog.lpt** and rename it **myway.wpt**.

If closed, MacENC's Navigation and Overview windows can be reopened via the **Window** menu.

Command-R toggles back and forth between MacENC's "stacked" raster and vector chart displays.

To integrate tidal prediction data in MacENC, simply download and install a companion piece of Mac OS X open source software called Mr. Tides (http://homepage.mac.com/augusth/MrTides).

Make sure you have at least one Apple Mail account set up in account preferences before requesting a GRIB weather forecast in MacENC's **Weather** menu.

Its designer, Rich Ray, sees MacENC as a worthy successor to his initial effort, GPSNavX, which was the name of both the company and the product, began as a personal project in 2002, when Ray wanted to view BSB charts on his Mac with OS X. He created a basic native-OS X chart viewer and distributed it for free.

Over the next year he incorporated new features requested by his "customers," and the application grew from a simple viewer to a full-fledged navigation package. NOAA's distribution of free vector format charts in 2005 forced a major overhaul of GPSNavX, which led to MacENC. Although GPSNavX is still available, we chose to review MacENC because it surpasses GPSNavX in both its features and practical utility.

Getting Started

MacENC is only available as a Web download. Ray believes that packaged software—in terms of a printed box, disc, and hardcopy manual—is an outdated form of distribution. In his model, customers gladly forfeit a pretty package in exchange for receiving the software immediately with a credit card purchase from GPSNavX's website (www.gpsnavx.com). The cost is $139.95 and the raster-only GPSNavX costs $59.95. A detailed comparison of the two packages is posted on the company's website.

You will also need to either download or purchase charts, in raster or vector formats. MacENC works with all NOAA and U.S. Army Corps of Engineer charts, as well as Maptech's encrypted BSB4 and BSB5 charts. Installing chart files is straightforward: simply copy the BSB (raster) or S-57 (vector) files to your hard drive, preferably to a folder in your Documents or User folder. MacENC uses a feature called the Chart Manager that makes it very easy to install and manage charts. Chart Manager brings up a browser window, letting you select one or more folders with charts.

After copying the charts, simply double-click on a chart folder (raster format) or a catalog (vector format) and it loads into Chart Manager. With MacENC's "directory based" chart loading, you can easily direct the application to add charts or a chart folder from anywhere on your hard drive.

Although MacENC does not include a PDF manual, the Help menu is very good. Each chapter is clearly written with sufficient detail to take you step-by-step through topics such as connecting your GPS or registering your BSB5 files. You don't need an online connection to use this linked help file, unless you want to hot-link directly to the manufacturer's website. These help chapters can also be printed as an excellent substitute for a manual.

GPSNavX provides direct customer support by email or telephone. They were very responsive to all our inquiries and other users report that technical support is prompt and helpful. Better yet, we had no trouble with the installation or operation.

Look and Feel

MacENC uses a typical Mac graphical user interface with floating windows. Windows can be displayed transparently, so you can see your chart display through any data window, giving a "layered" effect. This is a particularly nice feature to use with the Waypoint window. Some windows also include a drawer option. For example, the GPS window has a pop-out drawer that shows your satellite data. Like most Mac applications, MacENC includes a number of keyboard and mouse shortcuts, described in a compendium in the Help menu.

MacENC has a few oddly non-Mac aspects. For instance, Chart Manager can never rise to be the top window and is always obstructed by the Navigation window and the Overview window, forcing you to resize it. Like all software programs, these little quirks become irritating if you use the program extensively and repeat the inefficiencies over and over.

Beware of overdoing the window displays: the screen can easily become cluttered with window layers. The fixed-size Overview window takes up quite a bit of screen real estate. We'd love this application on dual 23-inch Apple Cinema Displays: one for all the windows and one for the chart view.

MacENC is a bit weak on customization and preference settings overall. There are a number of customizable properties in its Preferences window, such as setting compass displays to magnetic or true, and wind to apparent or true. However, MacENC doesn't have as many knobs and dials as a Mac user is accustomed to. For example, multiple windows cannot be combined into a tabbed window to save precious screen real estate. And the common ability to save a screen arrangement as a workspace is conspicuously absent.

Working with Charts

One of MacENC's nicest features is that it smoothly and automatically transitions between charts in all settings and circumstances. It brings up the next chart automatically—even in planning mode. Many other applications only include this feature when the boat is underway and connected to a GPS.

Even more impressive, MacENC works simultaneously with both raster and vector format charts. It doesn't simply cascade these two displays; it links the charts, making the raster display "smarter" by reading the vector data behind it. For example, clicking on an object on a raster chart would normally have no result, because it is just a scanned paper chart. But using MacENC, that same click brings up the data contained in the corresponding vector chart. You get the combined benefits of raster imagery and vector data.

Users can navigate across charts using Outline mode or with a feature called Automatic. Outline mode is a standard feature in many charting and navigation applications, displaying thin lines showing chart outlines. Right-click the mouse and a list of charts spanning your area appears in a pop-up window. Select one of those choices to refresh the chart display with that chart. Automatic mode simplifies the process by choosing the best chart as you zoom, scroll, or pan.

MacENC's standard zooming feature is a bit cumbersome. Most charting packages pan with the mouse and zoom with the scroll wheel. In MacENC, the scroll wheel moves the chart vertically or side-to-side. This relegates zooming to the main menu or an awkward keyboard shortcut. A frequently-used tool such as zooming is better served by the scroll wheel.

A nice feature in MacENC is its ability to measure distance and bearing even in planning mode. Some of the applications we tested didn't have a simple measuring tool, only providing range and bearing information when navigating to a waypoint. In MacENC, by simply selecting the line tool in the toolbar, and clicking-and-dragging the mouse, you can see the distance and bearing to the dinghy dock from your foggy Maine mooring.

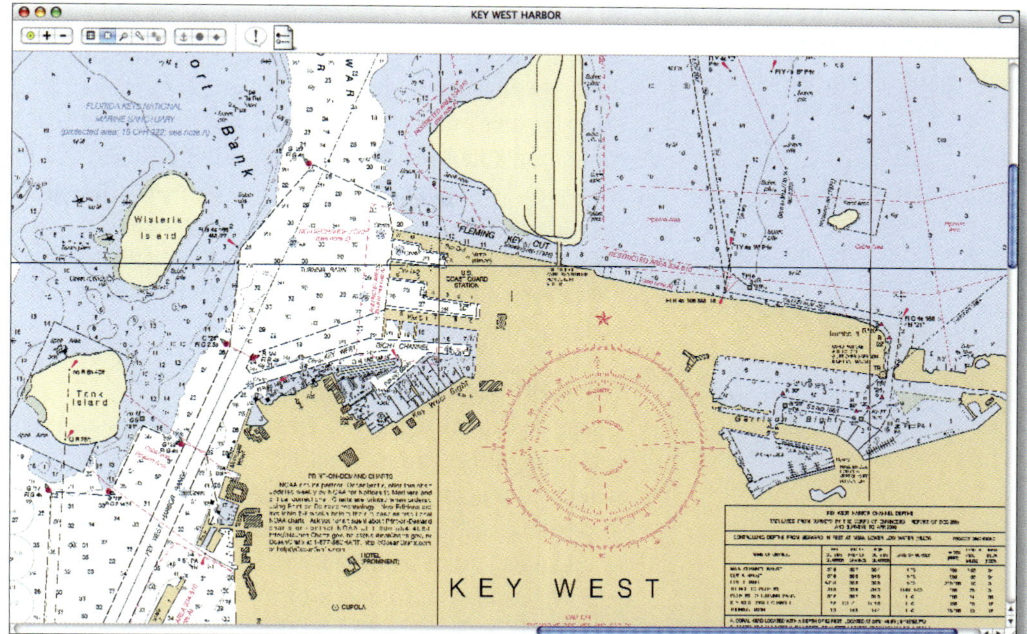

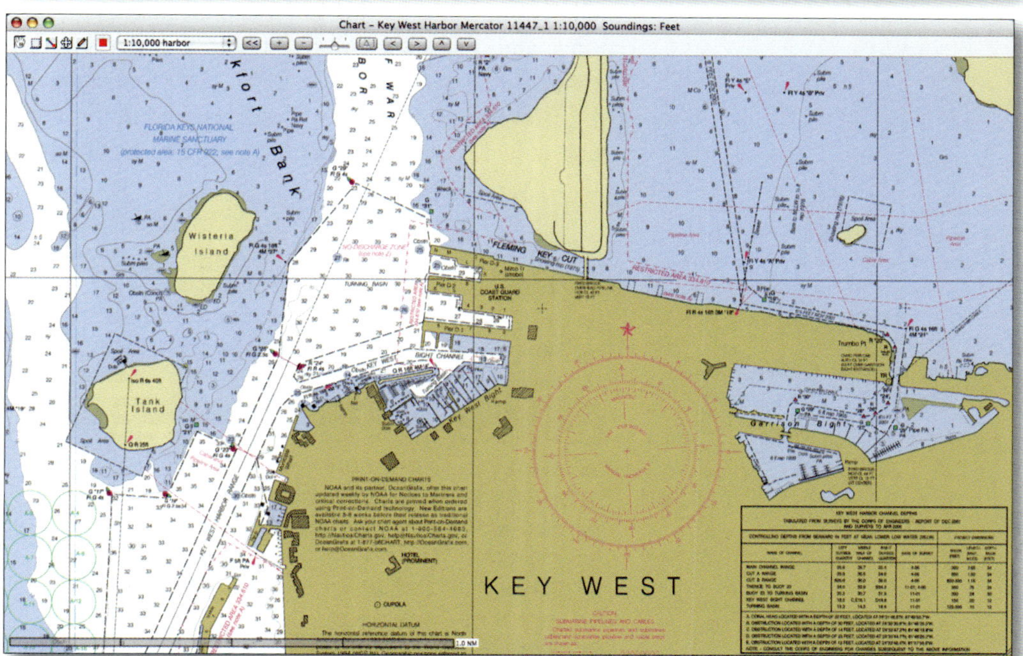

Quartz Rendering: *This OS X screen rendering technology is designed to deliver crisp graphics. The non-Quartz screenshot (top) demonstrates overall poor quality including broken lines and text as well as color shifts.*

One trade-off that comes with the ability to keep both raster and vector charts "in standby mode" is the program's refresh speed. Although it is not painfully slow, it is less responsive than many of the other applications we tested. A refresh on a laptop can take a few seconds and expect to see the Mac's "spinning beach ball" as you pan across chart files. We ran MacENC through its paces on several Macs and there is no question that more memory—2 GB or more—would be a welcome addition to your laptop. MacENC also performed better on our new Intel MacBook Pro.

Waypoints and Routes

Using dedicated windows, creating waypoints and routes is very straightforward in MacENC. For example, in the Routes window, you can easily draw from your waypoint collection to create a new route, copy a route, expand an existing route, or add a waypoint within legs of an existing route.

Ray has also put quite a bit of effort into data exchange. Waypoints and routes can be transferred to some GPS units, including select Garmin and Magellan models. Waypoints, routes, and tracks may all be imported or exported as CSV (comma-separated value) or GPX (GPS Exchange Format) files. However, note that any icons you choose for waypoints (such as fuel dock, anchorage, etc.) are not captured when ex-

porting to another computer.

One of our favorite simple features of MacENC is the keyword search field. This standard Macintosh tool adds powerful search capabilities to locate your waypoints and routes and for quickly displaying noncontiguous charts.

However, if you have hundreds of waypoints, MacENC's management of waypoints may be too simplistic. Its waypoint file is essentially a collection of all your waypoints. Although you can choose to manually "collect" subsets in named folders, unless you only travel in a limited geography or seldom enter waypoints, this process quickly blossoms into long, difficult-to-manage multiple lists. Most Mac applications leverage Apple's popular metaphor of "smart" folders to automate organizing and grouping tasks. We'd like to see Ray adopt this existing technology, providing a much more organized and efficient means to manage a coastal cruiser's large waypoint database.

Additional Features

MacENC integrates with many marine instruments, including GPS, depth sounder, heading sensor, autopilot, wind, speed, radar, and AIS receiver. MacENC can display a live overlay from an AIS receiver or a radar and is fully-featured in terms of setting proximity alarms.

Waypoint Search: *Use the keyword search field to quickly locate a chart. For example, typing* Miami *brings up all waypoints that contain this word. Select a waypoint and click* Scroll To. *Your waypoint appears, centered on the largest scale chart available (left).*

Chart Organization: *Cataloging charts by type, year, or geography makes loading and updating easy (right).*

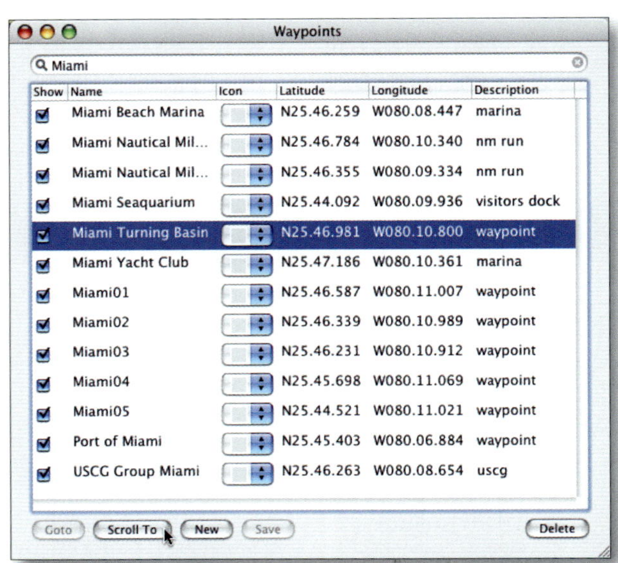

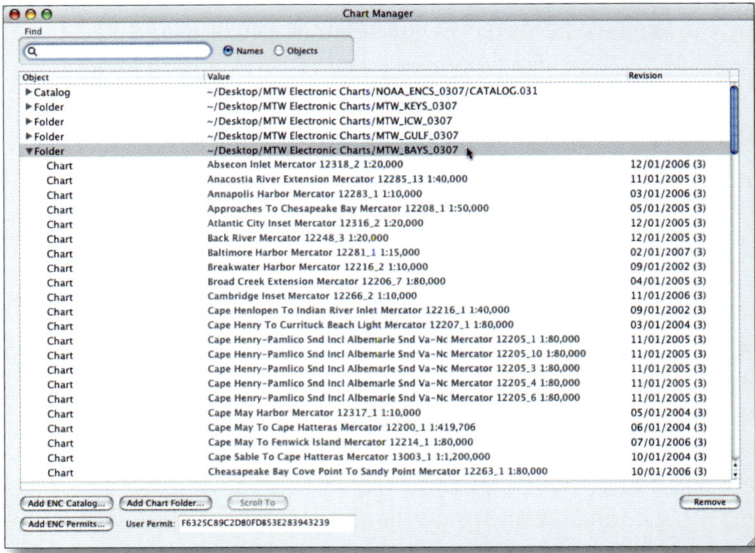

A particularly nice feature is the "Connect at Startup" option in File>Preferences. Many programs either automatically connect at startup or have you manually make the connection. MacENC's option to set connection preferences to automatic or manual lets you choose what to connect and when.

We had no trouble quickly downloading GRIB weather files from the Internet using MacENC. Simply choose Weather>Request GRIB to request surface wind, 500 millibar height, wave, surface temperature, or air temperature data. A GRIB data file arrives instantly as an email attachment in your inbox. Then select Weather>Open GRIB, choose the attachment file, and your weather data is overlaid on your chart display. There is no added cost.

MacENC shows tidal range data by incorporating an application called Mr. Tides. Right-clicking anywhere on a chart and choosing Mr. Tides in the contextual menu brings up a new window showing tidal prediction curves, in bright color-coding, for that day and location. You can advance to other dates or locations, or display a complete visual calendar of tides.

MacENC also recently integrated Google Earth, giving it the ability to show satellite imagery of your present chart position. Obviously this feature requires a fast online connection—Google Earth is a bandwidth hog—but it allows you to zoom and pan over a satellite image of your chart area and use Google Earth to locate map features. This free feature is a great alternative to more expensive cartography that includes aerial or satellite views, such as C-Map or Navionics cards.

Seeing your location on Google Earth is impressive, but what about displaying and saving your waypoints as a layer for Google Earth display? MacENC can easily transfer waypoints by saving a KML file to your desktop (Waypoints>Transfer>Export to Google Earth). You can then open the file in Google Earth.

If you have access to a large format color printer, MacENC also has the ability to save the chart as a TIFF graphics file and print high-resolution charts. MacENC can also print overlays of waypoints, routes, or tracks. There is very little control over this function so you may prefer a screenshot tool such as Snapz Pro X.

Another annotation feature is the pen tool, which lets you write notes on your charts. Although a mouse is a crude writing instrument—Mac's Sticky Notes or comments would work better for this purpose—it does allow you to scribble marks for fishing holes or sketch out route ideas.

For sailors, MacENC has a Sail-Timer feature that computes optimum sailing angles and target speeds to an active waypoint using wind speed, wind direction, and predicted performance. This functionality is usually reserved for more expensive e-charting applications.

Finally, for serious Mac tweaks (tech-weenies), MacENC can be extensively scripted, letting you write your own Apple-Script code to automate tedious tasks such as batch re-naming waypoints.

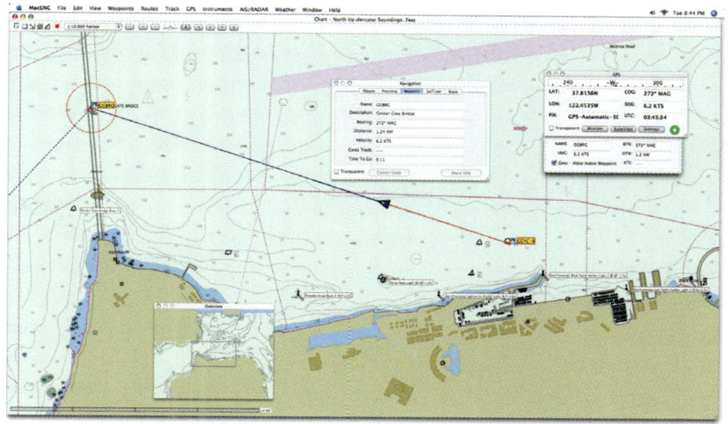

Follow a Route: *Leaving Golden Gate Yacht Club headed to the Golden Gate Bridge (course made good is 272° magnetic; speed over the ground is 6.2 knots; distance away is 1.2 nautical miles).*

Assessment

MacENC is an excellent choice if you already own a Macintosh laptop or are considering purchasing one. It's a mature, stable product that offers all the basic navigation features such as waypoints, routes, and tracks plus the ability to use both raster and vector charts.

It also has some advanced features, many of which typically only appear on more expensive PC applications. The ability to connect to AIS and radar, pull down GRIB weather files, and integrate with Google Earth to obtain satellite imagery all fit into this category.

There were some imperfections, but overall you'll be hard-pressed to find better ease of use and capability on either a Mac or PC at this price point. For Mac loyalists who do not want to switch to a PC when they step aboard, MacENC provides a solid option.

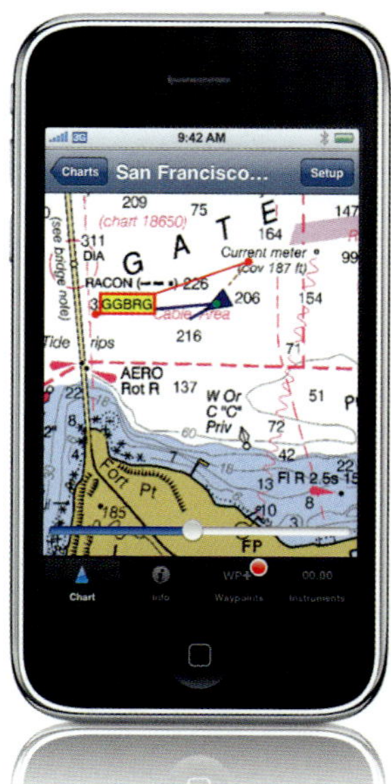

The Latest from GPSNavX: *Using iPhone's built-in location services (GPS, cell tower, or WiFi), iNavX displays NOAA RNCs and plots your position in real-time.*

Macintosh E-Charting Applications

GPSNavX (R)
www.gpsnavx.com

GPSy (R)
www.gpsy.com

MacENC (R and V)
www.gpsnavx.com

MacGPS Pro (R)
www.macgpspro.com

NavimaQ (R)
www.barcosoft.com

PassagePlus (R)
www.windvector.com

(R) Raster BSB support; (V) Vector S-57 support.

When connecting a GPS sensor, don't overlook the fact that Macs have different ports than PCs. If you already own a GPS sensor with a serial connection, you'll need to purchase a serial-to-USB adaptor. KeySpan's 19HS USB High Speed Serial Adapter costs $39 (www.keyspan.com). Alternatively, purchase a GPS sensor with a USB connection.

A screen capture tool is the best way to print chart excerpts showing your waypoints, routes, and annotations. A good choice for the Mac is Snapz Pro X (www.ambrosiasw.com), available as a $29 download.

For all things related to Macintosh computers and boating—including software questions—surf to macsailing.net (htttp://macsailing.net).

Parallels Desktop in Action: *An Intel Mac with virtualization software easily runs PC applications. The author's Macintosh shown above is loaded with many of the e-charting applications and utilities mentioned in this book. Examples are (top screenshot from left) SeaClear II, Coastal Explorer, Nobeltec Admiral MAX Pro, and TIKI Navigator Pro. An additional screenshot shows (clockwise from upper left) XM WxWorx On Water, NavSim BoatCruiser, GPSBabel, EarthNC, and Fugawi Marine ENC (bottom).*

NavimaQ
FEATURES AT A GLANCE

Each full-featured e-charting application was evaluated according to the following 81 features. Some ratings are binary, "yes" or "no." Others are more subjective, evaluated on a scale from 1 to 5 (5 being "Excellent"). All other comparisons are defined in the Legend.

Out of Box Experience

Packaging	2
Documentation	3
Software install	3
Load charts	3
Load supplemental data	N/A
GPS hookup	2
Technical support	3

User Interface

GUI metaphor	Mac
Screen space management	3
Program responsiveness	3
Chart display speed	2
Chart display quality	3
Customizable GUI	2
Customizable metrics	3
Split windows	No
Tabbed interface	No

Basic Features

One-button MOB	No
Waypoint creation	3
Route creation	3
Steer-to function	No
Track creation	3
Convert track to route	No
Waypoint and route management	2
Chart management	2
Search waypoint and route	2
Range/bearing tool	3
Chart annotation	No
Chart printing	1

Data Exchange and Networking

GPX support for waypoints	No
GPX support for routes	No
Waypoint exchange formats	None
Route exchange formats	None
Integrated GPS transfer	Yes
Multiple monitor support	No
Network application sharing	No

Cartography

Free U.S. rasters BSBs (RNCs)	Yes
Free U.S. vectors S-57s (ENCs, IENCs)	No
SoftCharts	Yes
CHS Canadian rasters	Yes
DNC	No
Seafarer	No
British Admiralty ARCS	No
International S-57s	No
International S-63s	No
Scan and geo-reference paper charts	No
Navionics cards	No
C-Map cards	No
C-Map CD/DVD	No
Satellite geo-referenced photos	No
Aerial informational nav photos	No
Topographic maps	No
Bathymetric data	No

Instruments and Sensors

GPS	Yes
Autopilot	Yes
AIS receiver	Yes
Wind	No
Depth	No
Water temperature	No
Radar	No
Heading sensor	No
Video camera	No

Advanced Features

Tides	3
Currents	No
Coast Pilot	No
POIs	No
Streets	No
Google Earth support	No
Advanced search	No
AIS	Yes
Buddy boat	No
Radar (ARPA/MARPA)	No
Radar (display overlay)	No
Fuel calculator	No
Transit calculator	No
Celestial calculator/Almanac	No
Auto route planning	No
Great Circle Route planning	No
Sail performance	No
GRIB weather integration	No
Weather options	No
Customizable bathymetric recorder	No

Legend

5	Excellent	CM	C-Map
4	Above average	NV	Navionics
3	Average	SC	SailCruiser
2	Below average	MT	Mr. Tides
1	Poor	PP	Plus Pack
CP	Chartplotter	C	CSV
Mac	Macintosh	E	Excel
NT	Non-traditional	G	GPX
Win	Windows	K	KML
N/A	Not applicable		

MacENC
FEATURES AT A GLANCE

Each full-featured e-charting application has been evaluated according to the following 81 features. Some ratings are binary, "yes" or "no." Others are more subjective, evaluated on a scale from 1 to 5 (5 being "Excellent"). Any other comparison is defined in the Legend.

Out of Box Experience

Packaging	N/A
Documentation	2
Software install	3
Load charts	3
Load supplemental data	N/A
GPS hookup	3
Technical support	3

User Interface

GUI metaphor	Mac
Screen space management	3
Program responsiveness	2
Chart display speed	2
Chart display quality	5
Customizable GUI	2
Customizable metrics	3
Split windows	No
Tabbed interface	No

Basic Features

One-button MOB	No
Waypoint creation	4
Route creation	3
Steer-to function	No
Track creation	3
Convert track to route	No
Waypoint and route management	3
Chart management	4
Search waypoint and route	4
Range/bearing tool	3
Chart annotation	2
Chart printing	4

Data Exchange and Networking

GPX support for waypoints	Yes
GPX support for routes	Yes
Waypoint exchange formats	GCK
Route exchange formats	GCK
Integrated GPS transfer	Yes
Multiple monitor support	No
Network application sharing	No

Cartography

Free U.S. rasters BSBs (RNCs)	Yes
Free U.S. vectors S-57s (ENCs, IENCs)	Yes
SoftCharts	Yes
CHS Canadian rasters	Yes
DNC	No
Seafarer	No
British Admiralty ARCS	No
International S-57s	Yes
International S-63s	Yes
Scan and geo-reference paper charts	No
Navionics cards	No
C-Map cards	No
C-Map CD/DVD	No
Satellite geo-referenced photos	No
Aerial informational nav photos	No
Topographic maps	No
Bathymetric data	No

Instruments and Sensors

GPS	Yes
Autopilot	Yes
AIS receiver	Yes
Wind	Yes
Depth	Yes
Water temperature	Yes
Radar	Yes
Heading sensor	Yes
Video camera	No

Advanced Features

Tides	MT
Currents	MT
Coast Pilot	No
POIs	No
Streets	No
Google Earth support	Yes
Advanced search	4
AIS	Yes
Buddy boat	No
Radar (ARPA/MARPA)	Yes
Radar (display overlay)	No
Fuel calculator	No
Transit calculator	No
Celestial calculator/Almanac	No
Auto route planning	No
Great Circle Route planning	No
Sail performance	3
GRIB weather integration	3
Weather options	No
Customizable bathymetric recorder	No

Legend

5	Excellent	CM	C-Map
4	Above average	NV	Navionics
3	Average	SC	SailCruiser
2	Below average	MT	Mr. Tides
1	Poor	PP	Plus Pack
CP	Chartplotter	C	CSV
Mac	Macintosh	E	Excel
NT	Non-traditional	G	GPX
Win	Windows	K	KML
N/A	Not applicable		

Part Five

E-Charting Resources

Appendix A, *Nautical Reference Library*, lists many of the free government publications, reference texts, and nautical calculators available on the Internet.

Appendix B, *Useful Web Links*, summarizes the web links by topic mentioned throughout the book.

The *Glossary of Terms* includes over 460 navigation, electronic charting, and general computing terms used in the main chapters.

Appendix A
Nautical Reference Library

Hundreds of nautical reference publications and calculators are available as free downloads on the Internet. Below is a sample of important U.S., Canadian, and World government publications, reference texts, and nautical calculators. The approximate hardcopy cost is included for comparison (in parentheses).

American Practical Navigator - Bowditch ($75)
Complete text and illustrations describing the principles of navigation, including piloting, electronic navigation, celestial navigation, mathematics, safety, oceanography, and meteorology.

Atlas of Pilot Charts Publications 105 to 109 ($165)
Pilot Charts depict averages in prevailing winds and currents, air and sea temperatures, wave heights, ice limits, visibility, barometric pressure, and weather conditions at different times of the year. The charts are intended to aid the navigator in selecting the fastest and safest routes with regard to expected weather and ocean conditions. The charts are not intended to be used for navigation.
- Pub. 105: South Atlantic Ocean ($30) (not yet available for download pending digital conversion)
- Pub. 106: North Atlantic Ocean (including Gulf of Mexico) ($45)
- Pub. 107: South Pacific Ocean ($30)
- Pub. 108: North Pacific Ocean ($30) (not yet available for download pending digital conversion)
- Pub. 109: Indian Ocean ($30)

Canadian Coast Guard Radio Aids to Navigation ($18.95)
Presents information on radio communication and radio navigational aid services provided in Canada. Also includes radio facilities of other government agencies that contribute to the safety of ships in Canadian waters.

Chart No. 1 ($9.95)
A description of the symbols, abbreviations, and terms that appear on nautical charts.

Distances Between Ports Pub. 151 ($19)
Tabulated distances between an alphabetical listing of departure ports, junction points, and arrival ports worldwide.

Distances Between U.S. Ports ($59)
Similar to Publication 151 (Distances Between Ports), but produced by NOAA and only covering U.S. port distances.

Federal Meteorological Handbook No. 1 (Free)
Twelve chapters and four appendices covering topics such as observation records, wind, visibility, sky conditions, temperature and dew point, and pressure coding. Includes a glossary and list of abbreviations and acronyms.

International Code of Signals Pub. 102 ($30)
Descriptions and/or graphic depictions of all signals for visual, sound, and radio communications.

Marine Weather Service Charts ($18.75)
Lists frequencies, schedules, and locations of stations disseminating National Weather Service information and additional weather information of interest to the mariner. Consists of 16 Marine Service Charts (MSCs):
- MSC-1 Eastport, ME to Montauk Point, NY ($1.25)
- MSC-2 Montauk Point, NY to Manasquan, NJ ($1.25)
- MSC-3 Manasquan, NJ to Cape Hatteras, NC ($1.25)
- MSC-4 Cape Hatteras, NC to Savannah, GA ($1.25)
- MSC-5 Savannah, GA to Apalachicola, FL ($1.25)
- MSC-6 Apalachicola, FL to Morgan City, LA ($1.25)
- MSC-7 Morgan City, LA to Brownsville, TX ($1.25)
- MSC-8 Mexican Border to Point Conception, CA ($1.25)
- MSC-9 Point Conception, CA to Point St. George, CA ($1.25)
- MSC-10 Point St. George, CA to Canadian Border ($1.25)
- MSC-11/12 Great Lakes ($1.25)
- MSC-13 Hawaiian Waters ($1.25)
- MSC-14 Puerto Rico and Virgin Islands ($1.25)
- MSC-15 Alaskan Waters ($1.25)
- MSC-16 Guam and the Northern Mariana Islands ($1.25)

Mariner's Guide for Hurricane Awareness (Free)
Seventy-page document covering topics such as storm categories, observations at sea, hurricane motion and monitoring, and hurricane evasion. Includes a risk analysis checklist and a tracking chart.

MF and HF Channels (Free)
List of middle and high frequency marine radiotelephone channels, including information on single sideband radiotelephone channels.

Nautical Calculators (Free)
A collection of 34 calculators for marine navigation, including celestial navigation, conversions, distance, log and trig, sailings, time zones, and weather data. Each calculator is a self-contained Java script program that can be used offline.

Nautical Chart User's Manual (Out of Print)
Over 300 pages covering the details of marine cartography, with topics such as layout, scales, projections, topography, hydrography, navigation aids, landmarks, and routes.

Navigation Rules-International and Inland ($14)

Published by the U.S. Coast Guard, the official rule book (CG169) required on all vessels over 12 meters. Contains the International Regulations for Preventing Collisions at Sea (COLREGS) and the Inland Navigation Rules for all inland waters including the Great Lakes.

NGA List of Lights ($343)

Similar to the U.S. Coast Guard Light List, contains information on lights and other aids to navigation under the authority of foreign governments by geographic region. Consists of seven volumes:

Pub. 110: Greenland, The East Coast of North and South America (Excluding Continental United States Except the East Coast of Florida) and the West Indies ($50)

Pub. 111: West Coasts of North and South America, Australia, Tasmania, New Zealand, and the Islands of the North and South Pacific Oceans ($49)

Pub. 112: Western Pacific and Indian Oceans Including the Persian Gulf and Red Sea ($60)

Pub. 113: The West Coasts of Europe and Africa, the Mediterranean Sea, Black Sea, and Azovskoye More (Sea of Azov) ($51)

Pub. 114: British Isles, English Channel and North Sea ($42)

Pub. 115: Norway, Iceland, and Arctic Ocean ($46)

Pub. 116: Baltic Sea with Kattegat, Belts and Sound and Gulf of Bothnia ($45)

NOAA Coast Pilots ($252)

Covers supplemental information difficult to portray on a nautical chart, including information on ports and marinas, channel descriptions, anchorages, bridge and cable clearances, currents, and tide and water levels. Consists of nine volumes:

Volume 1: Atlantic Coast: Eastport, ME to Cape Cod, MA ($28)

Volume 2: Atlantic Coast: Cape Cod, MA to Sandy Hook, NJ ($28)

Volume 3: Atlantic Coast: Sandy Hook, NJ to Cape Henry, VA ($28)

Volume 4: Atlantic Coast: Cape Henry, VA to Key West, FL ($28)

Volume 5: Gulf of Mexico, Puerto Rico, and Virgin Islands ($28)

Volume 6: Great Lakes ($28)

Volume 7: Pacific Coast: California, Oregon, Washington, Hawaii, and Pacific Islands ($28)

Volume 8: Pacific Coast Alaska: Dixon Entrance to Cape Spencer ($28)

Volume 9: Pacific Coast Alaska: Cape Spencer to the Beaufort Sea ($28)

NOAA Hurricane Basics (Free)

Nineteen-page document covering background on hurricane weather reports, including Saffir-Simpson scale and forecasting information.

NWS Observing Handbook No. 1 (Free)

Field guide format to observation and recording of meteorological data at sea.

Radar Navigation and Maneuvering Board Manual Pub. 1310 ($45)

Fundamentals of shipboard radar, radar operation, collision avoidance, navigation by radar, and a description of vessel traffic systems in U.S. waters.

Radio Navigation Aids Pub. 117 ($50)

List of selected worldwide radio stations broadcasting navigational warnings, time signals, or medical advice; communication traffic for distress, emergency, and safety; and long range navigational aids.

Sailing Directions Enroute ($1,320)

Detailed coastal and port approach information, supplementing the largest scale chart of the area. Contains information about the coastal weather, currents, ice, dangers, features, and ports for each geographic area. Consists of 38 publications:

Pub. 123: Southwest Coast of Africa ($20)

Pub. 124: East Coast of South America ($40)

Pub. 125: West Coast of South America ($35)

Pub. 126: Pacific Islands ($42)

Pub. 127: East Coast of Australia and New Zealand ($45)

Pub. 131: Western Mediterranean ($45)

Pub. 132: Eastern Mediterranean ($35)

Pub. 141: Scotland ($45)

Pub. 142: Ireland and the West Coast of England ($35)

Pub. 143: West Coast of Europe and Northwest Africa ($40)

Pub. 145: Nova Scotia and the St. Lawrence ($45)

Pub. 146: Newfoundland, Labrador and Hudson Bay ($48)

Pub. 147: Caribbean Sea, Bermuda, Bahamas, and Islands of the Caribbean Sea ($30)

Pub. 148: Caribbean Sea ($30)

Pub. 153: West Coasts of Mexico and Central America ($25)

Pub. 154: British Columbia ($40)

Pub. 155: East Coast of Russia ($30)

Pub. 157: Coasts of Korea and China ($30)

Pub. 158: Japan, Volume 1 ($35)

Pub. 159: Japan, Volume 2 ($33)

Pub. 161: South China Sea and the Gulf of Thailand ($24)

Pub. 162: Philippine Islands ($46)

Pub. 163: Borneo, Jawa, Sulawesi, and Nusa Tenggara ($46)

Pub. 164: New Guinea ($40)

Pub. 171: East Africa and the South Indian Ocean ($35)

Pub. 172: Red Sea and the Persian Gulf ($41)

Pub. 173: India and the Bay of Bengal ($36)

Pub. 174: Strait of Malacca and Sumatra ($30)

Pub. 175: North and West Coast of Australia ($31)

Pub. 181: Greenland and Iceland ($31)

Pub. 182: North and West Coasts of Norway ($32)

Pub. 183: Northern Coast of Russia ($22)

Pub. 191: English Channel ($25)

Pub. 192: North Sea ($35)

Pub. 193: Skagerrak and Kattegat ($30)

Pub. 194: Baltic Sea: Southern Part ($40)

Pub. 195: Gulf of Finland and Gulf of Bothnia ($22)

Pub. 200: Antarctica ($26)

Sailing Directions Planning Guides ($168)

Contains relevant physical, political, industrial, navigational, and regulatory information on countries adjacent to a particular ocean basin. Consists of five volumes:

Pub. 120: Pacific Ocean and Southeast Asia ($45)
Pub. 140: North Atlantic Ocean; Baltic, North and Mediterranean Seas ($45)
Pub. 160: South Atlantic Ocean and Indian Ocean ($35)
Pub. 180: The Arctic Ocean ($17)
Pub. 200: Antarctica ($26)

Sight Reduction Tables Pub. 229 ($119.70)

Tables necessary for celestial navigation. Consists of six volumes:

Volume 1: Latitudes 00º to 15º Inclusive ($19.95)
Volume 2: Latitudes 15º to 30º Inclusive ($19.95)
Volume 3: Latitudes 30º to 45º Inclusive ($19.95)
Volume 4: Latitudes 45º to 60º Inclusive ($19.95)
Volume 5: Latitudes 60º to 75º Inclusive ($19.95)
Volume 6: Latitudes 75º to 90º Inclusive ($19.95)

State Chart Catalogs (Free)

Separated by state, a list and illustration of NOAA charts by number and location.

USCG Light Lists ($336)

Published by the U.S. Coast Guard, a list of lights, sound signals, buoys, daybeacons, and other aids to navigation for U.S. waters. Consists of seven volumes:

Volume 1: Atlantic Coast: St. Croix River, ME to Shrewsbury River, NJ ($50)
Volume 2: Atlantic Coast: Shrewsbury River, NJ to Little River, SC ($50)
Volume 3: Atlantic Coast: Little River, SC to Ecofina River, FL, including Puerto Rico and USVI ($51)
Volume 4: Gulf of Mexico: Ecofina, FL to Rio Grande River, TX ($50)
Volume 5: Mississippi River System ($39)
Volume 6: Pacific Coast and Pacific Islands ($46)
Volume 7: Great Lakes ($50)

US VHF Channels (Free)

Table of VHF Channels and each channel's communication use.

World Port Index Pub. 150 ($41)

Contains the location, characteristics, known facilities, and available services of major ports and shipping facilities throughout the world.

Worldwide Marine Radiofacsimile Broadcast Schedules ($20)

Includes weatherfax frequencies and schedules for fax stations around the world.

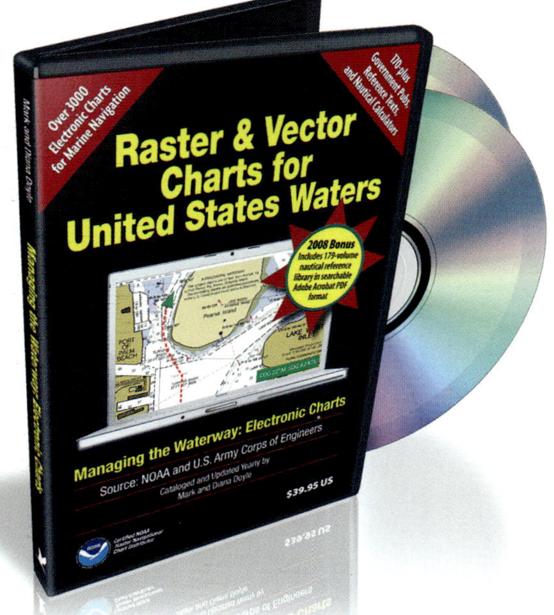

Onboard Digital Resource: Managing the Waterway: The Complete Electronic Chart and Nautical Reference Library ($39.95) is a 2-DVD set containing nearly 200 of the government publications, reference texts, and nautical calculators listed here as well as over 3,000 NOAA and USACE raster and vector charts. Updated yearly, the set is available from marine retailers or directly at www.managingthewaterway.com.

Useful Web Links

The following 162 websites are mentioned within the book. For convenience, they are grouped and listed by topic, in the approximate order of the chapters. Because the locator for a web page can change over time, an URL listed may not open. If this occurs, try the domain name only, omitting the server path (e.g., www.managingthewaterway.com instead of www.managingthewaterway.com/icwguide.htm).

Hardware and Accessories

Computer Accessories

C-Map USB Multimedia Reader
www.c-map.com

Belkin Cables
www.belkin.com/cables

Hoodman Corporation Screen Hoods
www.hoodmanusa.com

KeySpan USB High Speed Serial Adapter
www.keyspan.com

Navionics Multi Card Reader
www.navionics.com

Sony Digital Camera GPS Geo-Tagger
www.sony.com

Cellular, SSB and Satellite Internet

3G Cellular Network Information
www.evdoinfo.com

Globalstar
www.globalstar.com

iDirect
www.idirect.net

Inmarsat
www.inmarsat.com

Iridium Satellite
www.iridium.com

Port Networks
http://portnetworks.com

Shakespeare CruiseNet
www.shakespeare-marine.com

Syrens Onboard WiFi
www.syrens-at-sea.com

ThurayaMarine
www.thuraya.com

Antennas and Boosters

Digital Antenna
www.digitalantenna.com

WaveRV marine antenna
www.radiolabs.com

Rugged Computers, Cases, and Hoods

ARMOR Mobile Rugged Computers
www.drsarmor.com

Dell Latitude Rugged Laptop
www.dell.com

Itronix Rugged Computers
www.itronix.com

Motorola Rugged Notebook
www.motorola.com

OtterBox Waterproof Cases
www.otterbox.com

Panasonic Toughbook
www.panasonic.com

Mini Computers

mp3Car
www.mp3car.com

Logic Supply 12-Volt Computers
www.logicsupply.com

Simplified Innovation
www.simplifiedinnovation.com

Apple Mac mini
www.apple.com/macmini

Tablet Computers

ARMOR Mobile Rugged Computers
www.drsarmor.com

Faria Marine Tablets
www.faria-instruments.com/marine.php

Instruments and Sensors

USB GPS Sensors

USGlobalSat USB GPS
www.usglobalsat.com

Garmin USB GPS
www.garmin.com

AIS Stand-Alone Software

ShipPlotter
www.shipplotter.com

Wireless Marine Instruments

Tacktick Wireless Marine Instruments
www.tacktick.com

Electronic Charts

U.S. Chart Catalogs

U.S. Catalog of Charts & Publications
http://nauticalcharts.noaa.gov/mcd/ccatalogs.htm

State Chart Catalogs
http://nauticalcharts.noaa.gov/mcd/ccatalogs.htm#state

International Chart Catalogs

IHO List of Available ENC Cells
http://services.ecc.as/ihocc/public/

International Centre for ENCs Worldwide ENC Catalog
www.ic-enc.org

U.S. Free Chart Downloads

Army Corps of Engineers Inland Waterways Chart Downloads
www.tec.army.mil/echarts

ChartServer Downloads for NOAA RNCs and ENCs
http://chartmaker.ncd.noaa.gov

Digital Nautical Chart Downloads
www.nga.mil/portal/site/dnc

Maptech Chart Download Portal
www.freeboatingcharts.com

NOAA and NGA Online Chart Ordering
www.naco.faa.gov

Select International Hydrographic Offices

U.K. Hydrographic Office
www.ukho.gov.uk

Australian Hydrographic Service
www.hydro.gov.au

Canadian Hydrographic Service
www.charts.gc.ca

French Hydrographic Office
www.shom.fr

Land Navigation New Zealand
www.linz.govt.nz/audiences/mariners/index.html

Cartography Vendors

C-Map
www.c-map.com

Explorer Charts of the Bahamas
www.explorercharts.com

Garmin
www.garmin.com

Maptech
www.maptech.com

MapMedia
www.mapmedia.com

Nautical Data International (NDI)
www.digitalocean.ca

Navionics
www.navionics.com

Nobeltec
www.nobeltec.com

NV-Verlag Charts of Europe and the Caribbean
www.nv-verlag.de

Wavey Line Charts of the Caribbean
www.waveylinepublishing.com

Download Tests and Calculators

Intel Broadband Speed Test
www.intel.com/cd/personal/computing/emea/
eng/299414.htm

Speakeasy Speed Test
www.speakeasy.net/speedtest

Speedtest.net
www.speedtest.net

Numion Bandwidth Calculators
www.numion.com/calculators

E-Charting Software

E-Charting Freeware

CARIS Easy ENC
www.caris.com/products/easy-view

HydroSERVICE dKart Look
www.hydroservice.no

NOAA's On-Line Chart Viewer
http://ocsdata.ncd.noaa.gov/OnLineViewer

SevenCs SeeMyDENC
www.sevencs.com/?page=123

Sping SeaClear II
www.sping.com/seaclear

Full-Featured PC Software

B&G Deckman
www.sailmath.com/deckman_for_windows.htm

CherSoft Henry Navigation System
www.chersoft.co.uk

Des Newman's OziExplorer
www.oziexplorer.com

DigiBOAT Software-On-Board
www.digiboat.com.au

Expedition Navigation Software
www.iexpedition.org

Furuno MaxSea
www.maxsea.com

Global Navigation Software NavPak
www.globenav.com

Jeppesen Marine Nobeltec
www.nobeltec.com

Maptech Marine Software
www.maptech.com

NavGator Mariner
www.navgator.com

NavSim BoatCruiser
www.navsim.com

Northport Systems Fugawi Marine ENC
www.fugawi.com

Raymarine RayTech RNS
www.raymarine.com

Rose Point Navigation Coastal Explorer
www.rosepointnav.com

Seatrack Global Navigation Software
www.seatrack.co.uk

TIKI Navigator
www.tiki-navigator.com

Tridentnav Systems
www.tridentnav.se

Full-Featured Mac Software

Barco Software NavimaQ
www.barcosoft.com

GPSNavX MacENC
www.gpsnavx.com

James Associates MacGPS Pro
www.macgpspro.com

Macintosh GPS Communications GPSy
www.gpsy.com

Wind Vector PassagePlus
www.windvector.com

Internet-Based E-Charting Applications

EarthNC
http://earthnc.com

GeoGarage Google Earth Mashups
http://demo.geogarage.com/noaa

geowake Marina
http://marina.geowake.com

Tide and Current Utilities

Mr. Tides
http://homepage.mac.com/augusth/MrTides

XTide
www.flaterco.com/xtide

General Software

Adobe Acrobat Free Reader
www.adobe.com/downloads

Adobe Flash Player
www.adobe.com/downloads

Google Earth
http://earth.google.com

SnagIt Screen Capture Software
www.techsmith.com

Snapz Pro X Screen Capture Software
www.ambrosiasw.com

Mac PC Emulators

Apple Boot Camp
www.apple.com/support/bootcamp

Parallels Desktop
www.parallels.com/products/desktop

VMware Fusion
www.vmware.com/products/fusion

Networking and Data Exchange

Waypoint and GPS Helper Utilities

EasyGPS GPS Software
www.easygps.com

G7ToWin GPS Interface Software
www.gpsinformation.org/ronh/

GeoTrans Geographical Converter
http://earth-info.nga.mil/GandG/geotrans/

GooPs Google Earth GPS Utility
http://goopstechnologies.com

gps2geX Mac Google Earth GPS Utility
www.grandhighwizard.net/gps2gex.html

GPSBabel GPS Software
www.gpsbabel.org

Extending the Digital Metaphor

Downloadable Publications and Calculators

American Practical Navigator (Bowditch)
www.nga.mil/portal/site/maritime

NGA Nautical Publications
www.nga.mil/portal/site/maritime

NOAA Chart No. 1
http://nauticalcharts.noaa.gov/mcd/chartno1.htm

USCG Navigation Rules
www.navcen.uscg.gov/mwv/navrules/navrules.htm

U.S. Local Notice to Mariners
www.navcen.uscg.gov/lnm

Reference Sources

Geographic Names Information System Gazetteer
http://geonames.usgs.gov

GeoNet Name Server Online Gazetteer
http://earth-info.nima.mil/gns/html/index.html

Google Maps
http://maps.google.com

eNature Online Field Guides
www.enature.com/fieldguides

Mapquest
www.mapquest.com

Merck Medical Manuals
www.merck.com/mmpe

NavQuest
www.navquest.com

Weather Resources

National Weather Service Email/FTP Service
www.nws.noaa.gov/tg/ftpmail.htm

Saildocs Email GRIB requests
www.saildocs.com

Ugrib Free GRIB Viewer
www.grib.us

OCENS WeatherNet
www.ocens.com

National Data Buoy Center
www.ndbc.noaa.gov

XM Weather
www.xmwxweather.com

Sirius Marine Weather
www.sirius.com/marineweather

SkyMate Onboard Information Services
www.skymate.com

Xaxero SkyEye Software
www.xaxero.com

ClearPoint High Definition Weather
www.clearpointweather.com

Roffer's Ocean Fishing Forecasting Service
www.roffs.com

Weather Underground
www.wunderground.com

Accuweather
www.accuweather.com

Crown Weather
www.crownweather.com

National Weather Service
www.weather.gov

NWS Satellite Services Division
www.ssd.noaa.gov

HotSpots Charts
www.sstcharts.com

Rutgers University Satellite Images
http://marine.rutgers.edu/mrs/sat_
data/?nothumbs=0

User-Generated Sites

ActiveCaptain
www.activecaptain.com

Google Images
http://images.google.com

GPSWaypoint Repository
http://gpswaypoint.org

Panoramio Photos
www.panoramio.com

TravelByGPS Waypoint Repository
www.travelbygps.com

Wayhoo Geographic Coordinates
http://wayhoo.com

Wikipedia Encyclopedia
www.wikipedia.org

YouTube Videos
www.youtube.com

Forums

BoatUS.com
http://my.boatus.com/forum

Cruising Sailor
www.cruisenews.net/forum

GoBoating.com
www.goboating.com

Google Earth Blog
www.gearthblog.com

Living Aboard
www.livingaboard.com

macsailing.net
http://macsailing.net

Marine Trawler Owners Association
www.mtoa.net

Power Catamaran List
http://lists.samurai.com/mailman/listinfo/power-
catamaran

Rec.Boats.Cruising
http://groups.google.com/group/rec.boats.cruising

sailingtalks
www.sailingtalks.com

SailNet
www.sailnet.com

Seven Seas Cruising Association
http://www.ssca.org

TheBoaters.com
www.theboaters.com

The Hull Truth
www.thehulltruth.com

The Salty Southeast Cruiser's Net
www.cruisersnet.net

TopoGrafix Forums
http://forums.topografix.com

Trawlers-and-Trawlering List
http://lists.samurai.com/mailman/listinfo/trawlers-
and-trawlering

Glossary of Terms

This glossary contains over 460 navigation, electronic charting, and general computing terms. All abbreviated or acronym terms are listed in their full form and by their diminutive.

A

Abeam: To the side of a vessel (perpendicular to its heading).

AC: See Alternating Current.

Accuracy: The level of match between a GPS-measured position, time, and/or velocity and the true value.

Active Leg: The segment of a route (leg) currently being traveled.

Admiralty Raster Chart Service (ARCS): Worldwide raster charts supplied by the British Admiralty, Britain's hydrographic office.

Aerial Photo: Photograph taken of the ground where the camera is located on an aircraft, helicopter, or other low-altitude object. (Compare with Satellite Photo.)

Aid to Navigation: See Navigational Aid.

Air Card: A hardware device that connects a computer to a cellular network for Internet access.

AIS: See Automatic Identification System.

Alarm: An indicator or warning.

Alarm Radius: The distance set from a point (such as a waypoint or anchor location), within which an alarm will sound and/or a note will be displayed warning of entry into (in the case of waypoints) or outside of (in the case of an anchor watch) the proximity circle.

Almanac: An annual publication containing tabular information on a topic, often arranged according to the calendar

Almanac Data: Information transmitted by each satellite that allows a GPS receiver to acquire satellites rapidly upon startup.

Alphanumeric: A computer character that is either a letter or a number.

Alternating Current (AC): An electrical current whose magnitude and direction vary cyclically. The form in which electricity is delivered to businesses and residences in the U.S. as 120 volts. (Compare with Direct Current.)

American Practical Navigator: A classic reference text describing the principles of navigation, including piloting, electronic navigation, celestial navigation, mathematics, safety, oceanography, and meteorology. Named after Nathaniel Bowditch, the author of the first edition in the 1800s. Also called Bowditch.

Analog: An indication that is presented on a continuous scale rather than with binary digits. (Compare with Digital.)

Anchor Watch: An alarm function that warns if the vessel moves outside a preset proximity circle.

Anemometer: An instrument that measures wind speed and direction.

Annotation: A text box linked to a location on an electronic chart. Also called a note.

Application: See Software.

Approaching Waypoint: An alarm function that warns when the vessel is within a preset distance of a waypoint.

ARCS: See Admiralty Raster Chart Service.

ARPA: See Automatic Radar Plotting Aid.

Arrival Circle: A hypothetical radius surrounding a destination point, used to detect an approaching waypoint.

Arrival Radius: The distance from a waypoint location used by the GPS to mark that a waypoint has been reached, shifting navigation to the next leg in an active route.

Artifact: Degradation associated with a process. For example, a computer image may show display artifacts caused by the software code rather than actual data.

ASCII: American Standard Code for Information Interchange. A simple computer format for representing alphanumeric information such as waypoint names, latitude, and longitude.

Atlas: A book of maps or charts. (Compare with Chartkit and Gazetteer.)

Atomic Clock: A precise clock, used by GPS satellites, that works using the decay of the elements cesium or rubidium.

ATON: See Navigational Aid.

Attribute: A characteristic of an object, either qualitative or quantitative. An example is bridge height.

Automatic Identification System (AIS): A system for identifying, locating, and communicating between vessels by integrating an electronic navigation system with a GPS and a VHF transceiver system.

Automatic Radar Plotting Aid (ARPA): A radar that is ARPA-enabled can create tracks using radar contacts—calculating the tracked object's course, speed, and closest point of approach (CPA)—in order to warn of a potential collision.

Autopilot: A device that automatically steers a vessel

B

BA: See British Admiralty.

Bandwidth: In computing, the amount of data throughput available.

Bathymetric: The data of water depth, creating a representation of the topography of the sea bottom.

Bearing (BRG): The direction, in degrees measured from true or magnetic north, from present position to the next waypoint.

Blog: A website (usually maintained by an individual) that serves as an online journal with regular entries of text or images, commonly displayed in reverse chronological order. An abbreviation of weblog or web log.

Bluetooth: A short-range wireless connectivity standard used for computer peripherals.

Bowditch: See American Practical Navigator.

British Admiralty (BA): The hydrographic office of Great Britain.

BSB or BSB3: A digital file format for raster charts co-developed by Maptech and NOAA. Now the standard format for North American raster chart files.

BSB4, BSB5: Proprietary versions of BSB files requiring a user license.

C

Cache: Auxiliary computer memory from which high-speed retrieval is possible. Also used as a verb, meaning to store data in cache memory.

Canadian Coast Guard Radio Aids to Navigation: Government publication with information on radio communication and radio navigational aid services provided in Canada.

Canadian Hydrographic Service (CHS): The hydrographic office of Canada.

Card: Removable cartridge media containing chart cartography, commonly used in chartplotters. Also called a chip.

Card Reader: An electronic device that reads and writes data from portable flash memory storage devices called cards. Called a multimedia card reader if it accepts different card specifications, such as SD (Secure Digital), CF (CompactFlash), or MM (MultimediaCard).

Cartography: The making of maps or charts.

CDI: See Course Deviation Indicator.

CD-ROM: Compact Disc Read-Only Memory. A compact disc that contains up to 700 MB of data accessible by a computer.

Cell: The basic unit for the distribution of ENC (vector) data covering a geographic area. A cell rarely corresponds to the same area covered by the paper chart. (Compare with Panel.)

Cellular Network: A system of base stations providing mobile telephone support.

Channel: The frequency, receiver, and circuitry necessary to receive a radio signal from a transmitter.

Chart: A seagoing map showing depths and aids to navigation.

Chart Datum: A permanently established surface from which soundings or tide heights are referenced.

Chartkit: A book of nautical charts, often annotated with routes, marine facilities, and additional information. (Compare with Atlas and Gazetteer.)

Chart No. 1: Publication by NOAA with descriptions of the symbols, abbreviations, and terms that appear on U.S. nautical charts.

Chartplotter: A marinized hardware unit that displays cartridge-loaded electronic charts and optionally connects to navigation instruments such as a GPS receiver or autopilot.

Chip: See Card.

CHS: See Canadian Hydrographic Service.

Circle of Position (COP): A circle on the surface of the Earth surrounding a known object at fixed distance (radius). Used to compute global position.

Click: The action of quickly and singly depressing a computer mouse button. See also Double-Click.

Closest Point of Approach (CPA): The calculated minimum distance between two objects. Used for tracking and collision avoidance.

CMG: See Course Made Good.

Coast Pilots: A nine-volume annual government publication that supplements U.S. nautical charts with information such as channel descriptions, anchorages, bridge and cable clearances, currents, tide and water levels, prominent features, dangers, routes, traffic separation schemes, small-craft facilities, and federal regulations.

COG: See Course Over Ground.

Cold Start: Powering up a device, such as a GPS, when it needs to be initialized. (Compare with Warm Start.)

Collision Avoidance: Systems, such as radar or AIS, that track objects with the intent of warning of potential intersecting courses.

COLREGS: U.S. Coast Guard acronym for International Regulations for the Prevention of Collisions at Sea. Also known as the International Navigation Rules.

Comma-Separated Value (CSV): A file type that stores tabular data, separating fields with a comma.

Compass Rose: A printed circle showing the principal cardinal directions.

COM Port: Communications Port. The serial communication connection of an external device to a personal computer.

Coordinated Universal Time (UTC): A world standard for time, replacing Greenwich Mean Time (GMT) in 1986, based on atomic measurements rather than the rotation of the Earth. GPS uses UTC; GMT remains the standard time zone for the prime meridian and zero longitude.

Coordinates: A set of measurements that define the position of a point, such as latitude and longitude, time difference curves, or UTM.

COP: See Circle of Position.

Copy: To move a computer file from one location to another. Files can by copied between folders, drives, or external media such as CDs or DVDs. (Compare with Load.)

Course: The direction from a starting location to a destination point, measured clockwise in degrees from true or magnetic north.

Course Deviation Indicator (CDI): The display of lateral position in relation to a track. When used with a GPS it shows actual distance left or right of the programmed courseline.

Course Line: The line placed on a chart representing the ideal path between two waypoints.

Course Made Good (CMG): For each leg of a track, the resultant direction from a vessel's point of departure.

Course Over Ground (COG): The actual path of a vessel's course, typically an irregular line.

Course To Steer (CTS): The heading required in order to efficiently reach a destination.

Course-Up: A screen display orientation that shows the direction of the vessel's course upward. Similar to Heading Up but less sensitive to vessel course deviations. (Compare with Head-Up and North-Up.)

CPA: See Closest Point of Approach.

CPU: Central Processing Unit. The "brains" of the computer that executes programs. Also called the processor.

CRADA: Cooperative Research and Development Agreement. An agreement between a private company and a government agency to work together on a project in order to speed the commercialization of a federally-developed technology.

Crash: A colloquialism for an undesired computer system shut-down, usually requiring a restarting of the computer.

Cross-Track Error (XTE): The lateral distance in either direction (port or starboard) that the vessel is off course.

CSV: See Comma-Separated Value.

CTS: See Course To Steer.

Cursor: The indicator, typically an arrow icon, that the user moves around a computer screen (via a mouse, trackball, or arrow keys) to interact with the graphical user interface.

D

Datum (geodetic): A set of parameters specifying the reference coordinate system used for calculating points on the earth.

Datum (horizontal or vertical): A permanently established surface from which soundings or tide heights are referenced.

Datum Shift: The difference between the coordinates of a point using two separate references of measurement.

DC: See Direct Current.

Dead Reckoning (DR): Originally from "Deduced Reckoning," the deduction of one's position calculated from the course steered and the speed through water rather than by obtaining a fix.

Defense Mapping Agency (DMA): The U.S. agency created to consolidate mapping functions previously dispersed among the military services. Subsequently became part of NIMA and now part of the NGA.

Degree: A unit of measurement of angles, 1/360th of the circumference of a circle.

Demoware: See Trial Software.

Department of Defense (DoD): The U.S. agency that controls and manages the Global Positioning System (GPS).

Depth Sounder: An electronic instrument that determines the depth of water beneath a vessel.

Deviation: The difference between the actual magnetic direction and the compass, caused by local objects that interfere with magnetic compass readings.

DGLONASS: The differential system for GLONASS.

DGPS: The differential system for GPS.

Differential System: An augmentation system where radionavigation signals are monitored at a known reference position in order to make and broadcast local corrections.

Digital: Information that is stored in binary form, either as one/zero or plus/minus. (Compare with Analog.)

Digital Nautical Chart (DNC): The National Geospatial-Intelligence Agency's trademarked name for its vector-format chart data files. An alternative vector format to ENCs.

Digital Selective Calling (DSC): A device that has DSC capability is able to "dial" and "ring" other DSC-equipped radios, including instantly sending an automatically formatted distress call.

Dilution of Precision (DOP): The geometric contribution, such as by position or clock constraints, to the uncertainty of a GPS position fix.

Direct Current (DC): An electrical current with continuous and unidirectional flow of electric charge. The system used in cars and boats, typically at 12 volts. (Compare with Alternating Current.)

Display: A visual presentation of data.

Distance: The length between two points. This length can be measured in straight line (rhumb line) or Great Circle (over the Earth) terms.

Distance Off: A measure from a landmark to set an alarm or fix current position.

Distances Between Ports: A government publication tabulating distances between an alphabetical listing of departure ports, junction points, and arrival ports worldwide.

Distances Between U.S. Ports: Similar to Distances Between Ports, but produced by NOAA and only covering U.S. port distances.

Distance To Go (DTG): The distance between the present position and a destination.

DMA: See Defense Mapping Agency.

DNC: See Digital Nautical Chart.

Docking Station: A simplified way, using a permanent platform and readily accessible cables, of plugging in a portable laptop computer to frequently-used computer peripherals.

DoD: See Department of Defense.

Dongle: A small hardware device, also called a hardware key, that connects to a computer to authenticate a license and prevent illegal copying.

DOP: See Dilution of Precision.

Double-Click: The action of quickly depressing a computer mouse button twice in succession.

Download: To retrieve data from another computer, including from a server or from the Internet. (Compare with Upload.)

DR: See Dead Reckoning.

Drag-and-Drop: The action of depressing a computer mouse button, holding it down while sliding the mouse, then releasing the mouse button.

Driver: A computer program allowing higher-level computer programs to interact with a device. Also called a device driver or software driver.

DR Position: The point the vessel would have reached in the absence of current or wind. 

DSC: See Digital Selective Calling.

DTG: See Distance To Go.

DVD-ROM: Digital Versatile Disc Read-Only Memory. A storage media format that looks like a compact disc (CD) but stores 4.7 GB, six times as much data as a CD.

E

ECDIS: Electronic Chart Display and Information System. An Electronic Charting System that satisfies international standards set by the IHO for commercial navigation, such as the use of internationally-approved vector cartography.

E-Charting: A colloquial term for displaying charts, and optionally navigating, using a personal computer.

ECS: Electronic Charting System. A generic term for a system comprised of hardware devices integrating a personal computer or dedicated chartplotter with navigation software, chart data files, and additional marine electronics.

Edge-Matched: The scaling of displayed electronic charts to align adjacent charts edges.

Electronic Chart: The encoding of a nautical chart as a digital file to be viewed on an electronic display such as a computer or chartplotter.

Electronic Navigational Chart (ENC): *NOAA ENC* is the trademark name for its vector-format chart data files.

Elevation: The distance above or below mean sea level.

eLoran: Enhanced Loran System. An updated version of Loran, a ground-based radionavigation system that uses multiple low frequency radio transmitters to determine the location of the receiver.

ENC: See Electronic Navigational Chart.

Envoy: The use of an intermediary device to transfer data between two devices, such as using a chartplotter to transfer data between computers.

Ephemeris: A list of the precise positions of the GPS satellites as a function of time.

EPIRB: Emergency Position-Indicating Radio Beacon. Transmitters that aid in the detection and location of a vessel or person in maritime distress.

Equator: A great circle that is equidistant from the north and south poles.

Estimated Time En Route (ETE): The projected time remaining to a destination based on current speed and course.

Estimated Time of Arrival (ETA): The projected time of arrival at a destination based on current speed and course.

ETA: See Estimated Time of Arrival.

ETE: See Estimated Time En Route.

Ethernet: A standard for connecting computers or devices to form a local area network.

EVDO: Evolution-Data Optimized. A telecommunications standard for the wireless transmission of data through radio signals, typically for broadband Internet access.

Exchange: The transfer of data across computers or software systems. (Compare with Interchange.)

F

Fathom: A nautical measurement, equal to six feet, used to measure depth.

Feature: The representation of a real world object, such as by a symbol on a chart.

Federal Meteorological Handbook No. 1: A U.S. government publication covering topics such as observation records, wind, visibility,  sky conditions, temperature and dew point, and pressure coding.

FireWire: Apple's brand name for a high-speed serial bus interface standard used for connecting digital audio or video to a personal computer.

Fix: The known position on the surface of the Earth based on the intersection of position lines, obtained or calculated either by hand or electronically.

Flash Drive: A data storage device containing flash memory that has no moving parts and does not require batteries or a power supply to retain stored data. See also Thumb Drive.

Fluxgate Compass: An electronic device used to measure magnetic direction. 

Frequency: The number of cycles of radio energy per second (measured in KHz or MHz).

FTP: File Transfer Protocol. A network protocol to transfer data from one computer to another through a network such as the Internet.

Full-Featured Application: A software application that displays charts; performs planning, navigation, and piloting functions; and integrates with a vessel's marine electronics.

G

GARMN (or GRMN) Protocol: A proprietary protocol used by Garmin for communicating with its GPS devices.

Gazetteer: Printed or digital documents of place names with information such as geographic coordinates, elevation, and features. (Compare with Atlas and Chartkit.)

GB: See Gigabyte.

Geodetic: Relating to the size and shape of the Earth.

Geodetic Datum: A mathematical model to best fit the Earth's geometry and used to establish the origin of datum.

Geographic Information System (GIS): A computer system capable of assembling, storing, manipulating, and displaying geographically referenced information.

GEO/NOS: A digital file format for raster charts developed by SoftChart. This format has been largely superseded by the BSB file format.

Geo-referencing: The process of scaling, rotating, translating and de-skewing an image to match a particular size and position using a common reference system such as latitude and longitude. Necessary to establish a relation between raster and vector images and to allow data from different sources to be combined and overlaid.

GHz: Gigahertz. A measure of radio frequency or clock speed of a computer equal to one billion cycles per second.

Gigabyte: A unit of information equal to one billion bytes, where a byte consists of eight binary digits.

GIS: See Geographic Information System.

Global Maritime Distress and Safety System (GMDSS): A global satellite and land-based communications service that provides maritime safety information.

Global Navigation Satellite System: A positioning system of the European Union that would incorporate GPS, GLONASS, and other space- and ground-based positioning.

Global Positioning System (GPS): A space-based, radio-positioning navigation system maintained by the U.S. government.

GLONASS: A space-based, radio-positioning navigation system maintained by the Russian Federation.

GMDSS: See Global Maritime Distress and Safety System.

GMT: See Greenwich Mean Time.

GNSS: See Global Navigation Satellite System.

GoTo: A route consisting of one leg, with present position at the start of the route and a single waypoint as the destination.

GPS: See Global Positioning System.

GPS Receiver: An electronic receiver that decodes GPS satellite broadcasts and displays position, course, and speed.

GPS Sensor: A GPS receiver, without control buttons or a display, that outputs raw GPS data to a display unit such as a chartplotter.

GPX: GPS Exchange Format. A data format standard to transfer GPS data such as waypoints and routes.

Grabber Hand: A common computer tool where the cursor displays as a small hand icon that can grab and move objects by dragging-and-dropping with the mouse.

Graphical User Interface (GUI): The addition of icons and images, complimenting text, allowing a person means to interact with a computer.

Great Circle: The shortest line between two points on a sphere such as the Earth, formed by an arc over the surface.

Greenwich Mean Time (GMT): The mean solar time at the Greenwich meridian. Replaced by UTC in 1986.

Greenwich Meridian: The meridian of longitude, passing through Greenwich, England, that is the prime meridian with longitude 0°.

GRIB: Gridded Binary. A data format standard used in meteorology to store historical and forecasted weather data.

Grid: A pattern of regularly spaced horizontal and vertical lines forming zones on a map and used as a reference to establish location.

GUI: See Graphical User Interface.

H

Handshake: An exchange of standardized signals between devices that regulate the transfer of data.

Hardware: The physical components of a computer system.

HDG: See Heading.

Heading (HDG): The direction in which a vessel is pointed, expressed in degrees from north at zero degrees clockwise through 360 degrees.

Heading Sensor: A gyro compass/electro-magnetic device that measures a vessel's heading, pitch and roll, and rate of turn.

Head-Up: A screen display orientation that shows the vessel's heading upward. (Compare with Course-Up and North-Up.)

Highway Screen: Shows a "highway view" of the course over ground, used to graphically illustrate cross-track error

HO: See Hydrographic Office.

Horizontal Datum: See Chart Datum.

Hydrographic Office (HO): The government organization responsible for conducting marine surveys and publishing nautical charts of that country's coastal waters.

Hydrography: The science of surveying and charting bodies of water.

I

Icon: A symbol or graphic displayed by a software application, often one of several for selection.

IENC: See Inland Electronic Navigation Chart.

IHO: See International Hydrographic Office.

IMO: See International Maritime Organization.

Indicator: An alarm or warning.

Initialization: The first time a GPS receiver orients itself to its current location.

Inland Electronic Navigation Chart (IENC): Vector-format digital charts of the U.S. Inland Waterway System, created by the U.S. Army Corps of Engineers.

Inmarsat: See International Mobile Satellite Organization.

Interchange: The transfer of data within a computer or software system. (Compare with Exchange.)

Interface: An electrical or communications exchange connecting hardware or software so they can be operated jointly or communicate with each other.

International Code of Signals: A U.S. government publication with descriptions and/or graphic depictions of all signals for visual, sound, and radio communications.

International Convention for Safety of Life at Sea (SOLAS): An agreement among governments to ensure that a ship is fit for the service for which it is intended in terms of safety of life at sea.

International Hydrographic Office (IHO): The international governmental agency responsible for coordinating nautical charts across national hydrographic offices.

International Maritime Organization (IMO): The specialized agency of the United Nations responsible for measures to improve the safety of international shipping and to prevent marine pollution from ships.

International Mobile Satellite Organization (Inmarsat): The international organization that operates a system of geostationary satellites providing mobile communication.

International Navigation Rules: International regulations for preventing collisions at sea. Also known as COLREGS.

Internet: A global system of interconnected computer networks (a "network of networks") that shares electronic mail, data files, and the hypertext documents of the World Wide Web.

K

Kbps: Kilobits per second. A measure of data transmission speed equal to one thousand bits per second.

KHz: Kilohertz. A measure of radio frequency or clock speed of a computer equal to one thousand cycles per second.

L

LAN: See Local Area Network.

Latitude: The geographic distance north or south of the Equator measured in degrees, minutes, and seconds (or fractions of a minute).

LCD: See Liquid Crystal Display.

Leg: A portion of the route consisting of a starting waypoint and a destination waypoint.

Line of Position (LOP): A straight or curved line, obtained from visual, electronic, or celestial sources, along which your position is assumed to be located.

Liquid Crystal Display (LCD): A screen that displays information by applying an electric field to liquid crystal molecules.

Listener: A device that is listening for coded information from another device. An example is an autopilot. (Compare with Talker.)

List of Lights: A publication issued by a government tabulating navigational lights and their characteristics.

List of Radio Signals: A publication issued by a government tabulating information on radio services useful to a navigator such as coast and port radio stations and radio direction finding stations.

LNM: See Local Notice to Mariners.

Load: To tell a software application where particular files are located. Chart data files are loaded either by copying files to a designated folder ("smart folder" loading) or by pointing the application to the set of chart files ("directory" loading). (Compare with Copy.)

Local Area Network (LAN): A computer network covering a small area such as a home, office, school, or boat.

Local Notice to Mariners (LNM): A periodic notice issued by the U.S. government regarding changes in aids to navigation and other updates to issued charts.

Longitude: The east-west geographic coordinate measurement, ranging from 0° at the Prime Meridian to the International Date Line (+180° eastward and -180° westward), measured in degrees, minutes, and seconds (or fractions of a minute).

LOP: See Line of Position.

Loran: Long Range Radio Navigation. A ground-based radionavigation system that uses multiple low frequency radio transmitters to determine the location of the receiver.

M

Magnetic Direction: Direction expressed in 360 degrees from magnetic north.

Magnetic North: The direction a compass needle shows as north.

Man Overboard (MOB): In e-charting, denotes an emergency position-based marker that documents the location where a person has fallen off a vessel and is in need of rescue.

Map Display: A GPS display screen of an overhead or bird's-eye view of current position.

Mariner's Guide for Hurricane Awareness: A U.S. government publication covering topics such as storm categories, observations at sea, hurricane motion and monitoring, and hurricane evasion.

Marine Weather Service Charts (MSCs): A U.S. government publication listing frequencies, schedules, and locations of stations disseminating National Weather Service information.

Maritime Mobile Service Identity (MMSI): A series of nine digits transmitted over the radio that uniquely identify a vessel or station, similar to a telephone number linked to a vessel's radio.

Maritime Safety Information (MSI): Navigational and meteorological warnings and other urgent safety messages broadcast to ships.

Mark: A non-navigational geo-reference or a point not on a navigational route.

MARPA: Mini Automatic Radar Plotting Aid. A radar that is MARPA-equipped monitors a selected target to provide Closest Point of Approach (CPA) and Time to Closest Point of Approach (TCPA) data. Used for collision avoidance.

Mashup: A web application that combines data from more than one source into a single integrated tool.

MB: Megabyte. A unit of information equal to one million bytes.

Mbps: Megabits per second. A measure of data transmission speed equal to one million bits per second.

Menu: A list of commands or options displayed on the computer screen by a software application.

Mercator Projection: A cartographic projection system designed for mariners which accurately represents direction from north. Visualized by wrapping a cylinder around the Earth and aligned with the poles. (Compare with Polyconic Projection.)

Meridian: A line of longitude.

Meter: The fundamental unit of length in the metric system, equal to approximately 39.37 inches.

MF and HF Channels: Government-published list of middle and high frequency marine radiotelephone channels, including information on single sideband radiotelephone channels.

MFD: See Multifunction Display.

MHz: Megahertz. A measure of radio frequency or clock speed of a computer equal to one million cycles per second.

Mile: Also called statute mile, a measure of distance equal to 5,280 feet. (Compare with Nautical Mile.)

Minute: A measure of time equal to one sixtieth of an hour or a measure of distance equal to one sixtieth of a degree.

MMSI: See Maritime Mobile Service Identity.

MOB: See Man Overboard.

Mouse: A peripheral device used to move the cursor on a computer screen.

MSCs: See Marine Weather Service Charts.

MSI: See Maritime Safety Information.

Multifunction Display (MFD): A computer display that can show multiple views of graphical information simultaneously.

N

NAD-27: North American Datum 1927.

NAD-83: North American Datum 1983.

Nanosecond: One billionth of a second.

National Geospatial-Intelligence Agency (NGA): U.S. government office, part of the Department of Defense, with the mission of collecting, analyzing, and distributing geospatial intelligence in support of national security. Producer of Digital Navigation Charts (DNCs).

National Imagery and Mapping Agency (NIMA): The U.S. agency that preceded the National Geospatial-Intelligence Agency (NGA), responsible for geospatial intelligence for national security.

National Marine Electronics Association (NMEA): A U.S. standards committee that sets the data message structure, contents, and protocols that allow marine electronics to communicate with each other.

National Oceanic and Atmospheric Administration (NOAA): The federal agency responsible for oceanic and atmospheric data that includes the National Ocean Service and Office of Coast Survey.

National Ocean Service (NOS): An agency within NOAA that includes the Office of Coast Survey (OCS).

National Weather Service (NWS): The U.S. agency within NOAA that provides weather, hydrologic, and climate forecasts and warnings for the U.S. and its adjacent waters and ocean areas.

Nautical Chart User's Manual: A U.S. government publication covering the details of marine cartography, with topics such as layout, scales, projections, topography, hydrography, navigation aids, landmarks, and routes.

Nautical Mile (NM): An international distance measure (1852 meters or 1.151 miles) that equals one minute of latitude (but not one minute of longitude). (Compare with Statute Mile.)

Navaid: See Navigational Aid.

Navigation: The act of planning the course or heading from one location to another and monitoring position en route.

Navigational Aid: A visual, acoustical, or radio device external to a vessel that assists in the determination of a vessel's position and its safe course. Also called an aid to navigation, nav aid, or abbreviated as ATON.

Navigation Rules-International and Inland: The official rule book, published by the U.S. Coast Guard, required on all vessels over 12 meters. Contains the International Regulations for Preventing Collisions at Sea (COLREGS) and the Inland Navigation Rules.

NAVSTAR: Navigation Satellite Timing And Ranging. The original constellation of satellites that transmitted microwave signals to determine the receiver's location, speed, and direction. The precursor to GPS.

Nest: The placement of objects, such as a folder within a folder, in a hierarchical arrangement, resulting in the confusion of an application's directory.

NGA: See National Geospatial-Intelligence Agency.

NGA List of Lights: Similar to the U.S. Coast Guard Light List, a seven-volume set with worldwide information on lights and other aids to navigation.

NIMA: See National Imagery and Mapping Agency.

NM: See Nautical Mile.

NMEA: See National Marine Electronics Association.

NMEA 0183: An industry standard for electrical and data specifications, set by the National Marine Electronics Association (NMEA), for communication between marine electronic devices.

NMEA 2000: A revised data communications protocol, replacing NMEA 0183, that allows up to 50 marine devices to be connected.

NOAA: See National Oceanic and Atmospheric Administration.

Node: A computer, peripheral device, or marine instrument attached to a network.

North-Up: A screen display orientation that shows the north direction upward, corresponding to the usual orientation of a chart. (Compare with Course-Up and Head-Up.)

NOS: See National Ocean Service.

Notice to Mariners (NtM): A periodic notice issued by the government regarding changes in aids to navigation and other updates to issued charts.

NtM: See Notice to Mariners.

NWS: See National Weather Service.

O

Object: An identifiable set of information, associated with attributes and which may be related to other objects. An example is a bridge.

Object Class: A set of objects on a chart with the same characteristics, such as the object class of lit aids, of daybeacons, or of anchorage areas.

OCS: See Office of Coast Survey.

Office of Coast Survey (OCS): The hydrographic office of the United States, responsible for producing and maintaining the nautical charts of the coastal waters of the U.S. and its territories.

ONF: See Open Navigation Format.

Open Content Website: An Internet site where the content is created by users collaboratively contributing and editing posted information.

Open Navigation Format (ONF): A proprietary protocol used by Nobeltec for data exchange.

Operating System (OS): The software component of a computer system that is responsible for the management and coordination of activities and the sharing of the resources of the computer. Examples include Windows XP, Vista, OS X, and Linux.

OS: See Operating System.

OS X: A computer operating system for Macintosh computers, released in 2001, that replaced the "Classic" Mac (OS 9 and earlier) operating system. Pronounced "oh-es-ten."

Overscale: To display a chart at a scale larger than the compilation scale, resulting in pixelization. (Compare with Underscale.)

P

Package: See Software.

Pan: To swing in a horizontal plane for a panoramic effect or to follow a subject (as in to pan over a chart display).

Panel: The basic unit for the distribution of RNC (raster) data covering a geographic area. Typically consists of a main panel, extensions, and insets at different scales. A panel corresponds to the same area covered by the paper chart. (Compare with Cell.)

Parallel: A line of latitude.

Parallel Communication: A method of sending multiple data signals over a communication link at one time. (Compare with Serial Communication.)

PC: An abbreviation often specifically referring to a personal computer running the Microsoft Windows operating system. (Compare with Personal Computer.)

Personal Computer: A computer whose price, size, and capabilities are such that it is intended to be used by an individual directly, without an intervening computer operator.

Pilot Charts: Government publications depicting averages in prevailing winds and currents, air and sea temperatures, wave heights, ice limits, visibility, barometric pressure, and weather conditions at different times of the year. Intended to aid the navigator in selecting the fastest and safest routes with regard to expected weather and ocean conditions.

Piloting: The process of navigating using charts, visual landmarks, aids to navigation, and/or radio, sound, or electronic information.

Pirate: The use or reproduction of another's work without permission, usually in violation of a patent or copyright.

Pixel: Picture element. A single data or display point of an image. The more pixels, the higher the resolution and definition.

Planner: A software application that displays charts and is capable of creating and storing waypoints and routes for navigation.

Point of Departure: The last fix obtained, marked on a chart as the beginning of the track until the next fix.

Points-of-Interest (POIs): A collection of specific locations that someone may find useful or interesting.

POIs: See Points-of-Interest.

Polar Diagram: A diagram showing a vessel's speed at different angles of sail in different wind strengths, used to determine target speed for any given wind speed and direction.

Pole: Either of the two locations (North Pole or South Pole) on the surface of the earth that are the northern and southern ends of the axis of rotation. The reference points of lines for latitude and longitude.

Polyconic Projection: A conical map projection obtained by rolling a cone tangent to the Earth at all parallels of latitude. (Compare with Mercator Projection.)

Port: In computing, the socket where a cable can be connected.

Position: A geographic location on Earth, typically described using latitude and longitude.

Position Format: The way a GPS receiver displays position on the screen, most commonly as degrees, minutes, and thousandths of a minute.

Prime Meridian: The zero meridian from which longitude is measured east and west.

Program: See Software.

Proprietary: Relating to ownership (as in proprietary charts, which are owned by a private company, versus public-domain charts).

R

RACON: A radar beacon, which when triggered by pulses from a vessel's radar, transmits a reply.

Radar: Originally Radio Detection and Ranging, now the word "radar." A system that transmits radio waves—reflected by a target and detected by the receiver—to collect information on the target's range, altitude, direction, or speed.

Radar Navigation and Maneuvering Board Manual: A U.S. government publication covering the fundamentals of shipboard radar, radar operation, collision avoidance, navigation by radar, and a description of vessel traffic systems in U.S. waters.

Radionavigation: The determination of position using the propagation properties of radio waves, such as Loran or GPS.

Radio Navigation Aids: A U.S. government list of selected worldwide radio stations broadcasting navigational warnings, time signals, or medical advice; communication traffic for distress, emergency, and safety; and long range navigational aids.

RAM: Random Access Memory. A type of computer data storage used in personal computers. In computer specifications, a measure of the computer's memory capacity.

Range: The distance to a selected target.

Raster: A rectangular pattern of parallel scanning lines.

Raster Chart: A digital chart in raster format is a scanned image of a paper chart. The data file is comprised of millions of pixels to reproduce a color-scanned image. (Compare with Vector Chart.)

Raster Navigational Chart (RNC): *NOAA RNC* is the trademark name for its raster-format chart data files. The North American standard for raster chart data.

RCDS: Raster Chart Display System. A proposed IMO standard that permits ECDIS equipment to operate using raster charts (in an RCDS) mode) in the absence of vector charts.

Relative Bearing: The bearing measured to an object from a boat referenced to the boat's heading.

Relative Motion Display: A display where the vessel shows as stationary while all other information moves in relation to the vessel. (Compare with True Motion Display.)

Reverse Engineer: The reproduction of another manufacturer's product by undertaking a detailed examination of its construction.

Reverse Route: A function that reverses the sequence of waypoints in a saved route so the user can navigate from the end point back to the starting point.

Rhumb Line: A line on the surface of the Earth that crosses all meridians at a constant angle. A rhumb line is generally not the shortest path on the surface of the Earth. (Compare with Great Circle Route.)

RNC: See Raster Navigational Chart.

Route: A collection of waypoints, connected in sequence, typically marking a vessel's intended course.

Route Planning: The pre-determination of course, speed, and waypoints for the waters to be navigated.

RS-232: A serial input/output (I/O) standard that allows for compatibility between data communication equipment.

Rubber Band: A colloquialism for stretching and dropping lines on a display in a process of connecting straight lines between points.

Ruggedized Laptop: Portable computers designed for harsher environmental conditions including temperature change, impact, and humidity.

S

S-52: An international format for vector digital charts, developed by the International Hydrographic Organization (IHO).

S-57: The updated international format for vector digital charts, developed by the International Hydrographic Organization (IHO).

S-63: A data security-enabled international format for vector digital charts, developed by the International Hydrographic Organization (IHO).

SA: See Selective Availability.

Safety Contour: A contour line, selected by the user based on the vessel's draft, used to separate safe and unsafe depths for anti-grounding alarms.

Safety Depth: The depth chosen by the user, such as the vessel's draft plus under-keel clearance, as the minimum sounding.

Saildocs: An email-based document-retrieval system for the delivery of text-based Internet documents either on request or by subscription.

Sailing Directions Enroute: Thirty-eight publications by region with coastal and port approach information, supplementing the largest scale chart of the area. Contains information about coastal weather, currents, ice, dangers, features, and ports for each geographic area.

Sailing Directions Planning Guides: Five volumes by region with relevant physical, political, industrial, navigational, and regulatory information on countries adjacent to a particular ocean basin.

Satellite Differential GPS (SDGPS): A Federal Aviation Administration DGPS system using satellites, primarily for aeronautical use.

Satellite Photo: Photograph of the earth or other planets made by means of satellites.

Satellite Status Display: An information screen that shows technical data about each satellite in view, including receiver channel numbers, satellite ID numbers, status of satellite tracking or searching, satellite elevations and azimuths, signal-to-noise ratios, and dilution of precision ratings.

SatNav: Satellite Navigation System. A navigation system, maintained by the U.S. Navy, that is no longer operational.

Scale: The ratio between the linear dimensions of a chart and the actual dimensions represented. A larger scale (such as 1:10,000) shows larger-appearing information but covers a smaller area; a smaller scale (such as 1:80,000) shows smaller-appearing information over a larger area.

Screenshot: An image of the display on a computer screen.

Scroll: The action (typically initiated by the mouse wheel or keyboard arrows), that moves a computer screen display up or down to reveal new material.

SDGPS: See Satellite Differential GPS.

Second: A measure of time or distance equal to one sixtieth of an minute.

Selective Availability (SA): A feature of GPS (disabled as of 2000) that intentionally introduces slowly changing random errors into publicly available signals, reducing GPS accuracy for security reasons.

SENC: See System Electronic Navigational Chart.

Serial Communication: The process of sending data sequentially. (Compare with Parallel Communication.)

Sight Reduction Tables: Six volumes, by 15 degrees of latitude, containing the tables necessary for celestial navigation.

Sirius: A satellite radio company. Commercial-free subscription radio, news, and weather is received via Internet or a Sirius radio device.

Situational Awareness: The knowledge and understanding of your surroundings critical to making decisions.

Skype: A software that allows users to make telephone calls over the Internet using voice-over-Internet-protocol (VoIP).

Software: A computer program that uses the capabilities of a computer to perform certain tasks. Also called application, package, or program.

SOG: See Speed Over Ground.

SOLAS: See International Convention for Safety of Life at Sea.

Soundings: Measurements of the depth of the water.

Speed Over Ground (SOG): The actual speed over the sea bottom.

Split Window: A single computer window that displays more than one image in a side-by-side or above-below arrangement.

State Chart Catalogs: Separated by state, a list and illustration of NOAA charts by number and location.

Statute Mile: See Mile.

System Electronic Navigational Chart (SENC): A transformation of the standard ENC data by a vendor, adapting the data for display in their proprietary system.

T

Tabbed Interface: A computer window that displays file-folder-like tabs allowing a user to flip through layers of window displays.

Talker: A device that is sending coded information to another device. An example is GPS. (Compare with Listener.)

Target: An object detected by a radar scan.

TCPA: See Time to Closest Point of Approach.

TD: Time Difference or Time Delay. Curve computed by a Loran system by comparing the time between a master and secondary signal. The intersection of curves determines a geographic point in relation to the position of the stations. Also called TD lines.

Thumb Drive: A tiny portable USB flash drive.

Tidal Datum: The height of the mean lower low water (MLLW) (U.S.) or the height of the lowest astronomical tide (LAT) (Canada and the U.K.).

Time to Closest Point of Approach (TCPA): The calculated time until the minimum distance between two objects. Used for tracking and collision avoidance.

Time to First Fix (TTFF): The time it takes to find the satellites after the GPS receiver is turned on (or when a GPS receiver has lost signal or has been moved over 300 miles from its last location).

Toolbar: On a computer graphical user interface, a strip of icons used to perform functions.

Track: A series of points (defined by a latitude, longitude, and time) on a GPS receiver that shows a history of the receiver's course.

Transceiver: A device that can both transmit and receive communications.

Transit: The alignment of two visual landmarks.

Transit (SatNav): See SatNav.

Transreflective: A liquid crystal display (LCD) that reflects most of the sunlight shining on the screen and automatically adjusts screen brightness depending on how much light shines on it, reducing the need to manually adjust screen brightness.

Trial Software: A marketing method for computer software in which the software is available free of charge to try before buying the full version. Also called Demoware.

TRN: See Turn.

True Direction: Direction expressed in 360 degrees from true north.

True Motion Display: A display where the vessel and each target moves while charted information remains fixed. (Compare with Relative Motion Display.)

True North: The direction that points directly to Earth's north pole.

TTFX: See Time to First Fix.

Turn (TRN): The degrees that must be added to or subtracted from the current heading to reach the course to the intended waypoint.

U

Underscale: To display a chart that is not the largest-scale navigational data available for that area. (Compare with Overscale.)

Universal Coordinated Time (UTC): A world standard for time, replacing Greenwich Mean Time (GMT) in 1986, based on atomic measurements rather than the rotation of the Earth. GPS uses UTC; GMT remains the standard time zone for the prime meridian and zero longitude. 

Universal Transverse Mercator (UTM): A worldwide coordinate projection system using north and east distance measurements from reference points. UTM is used primarily for U.S. Geological Survey topographic maps, not for marine navigation.

Update: Information that keeps a chart or nautical publication up-to-date, typically issued as Notice to Mariners.

Upload: To transfer data to another computer system. (Compare with Download.)

USACE: See U.S. Army Corps of Engineers.

U.S. Army Corps of Engineers (USACE): The federal agency responsible for surveying and charting the U.S. Inland Waterway System.

USB: Universal Serial Bus. A standard port connection used for computer peripherals such as keyboards, mice, digital cameras, printers, and flash drives.

USCG: See U.S. Coast Guard.

U.S. Coast Guard (USCG): A branch of the U.S. armed forces responsible for maritime security and law enforcement.

USCG Light Lists: Seven volumes published by the U.S. Coast Guard listing lights, sound signals, buoys, daybeacons, and other aids to navigation for U.S. waters.

User Interface: The hardware and operating software by which a user executes commands on a device and the device conveys information to the user.

U.S. VHF Channels: U.S. publication listing VHF channels and each channel's communication use.

UTC: See Universal Coordinated Time.

UTM: See Universal Transverse Mercator.

V

Variation: The angular amount by which magnetic north differs from true north at any location on Earth.

Vector: A line designating both direction and magnitude.

Vector Chart: A digital chart in vector format is created from a data file of chart features. The data file consists of a collection of geospatially  referenced points, lines, polygons, symbols, and areas. (Compare with Raster Chart.)

Velocity: A measure of speed and direction.

Velocity Made Good (VMG): The speed and direction along the course intended.

VHF: Very High Frequency. The radio waves used by marine radios.

Viewer: A software application that displays charts on a personal computer.

VMG: See Velocity Made Good.

Voice over Internet Protocol (VoIP): An optimized standard for the transmission of voice through the Internet.

VoIP: See Voice over Internet Protocol.

W

WAAS: See Wide Area Augmentation System.

Warm Start: Powering up a device, such as a GPS, when it is already initialized. (Compare with Cold Start.)

Warning: An alarm or indicator.

Waypoint: A set of latitude and longitude coordinates that identify a location used for navigation.

WCI: World Calibrated Image. A file format commonly used for user-scanned chart images.

Weblog: See Blog.

WEND: See Worldwide Electronic Navigational Chart Data Base.

WGS: See World Geodetic System.

WGS-84: The set of parameters for the World Geodetic System as of 1984 and currently the datum standard used to produce ENCs.

Wide Area Augmentation System (WAAS): A feature on GPS receivers to improve position accuracy, typically to within three meters, using a system of approximately 25 ground reference stations positioned across the U.S.

Widget: An element of a graphical user interface that displays a narrow topic of information in a small window or text box.

WiFi: Wireless Fidelity. A group of technical standards enabling the transmission of data over wireless networks.

WiFi Card: A hardware device that connects a computer to a wireless network.

Window: A framed area on a display screen for viewing information.

World Geodetic System (WGS): A global geodetic reference system develop by the U.S.  for satellite position fixing and endorsed by the IHO.

World Port Index: A U.S. government publication with the location, characteristics, and available services of major ports and shipping facilities throughout the world.

Worldwide Electronic Navigational Chart Data Base (WEND): A worldwide network of ENC data sets based on IHO standards.

Worldwide Marine Radiofacsimile Broadcast Schedules: Government publication with weatherfax frequencies and schedules for fax stations around the world.

World Wide Web (WWW): A system of interlinked hypertext documents accessed via the Internet. 

WWW: See World Wide Web.

X

XM: A subscription satellite radio service that includes XM WX Weather and XM Satellite Radio.

XTE: See Cross-Track Error.

Z

Zoom: A method of enlarging (referred to as zooming in) or reducing (zooming out) the graphic information displayed on a screen.

Index

About the Authors

*Mark and Diana Doyle are authors of the popular cruising guide and electronic chart series, **Managing the Waterway**. They are also Electronics Editors for Mad Mariner and have written numerous articles for professional and boating publications, including Power Cruising, Sea Magazine, and Latitudes & Attitudes.*

Mark retired from high-tech, specializing in color digital imaging with companies such as Kodak, Agfa, and Xerox. He is the founder of Bluewater Yacht Deliveries and holds a 100-ton USCG Master's License. His first roundtrip of the Atlantic Intracoastal Waterway was in 1987.

Diana is a former university professor with a Ph.D. from Yale. She holds a 50-ton USCG Master's License. A lifetime birder and naturalist, she now enjoys sharing her knowledge with cruisers though her writing.

They live with their son, Morgan, juggling time between water and land—surveying future cruising guides, leading seminars, boat show-hopping—and, occasionally, hunkering down in the frigid North to write articles, books & guides.

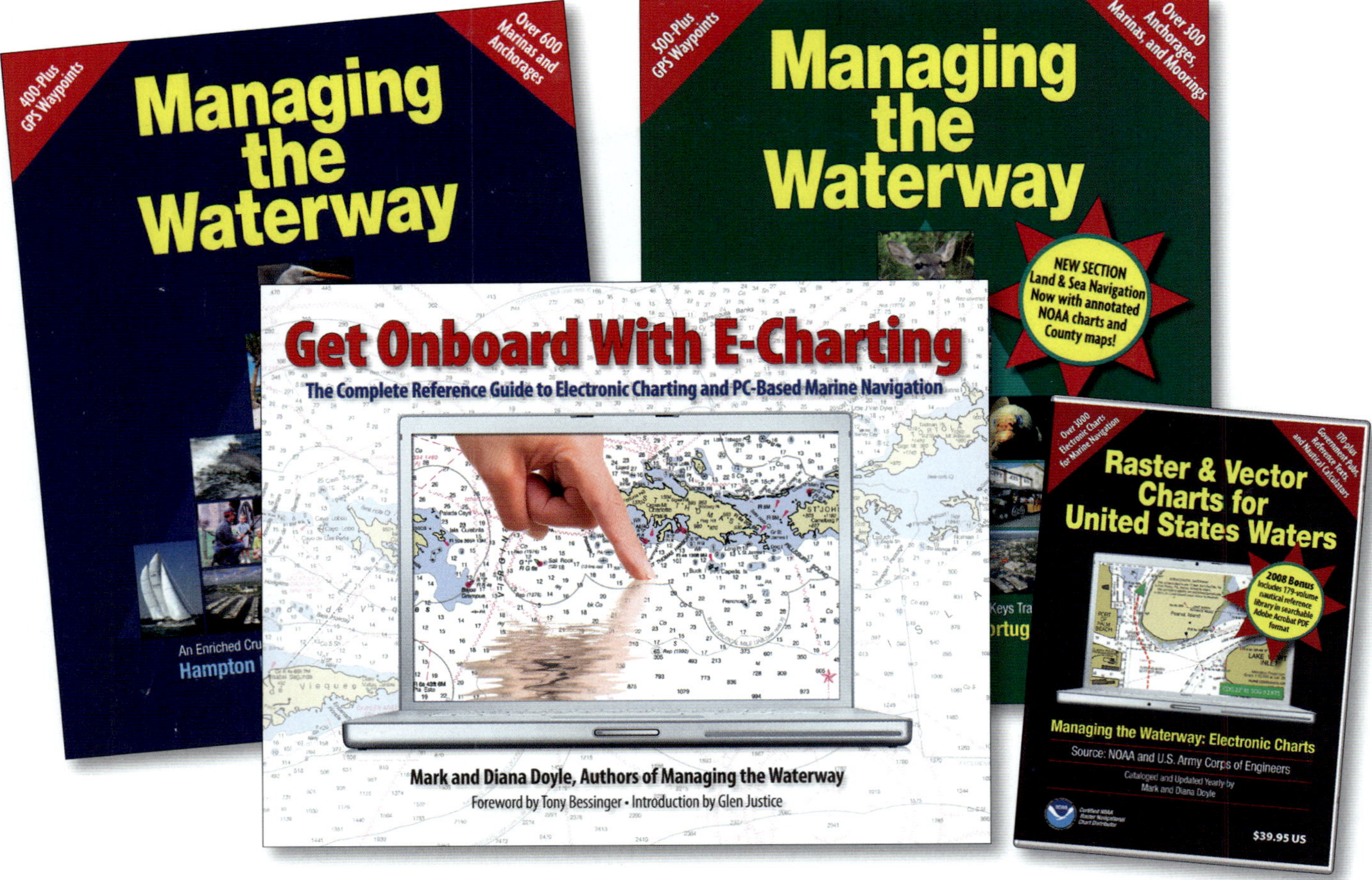

Managing the Waterway

Cruising Guides and Electronic Charting

www.managingthewaterway.com